Durability of Industrial Composites

Durability of Industrial Composites

Antonio Carvalho Filho

CRC Press
Taylor & Francis Group
Boca Raton London New York

CRC Press is an imprint of the
Taylor & Francis Group, an **informa** business

Cover photo: Continuous lamination of hoop-chop cylinders. Courtesy of NOV Fiber Glass Systems.

CRC Press
Taylor & Francis Group
6000 Broken Sound Parkway NW, Suite 300
Boca Raton, FL 33487-2742

First issued in paperback 2020

© 2019 by Taylor & Francis Group, LLC
CRC Press is an imprint of Taylor & Francis Group, an Informa business

No claim to original U.S. Government works

ISBN 13: 978-0-367-73298-1 (pbk)
ISBN 13: 978-1-138-33829-6 (hbk)

Library of Congress Cataloging-in-Publication Data

Names: Carvalho Filho, Antonio (Materials scientist), author.
Title: Durability of industrial composites / Antonio Carvalho Filho.
Description: Boca Raton : Taylor & Francis, a CRC title, part of the Taylor & Francis imprint, a member of the Taylor & Francis Group, the academic division of T&F Informa, plc, 2018. | Includes bibliographical references and index.
Identifiers: LCCN 2018024043 | ISBN 9781138338296 (hardback : acid-free paper) | ISBN 9780429441813 (ebook)
Subjects: LCSH: Composite materials -Service life. | Strength of materials.
Classification: LCC TA418.9.C6 C3257 2018 | DDC 620.1/18 -dc23
LC record available at https://lccn.loc.gov/2018024043

Visit the Taylor & Francis Web site at
http://www.taylorandfrancis.com

and the CRC Press Web site at
http://www.crcpress.com

Contents

PART II   Computation of Durability

Foreword

It ain't what you don't know that gets you into trouble, but what you know for sure and ain't so.

Mark Twain

Ignorance, more frequently than knowledge, begets unwavering confidence. Often times, seemingly impossible tasks are performed by people who know and fear little, rather than by those who know and fear much. This book is a loud example. Who, among the many experts, would consciously take on the abstruse issues involving the durability of industrial composites? The complexities involving this overwhelming problem are apparent in a multitude of factors. The enormous variety of laminate configurations, resin and fiber types, chemical environments and operating conditions, as well as the infinite variety of mechanical loadings – all of the above – act together to befuddle the issue of durability. The experts, who know much, recognize the enormous difficulty involved in this effort and dismiss it out of hand. Only a fearless non-expert would embrace such a daunting task.

In this book, I introduce several quantitative models that together predict the durability of any composite laminate operating in any chemical environment, subjected to any moisture, temperature or mechanical loading. I identify eight modes of long-term laminate failure. Of those, four are environment-dependent, and four are strictly load-dependent. The solutions of the environment-dependent problems are straightforward and require no outlandish hypotheses or new concepts. The load-dependent durability problems, however, require a host of new hypotheses and the rejection of old concepts. In this book, I propose several new concepts and assumptions to solve the load-dependent durability issues. To the best of my knowledge, these new concepts and assumptions are correct in spite of the ominous words of Mark Twain. Perhaps I am inviting trouble by asserting things that I know for sure, but may not be so. Consider, for example, the following five of the many bold proposals in this book.

- There are eight modes of long-term laminate failure, all governed by known mechanisms.
- The failure of the critical ply defines the durability of the entire laminate.
- The introduction of new concepts, such as the failure thresholds and the fatigue threshold.

- The outright rejection of old classical concepts, such as the HDB, Sb and Sc.
- The derivation of all regression equations from the experimental measurement of just one equation.

I have tried my very best to develop a set of compelling arguments in favor of the correctness of these and other new concepts and assumptions.

The current solutions to the load-dependent durability problems of composites have their basis on the classical Goodman diagrams. There are three difficulties with these diagrams. First, they are laminate and load specific, i.e., any change in the laminate construction or the applied loading requires the development of a new diagram. Second, they ignore the time variable. Third, the Goodman diagrams are restricted to rupture failures, and leave aside the important failure modes of infiltration, weeping and loss of stiffness. The unified equation, developed in Chapters 16, 17 and 18 of this book, takes a significant step forward and solves all load-dependent durability issues – any laminate and any loading – by explicitly taking into consideration the time variable. The lack of experimental data is the only current obstacle to the use of the unified equation. In Chapter 18, I discuss the experimental work needed to allow the full use of the unified equation. As explained there, this is a major effort requiring the experimental measurement of at least 55 regression equations. Such an expensive effort will probably not get started anytime soon, and the magnificent unified equation will remain useless for many years to come. To alleviate this difficulty, I have developed a simplified and approximate approach to allow the use of the unified equation in the absence of experimental data. The details are in the Appendix 18.1 of Chapter 18.

The concept of critical ply is central everywhere. The computation models developed in this book predict the durability of critical plies, not of laminates. The models themselves are straightforward, intuitive and easy to use. No complex mathematical apparatus is required in their derivation and application. In fact, the concepts developed and presented in this book derive from simple reinterpretation of well-known data published by many researchers. My role consisted essentially in collecting, examining and reinterpreting the existing data. In doing this, I have focused the critical plies, not the laminates themselves. I have tried to do for the durability of composites what James Clark Maxwell did for electromagnetism. Like Maxwell, I have collected many scattered and apparently unrelated experimental data into meaningful fundamental equations. I myself did no experiment. I have simply collected and reinterpreted the existing data.

In writing this book, I have put considerable effort on clarity. The text is perhaps wordy and sometimes certainly repetitive. This I have done on purpose, in the name of clarity. I have consciously sacrificed brevity and style to make sure the reader correctly understands the ideas and concepts as intended, with little risk of misinterpretation.

Antonio Carvalho Filho
Passos, MG, Brasil
December 11, 2017

Preface

The ambitious goal of this amazing book is nothing less than the *"quantification of the durability of any laminate in any environment under any loading"*. This is a huge goal, reaching far beyond anything in the current literature. Any claim to compute the durability of *"any laminate in any operating conditions and any loading"* is nothing short of outstanding.

My job as a reviewer consists essentially in verifying the usefulness of the book and checking the correctness of the assumptions proposed by the author.

I start with the usefulness, by checking the book coverage of topics of practical interest. I examined the eight modes of long-term failure proposed in Chapter 8 to check if they really cover all ageing scenarios. Apparently they do. I tried in vain to find a ninth mode of long-term failure. All ageing modes devised in my imagination related either to short-term failures – not the topic of this book – or to special cases of the eight proposed modes. The book seems to be complete and exhaustive, providing full coverage to all possible long-term modes of failure.

Still on the topic of usefulness, the author seems to have taken great pain in presenting a detailed description of each of the eight ageing processes, which is a great help to facilitate the understanding of composites. The academic researcher will design better experiments based on the clear descriptions of the ageing processes described in this book. The practicing engineer will benefit from the simple protocols proposed to quantify the laminate durability. Last, but not least, the young student will enjoy the opportunity to start his career in composites from a comprehensive and clear platform.

In summary, the usefulness of the book derives from its full coverage of the topic and the many solutions and insights presented and not found elsewhere. The book provides a multitude of fundamental insights and solutions to important engineering problems, way beyond the current methods.

Next, we check the validity of the hypotheses and their consistency with the mainstream ideas pertinent to composites technology. The analysis of the load-independent modes of failure presented by the author are straightforward and pose no concern regarding the correctness of the assumptions. However, in modelling the load-dependent durability, the author makes a few unheard of and controversial assumptions. Let us take a closer look at these non-mainstream deviations.

The author introduces a new unified equation to compute the load-dependent durability. The unified equation itself is not controversial and raises no objection. The controversy arises in the simplifications proposed to reduce the overwhelming amount of experimental data required by the unified equation. To reduce the experimental effort, the author proposes a multitude of daring and unusual assumptions. We have the following situation. The unified equation is positively mainstream, but impractical to use for lack of experimental data. To alleviate this, the author introduces several clever and unorthodox simplifications. The three most controversial assumptions are:

- The regression equations are independent of the operating temperature and moisture. At first glance, such a sweeping and bold simplification appears to be flat wrong.
- The regression lines start at the several failure thresholds – short-term ply strengths – and end at the same fatigue threshold. Why the same fatigue threshold for all plies, all resins and all modes of failure? This is hard to accept. To begin with, what is the fatigue threshold? There is no mention of this entity in the published literature.
- The failure of the critical ply defines the laminate durability. This central simplification shifts the difficult analysis of laminates to the simpler analysis of plies.

The correctness of the load-dependent durability models developed in this book rests on the veracity of the above assumptions. In my opinion these assumptions, although at first glance hard to accept, are in fact reasonable and even obvious from the compelling arguments developed by the author. My initial skepticism vanished on reading the argumentation presented in the book.

Having criticized the hypotheses, I proceed to highlight some of the strong and uncontroversial points of the book. First, the book provides a clear and consistent description of the long-term failure mechanisms and ageing processes of composite materials. All readers, skeptic or otherwise, will marvel and greatly benefit from the many precious insights and explanations of the ageing processes proposed by the author. Such powerful insights and explanations may well set the pathways for future research. Perhaps in time, as the readers and the technology mature, the hardest skeptics of today may change their minds and convert to the ideas in this book, if not in full, at least in part.

In conclusion, I highly recommend this book for immediate publication. The enormous mass of insights and innovative solutions contained in it, while in some cases departing from the mainstream thought, are nevertheless consistent with the accepted scientific methods. The solutions herein proposed promise new roads of development that may become the paradigm for future research. Furthermore, the presented material allows the systematic and accurate solution of durability problems outside the reach of the current knowledge. In this regard, this is a unique book, unmatched by any previous publication.

Prof. Antonio Marmo de Oliveira
Retired Full Professor of ITA (Instituto Tecnologico de Aeronautica)
S.P. Brasil

Acknowledgements

I am especially indebted to Carlos Marques, my intellectual partner of many years, for directing my efforts, challenging my early false starts and pointing my thoughts in the right direction. Carlos and I would often debate the durability of composites well into the night, trying to make sense of the available data and apparently conflicting reports. In doing this, we had no ambition or expectation of solving the problem ourselves. All we wanted and hoped for at that time was to get a glimpse of the factors affecting the durability and the ageing processes of composite materials. I wish Carlos were alive today, to see our early unpretentious discussions materialized in the present book.

I am also thankful to my colleagues Gabriel Gonzalez, Antonio Marmo de Oliveira, Francisco Xavier de Carvalho and Edouard Zurstrassen, for their continuous encouragement and many suggestions. Their expert criticism forced me to rewrite entire chapters and produce a better book.

Biography

Antonio Carvalho Filho is an electrical engineer with 47 years of experience in composites, starting in 1971 as technical support and market development specialist for a glass fiber manufacturer in South America. His activities in this period covered the entire customer base, including all molding processes and all markets. The upside of this apprenticeship was the wide exposure to composites as a versatile engineering and construction material. The downside was the diluted nature of the work, which added little in terms of technological development.

His real advancement in the knowledge and the technology of composites started 30 years later, in 2001, when fired from his job he started a consultancy firm dedicated to the analysis and design of industrial composite equipment. The fewer distractions and the well-focused nature of this new activity eventually paid off as a solid and deep understanding of composites in highly demanding service.

In his long involvement with composites, Carvalho has authored several technical articles, papers and books on a variety of topics, such as resin cure, structural analysis, molding processes, costing, and others, all directed to a Portuguese speaking audience.

Units

All quantities in this book, unless otherwise identified, enter the equations in the following units:

Quantity	unit
α – Coefficient of thermal expansion	1/C
β – Coefficient of hydric expansion	1/Δm
E – Young's modulus	kg/cm^2
[Q] – Ply stiffness matrix	kg/cm^2
Density	g/cm^3
t – Ply or laminate thickness	mm, cm
σ – Stress	kg/cm^2
[A] – Laminate stiffness matrix	kg/cm
[N^T] – Laminate thermal resultant	kg/cm
[N^H] – Laminate hydric resultant	kg/cm
[N^M] – Mechanical resultant	kg/cm
P – Pressure	kg/cm^2
D – Diameter	mm, cm
RT – Room temperature	C
OT – Operating temperature	C
PT – Stress-free temperature	C
PPT – Unassisted peak process temperature	C
PCT – Post-cure temperature	C
HDT – Heat distortion temperature	C
t – Time	sec, month, hour, year
Δe – Eroded or penetrated depth	mm
Loading frequency	Hz

Part I

Computation of Total Strains

The analysis of the load-dependent durability of composite materials requires the accurate knowledge of the total strain components on the critical plies. The Part I of this book, from Chapter 1 through Chapter 7, proposes a simple and easy protocol to compute the total strain components (mechanical + residual) on any ply embedded in circular cylindrical laminates. The reader familiar with the computation of strains in embedded plies can skip the next seven chapters and go directly to the Part II of the book, starting in Chapter 8.

1 Ply Properties

1.1 INTRODUCTION

Plies are the smallest macroscopic structural units recognizable in composite laminates. On a microscopic level, they are heterogeneous and exhibit a clearly distinct and discontinuous fibrous phase embedded in a continuous resin matrix. However, on a macroscopic scale, it is usual to smear the small diameter fibers to obtain essentially homogeneous plies. The smearing obliterates the fiber presence in the ply, while retaining their effect as a modifier of the otherwise isotropic matrix. The arrangement of the reinforcing fibers in a ply make take several patterns, the most common being their grouping in parallel continuous arrays (unidirectional or UD plies) and the random distribution of chopped strands (chopped fiber plies). Some plies consist exclusively of neat resin, with no reinforcement. Plies of resin reinforced with granular sand or woven fabrics are also used. In all cases, the fibers form a planar pattern that is repetitive and easily recognizable throughout the ply.

Plies are seldom, if ever, used by themselves in commercial applications. Most practical composites applications involve laminates composed of many distinct plies. The reinforcements discussed in this book are composed exclusively of glass fibers. This book does not discuss aramid and carbon fibers. The UD plies are orthotropic, meaning they have two orthogonal axes of symmetry identified by the digits 1 and 2. See Figure 1.1. The axis 1 coincides with the fiber direction. The axis 2 is transverse to the fibers. The axis 3, or z, perpendicular to the ply, is irrelevant to our discussions. The plies of sand, neat resin and chopped fibers are isotropic, meaning they have an infinite number of axes of symmetry. The properties of the macroscopically homogeneous plies vary with the fiber direction and arrangement, but are the same everywhere.

1.2 THE STANDARD PLIES

The concept of standard ply is rather intuitive. Most engineers think in terms of standard plies even if not aware. This section brings out and emphasizes this intuitive knowledge. The standard plies are idealized entities, the arbitrary product of our imagination, created especially for design purposes. Their composition represent as close as possible those of the real plies most commonly found in practice. The idealized standard plies are close, but never equal, to the real ones.

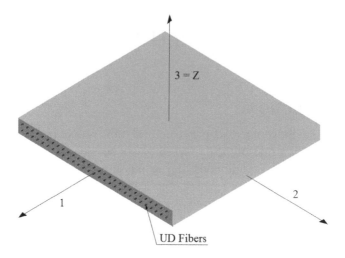

FIGURE 1.1 *Unidirectional UD ply and the local frame 1, 2, 3. The local axis 1 coincides with the fiber direction. The transverse axis 2 is perpendicular to the fibers. The properties of orthotropic UD plies vary with the direction. The isotropic plies of neat resin, or sand, or chopped fibers have the same properties in all directions.*

The four plies of commercial interest have the following standard compositions:

- Neat resin
- 30% by weight of chopped glass fibers
- 70% by weight of continuous UD glass fibers
- 80% by weight of sand

Table 1.1 lists the most important engineering properties characterizing a ply. Note that all properties, including the failure thresholds, refer to the local frame 1–2. The failure thresholds are in fact short-term ply strengths with new and more descriptive names. The choice of these new names correlates with the modes of failure they describe. The attentive reader will note that Table 1.1 contains several strengths – or thresholds – for the standard ply, one for each mode of failure. The values of the failure thresholds mentioned Chapter 8 derive from the published literature and may not be accurate. This, however, is not relevant, since our goal is to understand their concept. The acceptance of the concept would lead to the accurate measurements. Part II of this book discusses this in detail.

There are two ways to determine the properties of the standard plies. The first, and most common, consists in lab test measurements. As an alternative,

TABLE 1.1
This table lists the ply properties needed in engineering design. The reported properties apply to the local frame 1 – 2. The listed values refer to measurements at room temperature in a dry atmosphere.

Coefficients of thermal expansion

$$[\alpha] = \begin{bmatrix} \alpha_1 \\ \alpha_2 \\ 0 \end{bmatrix}$$

Coefficients of hydric expansion

$$[\beta] = \begin{bmatrix} \beta_1 \\ \beta_2 \\ 0 \end{bmatrix}$$

Elastic moduli

$$[E] = \begin{bmatrix} E_1 \\ E_2 \\ G_{12} \end{bmatrix}$$

Poisson's ratios

$$[v] = \begin{bmatrix} v_{12} \\ v_{21} \end{bmatrix}$$

Stiffness matrix

$$[Q] = \begin{bmatrix} Q_{11} & Q_{12} & 0 \\ Q_{12} & Q_{22} & 0 \\ 0 & 0 & Q_{66} \end{bmatrix}$$

Failure thresholds

$$\begin{bmatrix} T_i \\ T_w \\ T_s \\ T_r \\ T_0 \end{bmatrix} = \begin{bmatrix} Infiltration \\ Weep \\ Stiffness \\ Rupture \\ Fatigue \end{bmatrix}$$

the standard ply properties may derive from micromechanical models freely available in the literature. The lab tests work well in all cases. The micromechanical models apply to some ply properties, like the moduli and the coefficients of expansion, but cannot predict the failure thresholds. The determination of the failure thresholds require experimental lab measurements.

The first issue to address in the definition of the standard plies is the choice of the glass loading. This choice is arbitrary. To facilitate the discussions, this book establishes the standard glass loadings as close as possible to the most common loadings of real plies. The choice of the glass loading solves a long-standing issue of perennial discussion and disagreement between designers, fabricators and inspectors. The adoption of agreed upon standard plies – with fixed glass loadings and properties – eliminates this nightmare and brings the following benefits:

- The designer would do his work using standard plies of known properties. He would compute and specify the required number, type and orientation of plies in the standard laminate. Knowing the number of plies, the standard laminate thickness and the glass weights in kg/m² are easy to compute, if desired.

- The fabricator would execute the designer's specification with his mind set on the number of plies and unconcerned with the glass loading. The concept of standard plies facilitates the communication, understanding and execution at all levels for the personnel involved in design and fabrication.
- The inspector would also be pleased to check for nothing more than the presence of the specified plies. The use of standard plies and laminates eliminate the nagging discussions about laminate thickness, glass loadings and mechanical properties.

This chapter tabulates the properties of a few commercial standard plies. To minimize discrepancies between computations and reality, the glass loadings of the standard plies agree as much as possible with their "natural" or "most easily achievable" values. For example, the natural glass loading of UD plies is 70% by weight. Of course, this loading could be 65%, or 75%, but the most easily achievable average value is 70%. The plies of randomly distributed chopped fibers have a natural glass loading of 30% by weight. And so on. The fiber arrangement and the properties of the constituent materials determine the properties of the standard plies. The next section discusses the constituent materials.

The standard laminates are collections of standard plies firmly glued together. The standard and real laminates with the same construction – same number and arrangement of plies – have equivalent performances. The mechanical computations performed on standard laminates are valid for the real ones. This is a tremendous simplification, allowing the designer to perform calculations using the known properties of standard laminates, instead of the unknown properties of the real ones. The use of standard plies and laminates facilitates the communication between fabricators, inspectors and designers. The designer specifies the number and sequence of plies in a laminate, leaving thickness and glass content to the fabricator's discretion. The specification of the number and sequence of the plies – with no reference to laminate thickness and glass loading – is enough to ascertain the desired performance. The actual glass loading, laminate thickness and mechanical properties are irrelevant if the number, type and orientation of the plies are correct.

1.3   THE CONSTITUENT MATERIALS

The standard ply properties derive from the constituents listed in Table 1.2. The glass composition varies within a narrow range, which produces fibers of nearly equal properties. Unlike the fibers, the resin matrices may have widely different compositions and properties. However, the high fiber stiffness smooths out such differences, and we may assume the ply elastic

TABLE 1.2

Typical resin and fiber properties. The elongation at break of the resin affects mostly the failure thresholds and the cycle-dependent regression slopes.

Property	Glass fibers	Resin matrix
Elastic modulus (kg/cm²)	750 000	30 000
Poisson's ratio	0.20	0.30
Coefficient of thermal expansion (1/C)	5.5×10^{-6}	60.0×10^{-6}
Density (g/cm³)	2.62	1.20
Elongation at break (%)	--------	1.0 – 5.0
Water saturation (%)	0.0	0.3 – 1.2

properties independent of the resin. The only ply properties affected by the resin are (a) the water absorption Δm, (b) the failure thresholds and (c) the slope of the cycle-dependent regression equations. The type of resin does not affect the other standard ply properties, like the coefficients of thermal expansion, Poisson ratios, modulus of elasticity, etc.

1.4 STANDARD PLY PROPERTIES

As discussed, the properties of the standard plies – except the thresholds and the cycle-dependent slopes – are independent of the resin. This simplification facilitates the analysis, while having little effect on the computation accuracy. The resin effect will emerge in the Part II of this book, with the concepts of failure thresholds and regression lines. As an additional simplification, this book assumes the invariance of the standard ply properties with the operating temperature OT and moisture. These are reasonable assumptions, as long as the operating temperature OT is below the resin HDT and the water pick up Δm is small. The water saturation of commercial resins is indeed small (around 1.0%) and do not affect the ply properties much. For details, see Chapter 10.

As mentioned earlier, there are two ways to determine the standard ply properties. We will develop a few simple numerical examples to illustrate both methods. The derivation from the constituents is straightforward, using the micromechanical equations available in the published literature. The derivation from lab tests requires a simple correction discussed as a numerical example in Chapter 2.

Example 1.1

Derive the tensile modulus, the Poisson's ratio and the coefficient of thermal expansion of standard UD plies in the fiber direction 1.

Since all fibers are parallel and aligned in the principal ply direction 1, the tensile modulus derives from the simple law of mixtures

$$E_1 = \left(glass\right)_{vol} \times E_{glass} + \left(resin\right)_{vol} \times E_{resin} \qquad (1.1)$$

The glass and resin loadings by volume derive from the densities and standard weight fractions of the constituents. The standard weight fraction of glass in UD plies is 0.70. Table 1.2 lists the glass and resin densities.

$$\left(glass\right)_{vol} = \frac{\dfrac{0.70}{2.62}}{\dfrac{0.70}{2.62} + \dfrac{0.30}{1.20}} = 0.52$$

$$\left(resin\right)_{vol} = \frac{\dfrac{0.30}{1.20}}{\dfrac{0.70}{2.62} + \dfrac{0.30}{1.20}} = 0.48$$

Entering the above in Equation (1.1) we obtain

$$E_1 = 0.52 \times E_{glass} + 0.48 \times E_{resin}$$

Table 2 lists the elastic moduli of the glass and resin.

$$E_1 = 0.52 \times 750000 + 0.48 \times 30000 \approx 400000 \, kg/cm^2$$

The same reasoning applies to the Poisson's ratio

$$V_{12} = \left(glass\right)_{vol} \times V_{glass} + \left(resin\right)_{vol} \times V_{resin}$$

$$V_{12} = 0.52 \times 0.20 + 0.48 \times 0.30 = 0.25$$

The equation to compute the coefficient of thermal expansion is slightly more elaborate

$$\alpha_1 = \frac{\alpha_{glass} \times glass_{vol} \times E_{glass} + \alpha_{resin} \times resin_{vol} \times E_{resin}}{E_1}$$

$$\alpha_1 = \frac{5.5 \times 10^{-6} \times 0.52 \times 750000 + 60 \times 10^{-6} \times 0.48 \times 30000}{400000} = 7.5 \times 10^{-6}/^\circ C$$

Similar micromechanical equations exist to compute other ply properties.

Example 1.2

Derive the expressions linking the glass weight in kg/m² to the thicknesses of the standard ply in mm.

The applicable formulas are

$$\left[\frac{kg}{m^2}\right] = \left(glass\right)_{weight} \times area \times density \times thickness$$

$$density = \cfrac{1}{\cfrac{\left(glass\right)_{weight}}{2.62} + \cfrac{\left(resin\right)_{weight}}{1.20}}$$

For UD plies we have

$$\left(glass\right)_{weight} = 0.70$$

$$\left(resin\right)_{weight} = 0.30$$

Entering these in the above, we obtain

$$\left[\frac{kg}{m^2}\right]_{UD} = 0.70 \times 1 \times \cfrac{1}{\cfrac{0.70}{2.62} + \cfrac{0.30}{1.20}} \times t_{UD}$$

$$\left[\frac{kg}{m^2}\right]_{UD} = 1.5 \times t_{UD} \left(t_{UD} \text{ in mm}\right)$$

For plies of chopped glass we have

$$\left(glass\right)_{weight} = 0.30$$

$$\left(resin\right)_{weight} = 0.70$$

$$\left[\frac{kg}{m^2}\right]_{chop} = 0.30 \times 1 \times \cfrac{1}{\cfrac{0.30}{2.62} + \cfrac{0.70}{1.20}} \times t_{chop}$$

$$\left[\frac{kg}{m^2}\right]_{chop} = 0.43 \times t_{chop} \left(t \text{ in mm}\right)$$

For plies of sand we have

$$\left(sand\right)_{weight} = 0.80$$

$$\left(resin\right)_{weight} = 0.20$$

$$\left[\frac{kg}{m^2}\right]_{sand} = 0.80 \times 1 \times \frac{1}{\dfrac{0.80}{2.62} + \dfrac{0.20}{1.20}} \times t_{sand}$$

$$\left[\frac{kg}{m^2}\right]_{sand} = 1.69 \times t_{sand} \left(t \ in \ mm\right)$$

Table 1.3 summarizes these results

Example 1.3

Compute the standard thickness of a laminate with the following weight composition.
   UD fibers: 5.0 kg/m²; Chopped fibers: 1.2 kg/m²; Sand: 10.5 kg/m²

The laminate standard thickness is

$$t = t_{UD} + t_{chop} + t_{sand}$$

Using the relations from Table 1.3, we have

$$t = \frac{5.0}{1.35} + \frac{1.2}{0.43} + \frac{10.5}{1.69} = 12.71 \ \text{mm}$$

The above is the standard laminate thickness. The true thickness, measured on the actual laminate, is different.

TABLE 1.3
Glass weight and ply thickness for standard plies. The ply thicknesses are in mm.

Standard ply	Glass weight and ply thickness (mm)
70% by weight UD	$\left[\dfrac{kg}{m^2}\right]_{UD} = 1.35 \times t_{UD}$
30% by weight chopped fibers	$\left[\dfrac{kg}{m^2}\right]_{chop} = 0.43 \times t_{chop}$
80% by weight sand	$\left[\dfrac{kg}{m^2}\right]_{sand} = 1.69 \times t_{sand}$

Example 1.4

A thickness measurement prior to burnout of the laminate discussed in Example 1.3 revealed a thickness of 15.3 mm. This is 15.3–12.71 = 2.59 mm more than the standard value. Explain the discrepancy.

The excess thickness 15.3–12.71 = 2.59 mm corresponds to a fictitious ply of neat resin distributed throughout the actual laminate. The actual laminate is therefore equivalent to a standard laminate with the following standard ply construction.

- Standard UD ply of thickness $\dfrac{5.0}{1.35} = 3.70$ mm
- Standard chopped ply of thickness $\dfrac{1.2}{0.43} = 2.79$ mm
- Standard core ply of thickness $\dfrac{10.5}{1.69} = 6.21$ mm
- Ply of neat resin of thickness 2.59 mm.

The fictitious ply of neat resin is not really a ply. In fact, it spreads in the actual laminate.

1.5   THE STANDARD PLY MATRIX [Q]

The previous section illustrated the use of micromechanical equations to derive the engineering properties of standard plies. In this section, we will derive the standard ply matrices [Q] from those engineering properties. We do this because it is easier to work with ply matrices [Q] instead of engineering properties. The standard matrices [Q] tabulated in this chapter are very accurate.

The constitutive equation of plies under plane stress referred to the principal axes 1–2 is

$$
\begin{bmatrix} \sigma_1 \\ \sigma_2 \\ \tau_{12} \end{bmatrix} = \begin{bmatrix} Q_{11} & Q_{12} & 0 \\ Q_{21} & Q_{22} & 0 \\ 0 & 0 & Q_{66} \end{bmatrix} \times \begin{bmatrix} \varepsilon_1 \\ \varepsilon_2 \\ \gamma_{12} \end{bmatrix} \tag{1.2}
$$

The reader will note the ply constitutive Equation (1.2) presented in terms of stresses and moduli, like any homogeneous material. The concepts of stresses and moduli are valid for homogeneous plies. It is not possible to describe the non-homogeneous laminates, combining different plies, in terms of stresses and moduli.

The ply stiffness matrix [Q] in the local frame 1–2 is.

$$[Q]_{12} = \begin{bmatrix} Q_{11} & Q_{12} & 0 \\ Q_{21} & Q_{22} & 0 \\ 0 & 0 & Q_{66} \end{bmatrix} \text{kg/cm}^2$$

Terms of the engineering properties this

$$[Q]_{12} = \begin{bmatrix} \dfrac{E_1}{1-\upsilon_{12}\upsilon_{21}} & \dfrac{\upsilon_{21}E_1}{1-\upsilon_{12}\upsilon_{21}} & 0 \\ \dfrac{\upsilon_{12}E_2}{1-\upsilon_{12}\upsilon_{21}} & \dfrac{E_2}{1-\upsilon_{12}\upsilon_{21}} & 0 \\ 0 & 0 & G_{12} \end{bmatrix} \text{kg/cm}^2 \qquad (1.2A)$$

The following Tables 1.4 through 1.7 present the properties of a few select standard commercial plies. The data are valid for dry and wet plies operating below the resin HDT – heat distortion temperature. The next chapter discusses the laminates and their properties.

TABLE 1.4

The plies of neat resin are not critical for load-dependent failures and have no thresholds.

Coefficients of Thermal Expansion

$$\begin{bmatrix} \alpha_1 \\ \alpha_2 \\ \alpha_{12} \end{bmatrix} = \begin{bmatrix} 60.0 \\ 60.0 \\ 0 \end{bmatrix} \times 10^{-6}/^{0}C$$

Coefficients of Hydric Expansion

$$\begin{bmatrix} \beta_1 \\ \beta_2 \\ \beta_{12} \end{bmatrix} = \begin{bmatrix} 0.40 \\ 0.40 \\ 0 \end{bmatrix} 1/\Delta m$$

Modulus of elasticity

$$\begin{bmatrix} E_1 \\ E_2 \\ G_{12} \end{bmatrix} = \begin{bmatrix} 30000 \\ 30000 \\ 11500 \end{bmatrix} \text{kg/cm}^2$$

Poisson's ratio

$$\begin{bmatrix} v_{12} \\ v_{21} \end{bmatrix} = \begin{bmatrix} 0.30 \\ 0.30 \end{bmatrix}$$

Stiffness matrix

$$\begin{bmatrix} Q_{11} & Q_{12} & 0 \\ Q_{12} & Q_{22} & 0 \\ 0 & 0 & Q_{66} \end{bmatrix} = \begin{bmatrix} 33000 & 10000 & 0 \\ 10000 & 33000 & 0 \\ 0 & 0 & 11500 \end{bmatrix} \text{kg/cm}^2$$

Failure thresholds

$$\begin{bmatrix} T_i \\ T_w \\ T_s \\ T_r \\ T_0 \end{bmatrix} = \begin{bmatrix} — \\ — \\ — \\ — \\ — \end{bmatrix}$$

TABLE 1.5

Properties of standard chopped glass plies. The isotropic failure thresholds vary with the resin toughness.

Coefficients of thermal expansion

$$\begin{bmatrix} \alpha_1 \\ \alpha_2 \\ \alpha_{12} \end{bmatrix} = \begin{bmatrix} 25.0 \\ 25.0 \\ 0 \end{bmatrix} \times 10^{-6} / \,^{\circ}C$$

Coefficients of hydric expansion

$$\begin{bmatrix} \beta_1 \\ \beta_2 \\ \beta_{12} \end{bmatrix} = \begin{bmatrix} 0.25 \\ 0.25 \\ 0 \end{bmatrix} 1/\Delta m$$

Modulus of elasticity

$$\begin{bmatrix} E_1 \\ E_2 \\ G_{12} \end{bmatrix} = \begin{bmatrix} 70000 \\ 70000 \\ 27000 \end{bmatrix} kg/cm^2$$

Poisson's ratio

$$\begin{bmatrix} v_{12} \\ v_{21} \end{bmatrix} = \begin{bmatrix} 0.25 \\ 0.25 \end{bmatrix}$$

Stiffness matrix

$$\begin{bmatrix} Q_{11} & Q_{12} & 0 \\ Q_{12} & Q_{22} & 0 \\ 0 & 0 & Q_{66} \end{bmatrix} = \begin{bmatrix} 77,000 & 23,000 & 0 \\ 23,000 & 77,000 & 0 \\ 0 & 0 & 27,000 \end{bmatrix} kg/cm^2$$

Failure thresholds

$$\begin{bmatrix} T_i \\ T_w \\ T_s \\ T_r \\ T_0 \end{bmatrix} = \begin{bmatrix} 0.30\% - 0.50\% \\ 0.80\% - 1.00\% \\ - \\ 1.50\% - 2.50\% \\ 0.05\% \end{bmatrix}$$

TABLE 1.6

Properties of standard UD plies. The transverse failure thresholds vary with the resin toughness.

Coefficients of thermal expansion

$$\begin{bmatrix} \alpha_1 \\ \alpha_2 \\ \alpha_{12} \end{bmatrix} = \begin{bmatrix} 7.5 \\ 45.0 \\ 0 \end{bmatrix} \times 10^{-6} / \,^{\circ}C$$

Coefficients of hydric expansion

$$\begin{bmatrix} \beta_1 \\ \beta_2 \\ \beta_{12} \end{bmatrix} = \begin{bmatrix} 0.04 \\ 0.24 \\ 0 \end{bmatrix} 1/\Delta m$$

Modulus of elasticity

$$\begin{bmatrix} E_1 \\ E_2 \\ G_{12} \end{bmatrix} = \begin{bmatrix} 400000 \\ 100000 \\ 35000 \end{bmatrix} kg/cm^2$$

Poisson's ratio

$$\begin{bmatrix} v_{12} \\ v_{21} \end{bmatrix} = \begin{bmatrix} 0.25 \\ 0.07 \end{bmatrix}$$

Stiffness matrix

$$\begin{bmatrix} Q_{11} & Q_{12} & 0 \\ Q_{12} & Q_{22} & 0 \\ 0 & 0 & Q_{66} \end{bmatrix} = \begin{bmatrix} 400000 & 30000 & 0 \\ 30000 & 100000 & 0 \\ 0 & 0 & 35000 \end{bmatrix} kg/cm^2$$

Thresholds

$$\begin{bmatrix} T_i \\ T_w \\ T_s \\ T_r \\ T_0 \end{bmatrix} = \begin{bmatrix} 0.20\% - 0.30\% \\ 0.25\% - 0.40\% \\ 0.40\% - 0.60\% \\ 0.40\% - 0.60\% \\ 0.05\% \end{bmatrix}$$

TABLE 1.7

Properties of standard sand filled plies. The sand plies are not critical for load-dependent failures and have no failure thresholds.

Coefficients of thermal expansion

$$
\begin{bmatrix} \alpha_1 \\ \alpha_2 \\ \alpha_{12} \end{bmatrix} = \begin{bmatrix} 13.0 \\ 13.0 \\ 0 \end{bmatrix} \times 10^{-6} \ / \ ^{\circ}C
$$

Coefficients of hydric expansion

$$
\begin{bmatrix} \beta_1 \\ \beta_2 \\ \beta_{12} \end{bmatrix} = \begin{bmatrix} 0.28 \\ 0.28 \\ 0 \end{bmatrix} 1/\Delta m
$$

Modulus of elasticity

$$
\begin{bmatrix} E_1 \\ E_2 \\ G_{12} \end{bmatrix} = \begin{bmatrix} 60000 \\ 60000 \\ 23000 \end{bmatrix} kg/cm^2
$$

Poisson's ratio

$$
\begin{bmatrix} \nu_{12} \\ \nu_{21} \end{bmatrix} = \begin{bmatrix} 0.25 \\ 0.25 \end{bmatrix}
$$

Stiffness matrix

$$
\begin{bmatrix} Q_{11} & Q_{12} & 0 \\ Q_{12} & Q_{22} & 0 \\ 0 & 0 & Q_{66} \end{bmatrix} = \begin{bmatrix} 66000 & 20000 & 0 \\ 20000 & 66000 & 0 \\ 0 & 0 & 23000 \end{bmatrix} kg/cm^2
$$

Thresholds

$$
\begin{bmatrix} T_i \\ T_w \\ T_s \\ T_r \\ T_0 \end{bmatrix} = \begin{bmatrix} \text{Not applicable} \\ \text{Not applicable} \\ \text{Not applicable} \\ \text{Not applicable} \\ \text{Not applicable} \end{bmatrix}
$$

Example 1.5

Compute the stresses in the principal directions of standard UD plies subjected to the following strains.

$$
\begin{bmatrix} \varepsilon_1 \\ \varepsilon_2 \\ \gamma_{12} \end{bmatrix} = \begin{bmatrix} 0.015 \\ 0.002 \\ 0 \end{bmatrix}
$$

The ply stresses derive from the constitutive Equation (1.2)

$$
\begin{bmatrix} \sigma_1 \\ \sigma_2 \\ \tau_{12} \end{bmatrix} = \begin{bmatrix} Q_{11} & Q_{12} & 0 \\ Q_{21} & Q_{22} & 0 \\ 0 & 0 & Q_{66} \end{bmatrix} \times \begin{bmatrix} \varepsilon_1 \\ \varepsilon_2 \\ \gamma_{12} \end{bmatrix} kg/cm^2
$$

Table 1.6 lists the standard stiffness matrix [Q] of UD plies

$$
\begin{bmatrix} \sigma_1 \\ \sigma_2 \\ \tau_{12} \end{bmatrix} = \begin{bmatrix} 400000 & 30000 & 0 \\ 30000 & 100000 & 0 \\ 0 & 0 & 35000 \end{bmatrix} \times \begin{bmatrix} 0.015 \\ 0.002 \\ 0 \end{bmatrix}
$$

Developing the above, we have

$$\begin{bmatrix} \sigma_1 \\ \sigma_2 \\ \tau_{12} \end{bmatrix} = \begin{bmatrix} 6060 \\ 650 \\ 0 \end{bmatrix} \frac{kg}{cm^2}$$

Example 1.6

Use Equation (1.2A) to derive the stiffness matrix [Q] of UD plies.

$$[\mathcal{Q}]_{12} = \begin{bmatrix} \dfrac{E_1}{1 - v_{12}v_{21}} & \dfrac{v_{21}E_1}{1 - v_{12}v_{21}} & 0 \\ \dfrac{v_{12}E_2}{1 - v_{12}v_{21}} & \dfrac{E_2}{1 - v_{12}v_{21}} & 0 \\ 0 & 0 & G_{12} \end{bmatrix}$$

Entering the data from Table 1.6 in Equation (1.2A), we have

$$[\mathcal{Q}]_{12} = \begin{bmatrix} \dfrac{400000}{1 - 0.07 \times 0.25} & \dfrac{0.07 \times 400000}{1 - 0.07 \times 0.25} & 0 \\ \dfrac{0.30 \times 100000}{1 - 0.07 \times 0.25} & \dfrac{100000}{1 - 0.07 \times 0.25} & 0 \\ 0 & 0 & 35000 \end{bmatrix} kg/cm^2$$

$$[\mathcal{Q}]_{12} = \begin{bmatrix} 400000 & 30000 & 0 \\ 30000 & 100000 & 0 \\ 0 & 0 & 35000 \end{bmatrix} kg/cm^2$$

1.6   THE IMPORTANCE OF STRAINS

The total strains in the principal directions of the critical ply are essential to predict the load-dependent durability of laminates. The accurate knowledge of the total strains is so essential that we have divided this book in two parts to give this topic the attention it deserves. The first part introduces a simple and accurate method to compute the membrane strains of plies embedded in circular cylindrical laminates. The second part, starting in Chapter 8, deals with the durability issue. The reader familiar with strain computations may skip the next 6 chapters and go directly to the second part of the book, starting in Chapter 8.

Figure 1.2 shows a laminate and the global axes x and y. It is usual in the composites industry to take the x axis in the longitudinal direction of cylindrical laminates. The y axis is taken in the hoop, or circumferential, direction. When embedded in laminates, the UD plies can be laid up to take any orientation with respect to the global reference frame x − y. The angle "α" between the axis 1 and the global axis "x" defines the ply orientation. The distance "z" from the ply center of gravity to the laminate mid-surface defines the position. The distance "z" and the angle "α" completely define the position and orientation of embedded UD plies.

The standard laminate properties derive from those of the standard plies and are not the same as those of the actual laminate. Some manufacturers use too much resin and produce laminates thicker than the standard. Others do the opposite and produce thinner laminates. This discrepancy is not relevant, as long as the number and orientation of plies are those specified by the designer. Except for the very thin resin-rich liners in pipes and tanks for chemical service, the commercial laminates have no ply of neat resin. However, in spite of this, we have tabulated the matrices of plies of neat resin. Such matrices are useful to convert the standard laminate properties into the real properties of actual laminates. We will have more to say about this in the following chapters.

We conclude this chapter with a brief mention of the effects of the resin toughness and the operating conditions on the ply matrices.

From Table 1.8 we see that changes in resin toughness, operating temperature and moisture have no effect on the ply stiffness and coefficients of expansion. This is a tremendous design simplification. The failure thresholds, however, increase with increments in the resin toughness, as expected. Table 1.8 indicates that the operating temperature and moisture have no effect on the

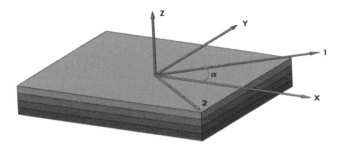

FIGURE 1.2   *The distance "z" from the laminate mid-surface defines the ply position. The angle "α" between the global "x" and the local "1" axes defines the ply orientation. The commercial laminates may combine several plies in any position and orientation. All commercial laminates are balanced and orthotropic.*

TABLE 1.8

The operating conditions have no effect on the ply properties. The resin toughness affects the failure thresholds and the regression slopes.

Ply property	Resin toughness	Operating temperature	Moisture
Stiffness	No effect	No effect	No effect
Termal expansion	No effect	No effect	No effect
Hydric expansion	No effect	No effect	No effect
Failure thresholds	Increases	No effect	No effect

failure thresholds, which is another important simplification. For details, see Chapter 10. The regression slopes also vary with the resin toughness. The following comments emphasize the importance of the failure thresholds.

- The infiltration threshold is a fundamental parameter in the design of corrosion barriers for chemical service. The allowable strains mentioned in product standards derive, mostly by intuition, from the infiltration threshold.
- The weep threshold is the most important design parameter for composite pipes and pressure vessels carrying non-aggressive fluids. The design criteria, qualification and quality control tests of all vessels, pipes and fittings complying with the standards AWWA C950, API 15 HR, ISO 14692 and ASME section X are for the most part based on the weep threshold.
- The stiffness threshold is the dominant long-term mode of failure for laminates in important structural service like wind blades and aircraft parts.

The Part II of this book carries a thorough discussion of the failure thresholds and the regression slopes.

2 Laminate Circularity

2.1 INTRODUCTION

The previous chapter presented the relevant properties of the most important commercial standard plies. The commercial laminates consist of stacks of one or more of those plies, each with their own set of properties. When embedded in laminates, the perfectly bonded plies deform without slipping, with the strains determined by their position and orientation in the laminate. The second part of this book will emphasize the fundamental importance of the ply strains in the study of load-dependent laminate durability.

Most industrial applications of composites involve thin circular cylindrical laminates under internal pressure. The circular cylinders subjected to internal pressure preserve their circular shape and all plies develop the same strain, regardless of their position in the laminate. The stresses, however, vary from ply to ply according to their stiffness and orientation. This prevents the laminate analysis in terms of stresses. The laminate constitutive equations make use of force and moment resultants in place of stresses.

In applications violating the circularity condition, as in underground pipes deflected diametrically by vertical loads, the plies develop bending strains that vary with their distance "z" from the laminate mid surface. The important bending strains are ignored in this book, since their presence complicates the analysis and throws no light on the subject of durability. This book deals only with cylinders in which the circularity is preserved. The preservation of circularity allows the analysis of asymmetrical laminates as if they were symmetrical, which is an important and welcome simplification.

There are many ways to embed plies in laminates. They can be placed near the laminate inner or outer surfaces, or near the mid surface, with their local axes 1–2 taking any orientation with respect to the global frame x – y. The laminate stiffness matrices are obtained from those of the plies, taking into account their orientation, thickness and position. Equation (2.1) shows the constitutive equation of laminates.

$$
\begin{bmatrix} N_x \\ N_y \\ N_{xy} \\ M_x \\ M_y \\ M_{xy} \end{bmatrix} =
\begin{bmatrix}
A_{xx} & A_{xy} & A_{xs} & B_{xx} & B_{xy} & B_{xs} \\
A_{yx} & A_{yy} & A_{ys} & B_{yx} & B_{yy} & B_{ys} \\
A_{sx} & A_{sy} & A_{ss} & B_{sx} & B_{sy} & B_{ss} \\
B_{xx} & B_{xy} & B_{xs} & D_{xx} & D_{xy} & D_{xs} \\
B_{yx} & B_{yy} & B_{ys} & D_{yx} & D_{yy} & D_{ys} \\
B_{sx} & B_{sy} & B_{ss} & D_{sx} & D_{sy} & D_{ss}
\end{bmatrix} \times
\begin{bmatrix} \varepsilon_x^0 \\ \varepsilon_y^0 \\ \gamma_{xy}^0 \\ \kappa_x \\ \kappa_y \\ \kappa_{xy} \end{bmatrix}
\tag{2.1}
$$

Compared to the ply constitutive equation discussed in Chapter 1. We see that the laminate Equation (2.1) uses force [N] and moment [M] resultants in place of stresses. Furthermore, the laminate matrices [A], [B] and [D] replace the ply matrix [Q]. The matrix [$\mathbb{C}^0$] represents the membrane strains of the laminate mid-surface. Likewise, the matrix [k] represents the changes of curvature of the mid-surface.

A full appreciation of Equation (2.1) requires familiarity with the fundamentals of classical laminate theory. The readers not familiar with this theory can refer to any of the many textbooks available in the literature. See ref. 1 and 2. We will not dwell on these fundamental details. We proceed directly with our discussions, lest we lose sight of the object of this book, which is to evaluate the durability of composites.

Equation (2.1) is general and holds for any laminate and loading. The designer knows the external force and moment resultants [N] and [M], as well as the laminate matrices [A], [B] and [D]. Equation (2.1) computes the unknown strains and changes in curvature. This book is not about laminate matrices and force resultants. These topics are touched upon just briefly, enough to allow the computation of the total ply strains. Our goal is the study of the durability of composites, not their technology.

The Part I of this book computes the total strains of plies embedded in circular laminates that maintain their circularity. The computation protocol is easy to understand, simple to use and very accurate. Its drawback is the exclusive and limited applicability to circular cylindrical laminates. In spite of this limitation, however, the model herein presented is of great help in conveying the essential ideas of total strain computation.

2.2 THERMAL LOADS

The embedded plies are not free to expand or contract when subjected to changes in temperature. Instead, they take the equilibrium position imposed by the laminate. At the equilibrium position, some of the plies are in tension and others in compression. The residual strains arise from the plies forced to comply with the laminate deformation, in spite of their own preferences. The force resultant obtained by adding the residual thermal forces in all plies is zero. The thermal resultants, however, are not zero. We will return to this point shortly.

The heat released in the process of resin cure raises the laminate temperature to a peak value PT. On reaching this peak temperature, the resin is still liquid and capable of flowing to eliminate thermal strains. The resin crosslinks at the peak temperature PT in a stress-free condition. The crosslinking congeals the plies and fix their positions in the laminate, which forces their subsequent expansion/contraction as a single body. We refer to the stress-free temperature as PT. The ply residual thermal strains develop as the laminate temperature changes from PT to the operating value OT. The laminate may shrink or expand in passing from PT to OT. If PT is higher than OT, it cools off and shrinks. Otherwise, it heats up and expands. For example, a room temperature cured laminate with a low stress-free temperature PT = 50C will expand when operating at OT = 80°C, and contract if operating at 20°C.

The laminate thermal equilibrium position stabilizes at the operating temperature OT. Equation (2.1) computes the mid-surface strains and the changes in curvature at this position with the thermal resultants defined as

$$\left[N^T \right] = \left[A \right] \times \left[\alpha \right] \times (OT - PT)$$

$$\left[M^T \right] = \left[B \right] \times \left[\alpha \right] \times (OT - PT)$$

Expanding the above, we have

$$
\begin{bmatrix} N_x^T \\ N_y^T \\ N_{xy}^T \end{bmatrix} = \begin{bmatrix} \sum \left(Q_{xx} t \alpha_x + Q_{xy} t \alpha_y + Q_{xs} t \alpha_{xy} \right) \\ \sum \left(Q_{yx} t \alpha_x + Q_{yy} t \alpha_y + Q_{ys} t \alpha_{xy} \right) \\ \sum \left(Q_{sx} t \alpha_x + Q_{sy} t \alpha_y + Q_{ss} t \alpha_{xy} \right) \end{bmatrix} \times (OT - PT)\ \text{kg/cm} \qquad (2.2)
$$

$$
\begin{bmatrix} M_x^T \\ M_y^T \\ M_{xy}^T \end{bmatrix} = \begin{bmatrix} \sum \left(Q_{xx} tz \alpha_x + Q_{xy} tz \alpha_y + Q_{xs} tz \alpha_{xy} \right) \\ \sum \left(Q_{yx} tz \alpha_x + Q_{yy} tz \alpha_y + Q_{ys} tz \alpha_{xy} \right) \\ \sum \left(Q_{sx} tz \alpha_x + Q_{sy} tz \alpha_y + Q_{ss} tz \alpha_{xy} \right) \end{bmatrix} \times (OT - PT)\ \text{kg·cm/cm} \qquad (2.3)
$$

The temperatures PT and OT are the same for all plies, since the thicknesses of commercial laminates are small. We ignore the temperature gradient across the thickness. In Equations (2.2) and (2.3) the "alphas" represent the coefficients of thermal expansion of each ply in the global system x − y, and "z" is the ply distance from the mid-surface. The thermal resultants depend on the ply properties and their thickness, position and orientation in the laminate. The temperature change takes as reference the stress-free peak temperature PT. This is an important detail. The reference temperature to compute the thermal resultants is the stress-free peak temperature PT, not the room temperature RT.

The mid surface strains and the changes in curvature at equilibrium are computed entering the thermal loadings from Equations (2.2) and (2.3) in the general Equation (2.1).

$$
\begin{bmatrix} N_x^T \\ N_y^T \\ N_{xy}^T \\ M_x^T \\ M_y^T \\ M_{xy}^T \end{bmatrix}
=
\begin{bmatrix}
A_{xx} & A_{xy} & A_{xs} & B_{xx} & B_{xy} & B_{xs} \\
A_{yx} & A_{yy} & A_{ys} & B_{yx} & B_{yy} & B_{ys} \\
A_{sx} & A_{sy} & A_{ss} & B_{sx} & B_{sy} & B_{ss} \\
B_{xx} & B_{xy} & B_{xs} & D_{xx} & D_{xy} & D_{xs} \\
B_{yx} & B_{yy} & B_{ys} & D_{yx} & D_{yy} & D_{ys} \\
B_{sx} & B_{sy} & B_{ss} & D_{sx} & D_{sy} & D_{ss}
\end{bmatrix}
\times
\begin{bmatrix} \varepsilon_x^e \\ \varepsilon_y^e \\ \gamma_{xy}^e \\ \kappa_x^e \\ \kappa_y^e \\ \kappa_{xy}^e \end{bmatrix}
\qquad (2.4)
$$

The reader will note the superscript "e" designating "equilibrium" in the strain and curvature matrices. The designer knows the laminate matrices [A], [B] and [D], as well as the thermal resultants. The laminate strains and changes in curvature at thermal equilibrium come from inverting Equation (2.4).

$$
\begin{bmatrix} \varepsilon_x^e \\ \varepsilon_y^e \\ \gamma_{xy}^e \\ \kappa_x^e \\ \kappa_y^e \\ \kappa_{xy}^e \end{bmatrix}
=
\begin{bmatrix}
A_{xx} & A_{xy} & A_{xs} & B_{xx} & B_{xy} & B_{xs} \\
A_{yx} & A_{yy} & A_{ys} & B_{yx} & B_{yy} & B_{ys} \\
A_{sx} & A_{sy} & A_{ss} & B_{sx} & B_{sy} & B_{ss} \\
B_{xx} & B_{xy} & B_{xs} & D_{xx} & D_{xy} & D_{xs} \\
B_{yx} & B_{yy} & B_{ys} & D_{yx} & D_{yy} & D_{ys} \\
B_{sx} & B_{sy} & B_{ss} & D_{sx} & D_{sy} & D_{ss}
\end{bmatrix}^{-1}
\times
\begin{bmatrix} N_x^T \\ N_y^T \\ N_{xy}^T \\ M_x^T \\ M_y^T \\ M_{xy}^T \end{bmatrix}
\qquad (2.5)
$$

Equation (2.5) computes the equilibrium strains and changes in curvature of the mid-surface of any laminate subjected to a temperature change from PT to OT. If OT is lower than PT, the thermal forces and moments are negative. Otherwise, they are positive.

We next simplify Equation (2.5) for the special case of circular cylindrical laminates.

2.3 THERMAL LOADS AND LAMINATE CIRCULARITY

Equations (2.2), (2.3) and (2.4) are applicable to any laminate and thermal loading. They compute the laminate warping and distortion that occur in response to changes in temperature. However, the cylindrical laminates of circular cross sections preserve their round shape and do not warp. They maintain their curvatures at any temperature. Mathematically, we describe the preservation of circularity as

$$
\begin{bmatrix} \kappa_x \\ \kappa_y \\ \kappa_z \end{bmatrix} = \begin{bmatrix} 0 \\ 0 \\ 0 \end{bmatrix} \tag{2.6}
$$

The preservation of circularity develops internal reactive moments computed by entering Equation (2.6) in the general Equation (2.4).

$$
\begin{bmatrix} N_x^T \\ N_y^T \\ N_{xy}^T \\ M_x^T + M_x^R \\ M_y^T + M_y^R \\ M_{xy}^T + M_{xy}^R \end{bmatrix} = \begin{bmatrix} A_{xx} & A_{xy} & A_{xs} & B_{xx} & B_{xy} & B_{xs} \\ A_{yx} & A_{yy} & A_{ys} & B_{yx} & B_{yy} & B_{ys} \\ A_{sx} & A_{sy} & A_{ss} & B_{sx} & B_{sy} & B_{ss} \\ B_{xx} & B_{xy} & B_{xs} & D_{xx} & D_{xy} & D_{xs} \\ B_{yx} & B_{yy} & B_{ys} & D_{yx} & D_{yy} & D_{ys} \\ B_{sx} & B_{sy} & B_{ss} & D_{sx} & D_{sy} & D_{ss} \end{bmatrix} \times \begin{bmatrix} \varepsilon_x^e \\ \varepsilon_y^e \\ \gamma_{xy}^e \\ 0 \\ 0 \\ 0 \end{bmatrix} \tag{2.7}
$$

In the above, we identify the reactive moments with the superscript "R". The inclusion of the reactive moments in Equation (2.7) eliminates the changes in curvature and assures the preservation of circularity. The thermal reactive moments develop every time cylindrical laminates of circular cross section undergo temperature changes.

In the laminate manufacturing process, the plies congeal in a stress-free condition at the peak temperature PT. As the temperature changes from PT to OT, the cylinder takes an equilibrium position while preserving its circular cross section. The preservation of circularity gives rise to the reactive moments just described and decouples the force and moment resultants, as seen by expanding Equation (2.7)

$$
\begin{bmatrix} N_x^T \\ N_y^T \\ N_{xy}^T \end{bmatrix} = \begin{bmatrix} A_{xx} & A_{xy} & A_{xs} \\ A_{yx} & A_{yy} & A_{ys} \\ A_{sx} & A_{sy} & A_{ss} \end{bmatrix} \begin{bmatrix} \varepsilon_x^e \\ \varepsilon_y^e \\ \gamma_{xy}^e \end{bmatrix} \tag{2.8}
$$

$$
\begin{bmatrix} M_x^T + M_x^R \\ M_y^T + M_y^R \\ M_{xy}^T + M_{xy}^R \end{bmatrix} = \begin{bmatrix} B_{xx} & B_{xy} & B_{xs} \\ B_{yx} & B_{yy} & B_{ys} \\ B_{sx} & B_{sy} & B_{ss} \end{bmatrix} \begin{bmatrix} \varepsilon_x^e \\ \varepsilon_y^e \\ \gamma_{xy}^e \end{bmatrix} \tag{2.9}
$$

$$
\begin{bmatrix} M_x^R \\ M_y^R \\ M_{xy}^R \end{bmatrix} = \begin{bmatrix} B_{xx} & B_{xy} & B_{xs} \\ B_{yx} & B_{yy} & B_{ys} \\ B_{sx} & B_{sy} & B_{ss} \end{bmatrix} \begin{bmatrix} \varepsilon_x^e \\ \varepsilon_y^e \\ \gamma_{xy}^e \end{bmatrix} - \begin{bmatrix} M_x^T \\ M_y^T \\ M_{xy}^T \end{bmatrix} \tag{2.9A}
$$

In the special cases of symmetrical laminates, the coupling matrix [B] = 0 and the reactive moments are

$$
\begin{bmatrix} M_x^R \\ M_y^R \\ M_{xy}^R \end{bmatrix} = - \begin{bmatrix} M_x^T \\ M_y^T \\ M_{xy}^T \end{bmatrix} \tag{2.9B}
$$

Equation (2.8) computes the laminate equilibrium strains using the matrix [A] alone, regardless of the matrix [B]. This is a very important simplification. The preservation of circularity allows the analysis of asymmetric laminates as if they were symmetric, i.e., by ignoring the matrix [B]. Equation (2.9A) indicates the reactive moments computed from the thermal moment resultants, the coupling matrix [B] and the membrane strains.

2.4 HYDRIC LOADS

The foregoing analysis is also applicable to hydric loadings. The only difference is the laminate expansion from absorbing water, instead of contraction from cooling down. In both cases, the laminate constraint prevent the embedded plies from taking their natural dimensions. In the thermal case, the plies cannot freely contract from PT to OT. In the hydric case, they cannot freely expand when saturated with water. The computation of the hydric forces and moment resultants is similar to the thermal case

$$
\begin{bmatrix} N_x^H \\ N_y^H \\ N_{xy}^H \end{bmatrix} = \begin{bmatrix} \sum\left(Q_{xx}t\beta_x + Q_{xy}t\beta_y + Q_{xs}t\beta_{xy}\right) \\ \sum\left(Q_{yx}t\beta_x + Q_{yy}t\beta_y + Q_{ys}t\beta_{xy}\right) \\ \sum\left(Q_{sx}t\beta_x + Q_{sy}t\beta_y + Q_{ss}t\beta_{xy}\right) \end{bmatrix} \times (\Delta m)\,\text{kg/cm} \tag{2.10}
$$

$$
\begin{bmatrix} M_x^H \\ M_y^H \\ M_{xy}^H \end{bmatrix} = \begin{bmatrix} \sum\left(Q_{xx}tz\beta_x + Q_{xy}tz\beta_y + Q_{xs}tz\beta_{xy}\right) \\ \sum\left(Q_{yx}tz\beta_x + Q_{yy}tz\beta_y + Q_{ys}tz\beta_{xy}\right) \\ \sum\left(Q_{sx}tz\beta_x + Q_{sy}tz\beta_y + Q_{ss}tz\beta_{xy}\right) \end{bmatrix} \times (\Delta m)\,\text{kg}\cdot\text{cm/cm} \tag{2.10A}
$$

Equations (2.10) and (2.10A) show the hydric resultants computed in the same way as in the thermal case. Note the water (or solvent) pick up – Δm – introduced in place of the temperature change $\Delta T = OT - PT$. Furthermore, the hydric expansion coefficients "betas" replace the thermal coefficients "alphas". The amount of water picked up by the resin, Δm, is assumed the same for all plies. As in the thermal case, the hydric gradient is zero. Developing the hydric equations as we did in the thermal case, we have

$$
\begin{bmatrix} N_x^H \\ N_y^H \\ N_{xy}^H \end{bmatrix} = \begin{bmatrix} A_{xx} & A_{xy} & A_{xs} \\ A_{yx} & A_{yy} & A_{ys} \\ A_{sx} & A_{sy} & A_{ss} \end{bmatrix} \begin{bmatrix} \varepsilon_x^e \\ \varepsilon_y^e \\ \gamma_{xy}^e \end{bmatrix}_{HID} \tag{2.11}
$$

$$
\begin{bmatrix} M_x^H + M_x^R \\ M_y^H + M_y^R \\ M_{xy}^H + M_{xy}^R \end{bmatrix} = \begin{bmatrix} B_{xx} & B_{xy} & B_{xs} \\ B_{yx} & B_{yy} & B_{ys} \\ B_{sx} & B_{sy} & B_{ss} \end{bmatrix} \begin{bmatrix} \varepsilon_x^e \\ \varepsilon_y^e \\ \gamma_{xy}^e \end{bmatrix}_{HID} \tag{2.12}
$$

$$
\begin{bmatrix} M_x^R \\ M_y^R \\ M_{xy}^R \end{bmatrix}_{HID} = \begin{bmatrix} B_{xx} & B_{xy} & B_{xs} \\ B_{yx} & B_{yy} & B_{ys} \\ B_{sx} & B_{sy} & B_{ss} \end{bmatrix} \begin{bmatrix} \varepsilon_x^e \\ \varepsilon_y^e \\ \gamma_{xy}^e \end{bmatrix}_{HID} - \begin{bmatrix} M_x^H \\ M_y^H \\ M_{xy}^H \end{bmatrix} \tag{2.12A}
$$

In symmetric laminates, the coupling matrix [B] = 0 and the reactive hydric moments are

$$
\begin{bmatrix} M_x^R \\ M_y^R \\ M_{xy}^R \end{bmatrix}_{HID} = - \begin{bmatrix} M_x^H \\ M_y^H \\ M_{xy}^H \end{bmatrix}_{HID}
\tag{2.12B}
$$

As in the thermal case, the preservation of the circular shape allows the computation of the laminate equilibrium hydric strains from Equation (2.11), ignoring the coupling matrix [B].

2.5   MECHANICAL LOADS

The constitutive equation of circular cylindrical laminates under mechanical loads comes from the general Equation (2.1) and the condition of circularity. The analysis is almost identical as in the thermal and hydric cases. The difference lies in the arbitrary nature of the external forces, moments and torques. The external mechanical load resultants are independent of the laminate construction. By contrast, the thermal and hydric resultants derive from the laminate construction.

The mechanical force resultants N act on the laminate mid-surface to generate membrane strains. The moment resultants M give rise to laminate bending.

2.6   MEMBRANE FORCES ON CIRCULAR CYLINDERS

The external mechanical forces acting on the circular cylinder are the axial resultant Nx, the circumferential resultant Ny and the torque resultant Nxy. The membrane forces preserve the circular shape. The laminate curvature does not change under membrane loads and the general Equation (2.1) reduces to

$$
\begin{bmatrix} N_x^M \\ N_y^M \\ N_{xy}^M \\ 0+M_x^R \\ 0+M_y^R \\ 0+M_{xy}^R \end{bmatrix}_{MEC} = \begin{bmatrix} A_{xx} & A_{xy} & A_{xs} & B_{xx} & B_{xy} & B_{xs} \\ A_{yx} & A_{yy} & A_{ys} & B_{yx} & B_{yy} & B_{ys} \\ A_{sx} & A_{sy} & A_{ss} & B_{sx} & B_{sy} & B_{ss} \\ B_{xx} & B_{xy} & B_{xs} & D_{xx} & D_{xy} & D_{xs} \\ B_{yx} & B_{yy} & B_{ys} & D_{yx} & D_{yy} & D_{ys} \\ B_{sx} & B_{sy} & B_{ss} & D_{sx} & D_{sy} & D_{ss} \end{bmatrix} \times \begin{bmatrix} \varepsilon_x^e \\ \varepsilon_y^e \\ \gamma_{xy}^e \\ 0 \\ 0 \\ 0 \end{bmatrix}_{MEC}
\tag{2.13}
$$

Equation (2.13) emphasizes the absence of external bending moments by writing "zeros" in their places. It includes the reactive moments required to preserve the circular shape, as in the hydric and thermal cases. Expansion of Equation (2.13) gives the mechanical strain components at the equilibrium position

$$
\begin{bmatrix} N_x^M \\ N_y^M \\ N_{xy}^M \end{bmatrix} = \begin{bmatrix} A_{xx} & A_{xy} & A_{xs} \\ A_{yx} & A_{yy} & A_{ys} \\ A_{sx} & A_{sy} & A_{ss} \end{bmatrix} \times \begin{bmatrix} \varepsilon_x^e \\ \varepsilon_y^e \\ \gamma_{xy}^e \end{bmatrix}_{MEC}
\tag{2.14}
$$

The reactive mechanical moments are:

$$
\begin{bmatrix} M_x^R \\ M_y^R \\ M_{xy}^R \end{bmatrix}_{MEC} = \begin{bmatrix} B_{xx} & B_{xy} & B_{xs} \\ B_{yx} & B_{yy} & B_{ys} \\ B_{sx} & B_{sy} & B_{ss} \end{bmatrix} \times \begin{bmatrix} \varepsilon_x^e \\ \varepsilon_y^e \\ \gamma_{xy}^e \end{bmatrix}_{MEC}
\tag{2.15}
$$

The preservation of circularity generate reactive moments that, like all reactive quantities, have the exact magnitudes to neutralize their cause. In symmetric laminates, the coupling matrix [B] = 0, and the reactive mechanical moments are zero.

$$
\begin{bmatrix} M_x^R \\ M_y^R \\ M_{xy}^R \end{bmatrix}_{MEC} = \begin{bmatrix} 0 \\ 0 \\ 0 \end{bmatrix}
$$

The preceding analysis produces two interesting results.

- First, the preservation of circularity eliminates the coupling matrix [B] from the computation of the membrane strains. This is a welcome simplification.
- Second, the membrane strains are independent of the reactive bending moments.

These apparently surprising results are in fact obvious when given some thought. If the circular shape is preserved, the coupling matrix [B] cannot affect the membrane strains. The matrix [B] is not relevant in the analysis of circular cylinders that maintain circularity. Furthermore, the preservation of circularity automatically takes care of the reactive moments.

2.7 CIRCULAR CYLINDERS DEFLECTED BY LATERAL LOADS

The foregoing analysis is valid for circular cylinders subjected to loads that preserve the circular shape. There are loads, however, that change the cylinder curvature. Examples of such loads are the vertical soil weight on underground pipes and the bending of the knuckle region of vertical storage tanks. The analysis of such cases require the general Equation (2.1).

The matrix [B] couples the force and moment resultants in Equation (2.1). This coupling is a big complicatoe and is the reason why, in engineering practice, the laminates are symmetrical or quasi-symmetrical. In fully symmetrical laminates, the coupling matrix [B] = 0 and Equation (2.1) reduces to

$$
\begin{bmatrix} N_x^M \\ N_y^M \\ N_{xy}^M \end{bmatrix} = \begin{bmatrix} A_{xx} & A_{xy} & A_{xs} \\ A_{yx} & A_{yy} & A_{ys} \\ A_{sx} & A_{sy} & A_{ss} \end{bmatrix} \times \begin{bmatrix} \varepsilon_x^e \\ \varepsilon_y^e \\ \gamma_{xy}^e \end{bmatrix}
$$

$$
\begin{bmatrix} M_x^M \\ M_y^M \\ M_{xy}^M \end{bmatrix} = \begin{bmatrix} D_{xx} & D_{xy} & D_{xs} \\ D_{yx} & D_{yy} & D_{ys} \\ D_{sx} & D_{sy} & D_{ss} \end{bmatrix} \times \begin{bmatrix} \kappa_x^e \\ \kappa_y^e \\ \kappa_{xy}^e \end{bmatrix}
$$

The last of the above equations compute the changes in curvature. The changes in curvature [k] and the ply position z allow the computation of the bending strains.

$$
\begin{bmatrix} \epsilon_x^b \\ \epsilon_y^b \\ \gamma_{xy}^b \end{bmatrix} = \begin{bmatrix} \epsilon_x^e \\ \epsilon_y^e \\ \gamma_{xy}^e \end{bmatrix} + \begin{bmatrix} k_x^e \\ k_y^e \\ k_{xy}^e \end{bmatrix} \times z
$$

The simplicity of the preceding equations justifies the strong incentive to specify symmetrical laminates in the presence of external bending moments. This book ignores the bending strains, since their inclusion would complicate the analysis while bringing no benefit to the understanding of the durability problem.

2.8   CONCLUSION

This chapter presented the constitutive equations of circular cylindrical laminates for the three loadings that appear in engineering practice, namely the hydric, the thermal and the mechanical loadings. The development assumed the following:

1. The laminate is stress-free when dry and at the peak temperature PT
2. The cylinder preserves its original circular shape
3. All plies operate at the same temperature OT. There is no thermal gradient
4. The resin in all plies pick up the same amount of water Δm. There is no hydric gradient
5. The laminate properties do not change with temperature or water pick up.
6. The operating temperature OT is less than the resin HDT.

The above assumptions are not overly restrictive. Not all commercial laminates are symmetrical, but they are all balanced. Furthermore, the laminate properties change very little for operating temperatures below the HDT – Heat Distortion Temperature – of the resin.

3 Computing the Ply Strains

3.1 INTRODUCTION

The ply total strains play a central role in the prediction of the load-dependent laminate durability. The first step in assessing the durability of laminates subjected to mechanical loadings is the accurate computation of the total strain components on the critical ply, i.e., the ply actually governing the failure. We will have more to say about the critical plies and the importance of their strains in the Part II of this book. For now, let us tackle the strain issue.

This chapter introduces a simple and accurate protocol to compute the total strains of plies embedded in circular cylindrical laminates subjected to thermal, hydric and mechanical loadings. The total strains are direct inputs to the unified equation used to compute the load-dependent durability. The published literature is highly competent in computing the ply strains from mechanical loadings, but less than satisfactory in the estimation of the thermal and hydric residual strains. In fact, the published literature is very poor in the computation of the hydric and thermal residual strains. The simple and accurate protocol developed in this chapter computes the total strains – mechanical + thermal + hydric – on plies embedded in circular cylindrical laminates. In Chapter 4, we will introduce the specific data of commercial laminates and put the generic protocol developed here to practical use. Finally, in Chapter 5, we compute the total ply strains of several numerical examples.

The protocol proposed in this book computes the ply strains before the stresses. The computation stops with the strains, since the laminate durability models make use of strains, not stresses. The ply stresses are not discussed much in this book. The total ply strains consist of three components.

$$
\begin{bmatrix} total \\ strain \end{bmatrix} = \begin{bmatrix} mechanical \\ strain \end{bmatrix} + \begin{bmatrix} residual\ thermal \\ strain \end{bmatrix} + \begin{bmatrix} residual\ hydric \\ strain \end{bmatrix} \quad (3.1)
$$

To facilitate the discussion and clarify the issue, we will develop the computation protocol separately for each component. The reader will realize the similarities in each case, the small differences that exist being for the most part obvious and intuitive. Each of the above strain components are first computed in the global laminate system x – y and subsequently rotated to the ply frame 1–2.

The protocol is applicable to circular cylindrical laminates subjected to membrane strains. As explained in the previous chapter, the equilibrium strains of such laminates do not depend on the coupling matrix [B]. This is a tremendous simplification. However, in spite of this simplification, the computation of all strains, for all plies, on the local and global frames, requires patience and attention. The prudent designer repeats the computations to avoid possible mistakes. We start with the thermal strains.

3.2   COMPUTING THE THERMAL STRAINS

The commercial laminates are laid up at room temperature, combining several plies of glass fibers impregnated with liquid resin. Following this lamination, the liquid resin gradually undergoes a curing process and takes the solid form of the final product. The cure process releases chemical energy that may substantially increase the laminate temperature. In addition to this cure heat, the manufacturer may also provide outside heating to expedite the process. In either case, the laminate heats up from the room temperature (RT) to a peak temperature (PT) before cooling down. On reaching the peak temperature (PT), the individual plies attain their maximum thermal expansion while the resin is still liquid and in a stress-free condition. The plies and the laminate remain at the peak temperature (PT) for a while as the resin cures and converts from liquid to solid. The important thing to consider here is the stress-free condition of the laminate at the temperature PT. After a while, the peak temperature (PT) starts to drop and, since the resin has already set into a solid, the laminate shrinks as a single body. This shrinkage induces thermal strains in the plies.

In the early phase of cure, when the temperature first reaches the peak temperature PT and the crosslinking has not advanced much, the still plastic resin flows and keeps the plies in a stress-free condition. Thereupon, the cure proceeds quickly and sets the resin into a rigid solid that congeals and immobilizes the plies in this stress-free condition. The thermal stresses develop as the temperature drops from PT and the plies shrink together as a solid laminate. All plies cool and shrink together, as a laminate, to take the same equilibrium dimensions. At the equilibrium temperature OT, the plies that shrink more come to rest in a stretched condition, while those that shrink less are compressed. This is because all plies shrink together as a laminate.

The stress-free temperature PT is determined from consideration of four temperatures.

- **Heat Distortion Temperature HDT.** This is the temperature at which the cured resin gains substantial molecular mobility that dissipates all residual stresses. The HDT sets the maximum possible value for the stress-free temperature. Therefore, the peak temperature PT cannot exceed the HDT and we have PT < HDT.
- **Peak process temperature PPT.** This is the highest temperature achieved in unassisted room temperature cure. In laminates that are not post-cured, the stress-free temperature is PT = PPT.
- **Post-cure Temperature PCT.** This is the highest temperature reached in laminates that are post-cured. When PCT < HDT we have PT = PCT. If PCT > HDT, then PT = HDT.
- **Operating Temperature OT.** Also known as working temperature. At OT, the laminate takes its equilibrium position and the plies develop their residual thermal strains.

The stress-free temperature falls between a minimum at PT = PPT in laminates that are not post-cured and a maximum at PT = HDT in laminates that are post-cured above the HDT. In laminates that are post-cured at temperatures less than the HDT, PT = PCT. See Figure 3.1.

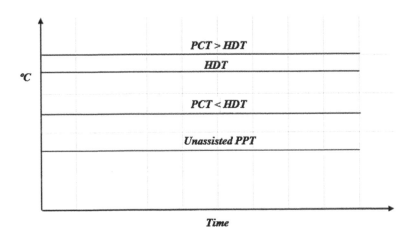

FIGURE 3.1 *The stress-free temperature PT ranges from a minimum unassisted peak process temperature PPT to a maximum limited by the resin HDT.*

Example 3.1

This numerical example clarifies the preceding discussion. Let us assume a laminate of vinyl ester resin with HDT = 105C. The unassisted peak process temperature is PPT = 70C.

- Suppose the laminate is not post-cured. The stress-free temperature in this case is PT = PPT = 70C.
- Suppose the laminate is post-cured for 1 hour at PCT = 90C. The stress-free temperature in this case is PT = PCT = 90C.
- Suppose the laminate is post-cured for 1 hour at PCT = 120C. The stress-free temperature in this case is PT = HDT = 105C.

In mathematical language, the above statements reduce to

PT = PCT < HDT.

Expressed in words, the stress-free temperature PT is equal to the post-cure temperature PCT limited by the resin HDT.

3.3   PROTOCOL TO COMPUTE THE THERMAL STRAINS

The equilibrium thermal strains of circular cylindrical laminates derive from the stiffness matrix [A] and the force resultants. The thermal force resultants and all matrices required in the computation refer to the global system. To do the computations, the ply matrices [Q] and [α] are rotated from their local frames 1–2 to the global frame x – y.

The protocol to compute the residual thermal strains is like follows.

1. Rotation of the ply matrices [Q] to the laminate frame.
2. Computation of the laminate matrix [A]
3. Rotation of the ply coefficients of thermal expansion to the global frame.
4. Computation of the laminate thermal resultants
5. Computation of the laminate thermal equilibrium strains
6. Rotation of the equilibrium strains to the local ply system 1–2.
7. Computation of the residual ply strains in the local system 1–2

The seven steps outlined above compute the residual thermal strains of plies embedded in circular cylindrical laminates. The rest of this section is dedicated to a full and detailed description of the entire procedure, often repeating things and concepts for the sake of clarity. The experienced reader will bear with this repetition.

1. *Rotation of the ply matrices [Q] from the local to the global frame.* The ply matrices [Q] in the local 1–2 and in the global x-y systems are.

$$[Q]_{12} = \begin{bmatrix} Q_{11} & Q_{12} & 0 \\ Q_{21} & Q_{22} & 0 \\ 0 & 0 & Q_{66} \end{bmatrix}_{LOCAL}$$

$$[Q]_{xy} = \begin{bmatrix} Q_{xx} & Q_{xy} & Q_{xs} \\ Q_{yx} & Q_{yy} & Q_{ys} \\ Q_{sx} & Q_{sy} & Q_{ss} \end{bmatrix}_{GLOBAL}$$

The matrices $[Q]_{12}$ are known for all plies, as explained in Chapter 1. The elements of the rotated matrices $[Q]_{xy}$ are

$$Q_{xx} = m^4 Q_{11} + n^4 Q_{22} + 2m^2 n^2 Q_{12} + 4m^2 n^2 Q_{66}$$

$$Q_{xy} = Q_{yx} = m^2 n^2 Q_{11} + m^2 n^2 Q_{22} + (m^4 + n^4)Q_{12} - 4m^2 n^2 Q_{66}$$

$$Q_{xs} = Q_{sx} = m^3 n Q_{11} - mn^3 Q_{22} + (mn^3 - m^3 n)Q_{12} + 2(mn^3 - m^3 n)Q_{66}$$

$$Q_{yy} = n^4 Q_{11} + m^4 Q_{22} + 2m^2 n^2 Q_{12} + 4m^2 n^2 Q_{66}$$

$$Q_{ys} = Q_{sy} = mn^3 Q_{11} - m^3 n Q_{22} + (m^3 n - mn^3)Q_{12} + 2(m^3 n - mn^3)Q_{66}$$

$$Q_{ss} = m^2 n^2 Q_{11} + m^2 n^2 Q_{22} - 2m^2 n^2 Q_{12} + (m^2 - n^2)^2 Q_{66}$$

Where "m" is the cosine and "n" is the sine of the angle formed by the global laminate axis "x" and the local axis "1" of each ply. The above computations include all UD plies. They are rather complex and full of opportunities for errors. The prudent designer will repeat the computations to eliminate possible mistakes. The orthotropic UD plies are the only ones requiring rotation. The isotropic plies require no rotation.

2. *Computation of the laminate matrix [A].* The elements of the matrix [A] come from the summation of the contributions from all plies. The expression below is self-explanatory.

$$[A] = \begin{bmatrix} \sum Q_{xx}t & \sum Q_{xy}t & \sum Q_{xs}t \\ \sum Q_{yx}t & \sum Q_{yy}t & \sum Q_{ys}t \\ \sum Q_{sx}t & \sum Q_{sy}t & \sum Q_{ss}t \end{bmatrix} \text{kg/cm} \qquad (3.2)$$

In the above, "t" is the thickness and $[Q]_{xy}$ are the matrix elements of each ply in the global system.

3. *Rotation of the ply coefficients of thermal expansion.* This rotation is required for the UD plies and is done like follows.

$$\alpha_x = m^2\alpha_1 + n^2\alpha_2$$

$$\alpha_y = n^2\alpha_1 + m^2\alpha_2 \qquad (3.3)$$

$$\alpha_{xy} = 2mn\left(\alpha_2 - \alpha_1\right)$$

Where again m = cos(α) and n = sin(α), α being the angle formed by the local fiber direction 1 and the longitudinal global axis x. The rotation applies to all UD plies.

4. *Computation of the thermal resultants.* The general equations to compute the thermal resultants are self-explanatory.

$$\begin{bmatrix} N_x^T \\ N_y^T \\ N_{xy}^T \end{bmatrix} = \begin{bmatrix} \sum\left(Q_{xx}t\alpha_x + Q_{xy}t\alpha_y + Q_{xs}t\alpha_{xy}\right) \\ \sum\left(Q_{yx}t\alpha_x + Q_{yy}t\alpha_y + Q_{ys}t\alpha_{xy}\right) \\ \sum\left(Q_{sx}t\alpha_x + Q_{sy}t\alpha_y + Q_{ss}t\alpha_{xy}\right) \end{bmatrix} \times\left(\text{OT} - \text{PT}\right)\text{kg/cm} \qquad (3.4)$$

Since all commercial laminates are balanced, Equation (3.4) reduces to

$$\begin{bmatrix} N_x^T \\ N_y^T \\ N_{xy}^T \end{bmatrix} = \begin{bmatrix} \sum\left(Q_{xx}t\alpha_x + Q_{xy}t\alpha_y + Q_{xs}t\alpha_{xy}\right) \\ \sum\left(Q_{yx}t\alpha_x + Q_{yy}t\alpha_y + Q_{ys}t\alpha_{xy}\right) \\ 0 \end{bmatrix} \times\left(\text{OT} - \text{PT}\right)\text{kg/cm} \qquad (3.4A)$$

Balanced laminates have no shear thermal resultant in the global system.

5. *Computation of the equilibrium thermal strains.* The thermal equilibrium takes place when the laminate cools down from PT to OT. The circularity condition guarantees the same equilibrium strains for all plies. The equilibrium strains come from the laminate matrix [A] and the thermal resultants [N^T].

$$
\begin{bmatrix} N_x^T \\ N_y^T \\ N_{xy}^T \end{bmatrix} = \begin{bmatrix} A_{xx} & A_{xy} & A_{xs} \\ A_{yx} & A_{yy} & A_{ys} \\ A_{sx} & A_{sy} & A_{ss} \end{bmatrix} \times \begin{bmatrix} \varepsilon_x^e \\ \varepsilon_y^e \\ \gamma_{xy}^e \end{bmatrix} \tag{3.5}
$$

Since all commercial laminates are balanced, Equation (3.5) reduces to

$$
\begin{bmatrix} N_x^T \\ N_y^T \\ 0 \end{bmatrix} = \begin{bmatrix} A_{xx} & A_{xy} & 0 \\ A_{yx} & A_{yy} & 0 \\ 0 & 0 & A_{ss} \end{bmatrix} \times \begin{bmatrix} \varepsilon_x^e \\ \varepsilon_y^e \\ \gamma_{xy}^e \end{bmatrix} \tag{3.5A}
$$

Equation (3.5A) immediately leads to

$$
\gamma_{xy}^e = 0
$$

The equilibrium thermal strains come from the inversion of Equation (3.5A)

$$
\begin{bmatrix} \varepsilon_x^e \\ \varepsilon_y^e \\ 0 \end{bmatrix} = \begin{bmatrix} A_{xx} & A_{xy} & 0 \\ A_{yx} & A_{yy} & 0 \\ 0 & 0 & A_{ss} \end{bmatrix}^{-1} \times \begin{bmatrix} N_x^T \\ N_y^T \\ 0 \end{bmatrix} \tag{3.6}
$$

In Equation (3.6), the superscript "e" designates equilibrium strains. The equilibrium strains take as reference the laminate dimensions at the stress-free peak temperature PT, not the room temperature RT.

Figure 3.2 shows a laminate with three plies of equal dimensions at the stress-free temperature PT. Also shown are the ply dimensions

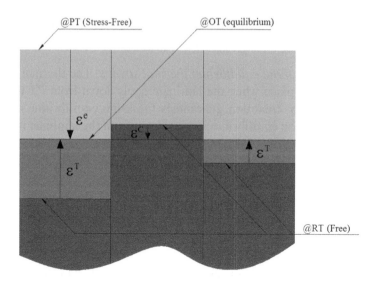

FIGURE 3.2 *The thermal strains are zero (stress-free condition) at the tempera-ture PT. The equilibrium strains are equal for all plies at the operating temperature OT. The shaded areas show the rest position of each ply if allowed to contract freely.*

if allowed to shrink freely from PT to OT. As already discussed, the embedded plies develop residual strains when forced to take the same equilibrium dimensions. The equilibrium strains are negative, since the laminate shrinks when cooling from PT to the operating temperature OT. The residual strains, however, can be positive or negative, as shown in Figure 3.2.

The preservation of circularity assures that all plies have the equi-librium strains computed in Equation (3.6). The equilibrium strains referred to the global system x − y are rotated to the local systems 1–2 of each ply.

6. *Rotation of the equilibrium strains.* The strain rotation involves the multiplication of the equilibrium strains by the matrix [T].

$$[T] = \begin{bmatrix} m^2 & n^2 & 2mn \\ n^2 & m^2 & -2mn \\ -mn & mn & m^2 - n^2 \end{bmatrix}$$

In the above, "m" is the cosine and "n" is the sine of the ply angle α. The equilibrium thermal strains referred to the local system 1–2 are:

$$
\begin{bmatrix} \varepsilon_1^e \\ \varepsilon_2^e \\ \dfrac{1}{2}\gamma_{12}^e \end{bmatrix} = \begin{bmatrix} m^2 & n^2 & 2mn \\ n^2 & m^2 & -2mn \\ -mn & mn & m^2 - n^2 \end{bmatrix} \times \begin{bmatrix} \varepsilon_x^e \\ \varepsilon_y^e \\ 0 \end{bmatrix} \tag{3.7}
$$

Multiplication by the matrix [T] rotates the strains from the global $x - y$ to the local 1–2 system. The equilibrium shear strain is not zero in the local frame. In fact, the equilibrium shear strain in the local system is

$$
\gamma_{12}^e = 2 \times mn \times \left(\varepsilon_y^e - \varepsilon_x^e \right)
$$

7. *Computing the residual thermal strains.* The residual thermal strains come from subtracting the free ply expansion from the equilibrium strains.

$$
\begin{bmatrix} \varepsilon_1^r \\ \varepsilon_2^r \\ \gamma_{12}^r \end{bmatrix} = \begin{bmatrix} \varepsilon_1^e - \alpha_1 (\mathrm{OT} - \mathrm{PT}) \\ \varepsilon_2^e - \alpha_2 (\mathrm{OT} - \mathrm{PT}) \\ \gamma_{12}^e \end{bmatrix} \tag{3.8}
$$

The residual thermal strains (superscript "r") exist because the embedded plies take an equilibrium dimension different from those they would take if allowed to shrink freely. The residual strains computed by Equation (3.8) take as reference the stress-free ply dimensions at the temperature PT. They can be negative or positive, depending on the thermal contraction of the ply.

Note: The residual thermal stresses are not required to predict the load-dependent durability. The residual stresses derive from the ply constitutive equation and the residual strains.

$$
\begin{bmatrix} \sigma_1^r \\ \sigma_2^r \\ \tau_{12}^r \end{bmatrix} = \begin{bmatrix} Q_{11} & Q_{12} & 0 \\ Q_{12} & Q_{22} & 0 \\ 0 & 0 & Q_{ss} \end{bmatrix} \times \begin{bmatrix} \varepsilon_1^r \\ \varepsilon_2^r \\ \gamma_{12}^r \end{bmatrix}
$$

3.4   THE PROTOCOL FOR HYDRIC STRAINS

The protocol to compute the hydric residual strains is essentially the same of Section 3.3. The difference is the thermal coefficients replacing their hydric counterparts. Furthermore, the water pickup Δm replaces the temperature drop $PT - OT$. The protocol to compute the hydric strains is like follows:

1. Rotation of the ply matrices [Q] to the global frame.
2. Computation of the matrix [A]
3. Rotation of coefficients of hydric expansion to the global frame.
4. Computation of the hydric resultants
5. Computation of the equilibrium strains in the global system $x - y$
6. Rotation of the equilibrium strains to the local system 1–2 of each UD ply
7. Computation of the residual hydric strains in the local system 1–2.

The computations are similar to those of the thermal case. What follows is a highlight of the differences. The hydric coefficients of expansion in the global system are.

$$\beta_x = m^2\beta_1 + n^2\beta_2$$

$$\beta_y = n^2\beta_1 + m^2\beta_2$$

$$\beta_{xy} = 2mn(\beta_2 - \beta_1)$$

The hydric force resultants are

$$\begin{bmatrix} N_x^H \\ N_y^H \\ N_{xy}^H \end{bmatrix} = \begin{bmatrix} \sum(Q_{xx}t\beta_x + Q_{xy}t\beta_y + Q_{xs}t\beta_{xy}) \\ \sum(Q_{yx}t\beta_x + Q_{yy}t\beta_y + Q_{ys}t\beta_{xy}) \\ \sum(Q_{sx}t\beta_x + Q_{sy}t\beta_y + Q_{ss}t\beta_{xy}) \end{bmatrix} \times (\Delta m) \tag{3.9}$$

In the above, the "betas" are the coefficients of hydric expansion and "Δm" is the amount of water picked up by the resin. The hydric equilibrium strains are

$$\begin{bmatrix} \varepsilon_x^e \\ \varepsilon_y^e \\ \gamma_{xy}^e \end{bmatrix} = \begin{bmatrix} A_{xx} & A_{xy} & A_{xs} \\ A_{yx} & A_{yy} & A_{ys} \\ A_{sx} & A_{sy} & A_{ss} \end{bmatrix}^{-1} \times \begin{bmatrix} N_x^H \\ N_y^H \\ N_{xy}^H \end{bmatrix}$$

The equilibrium equation for balanced laminates is

$$
\begin{bmatrix} \varepsilon_x^e \\ \varepsilon_y^e \\ 0 \end{bmatrix} = \begin{bmatrix} A_{xx} & A_{xy} & 0 \\ A_{yx} & A_{yy} & 0 \\ 0 & 0 & A_{ss} \end{bmatrix}^{-1} \times \begin{bmatrix} N_x^H \\ N_y^H \\ 0 \end{bmatrix} \tag{3.10}
$$

Entering (3.9) in (3.10) we have

$$
\begin{bmatrix} \varepsilon_x^e \\ \varepsilon_y^e \\ 0 \end{bmatrix} = \begin{bmatrix} A_{xx} & A_{xy} & 0 \\ A_{yx} & A_{yy} & 0 \\ 0 & 0 & A_{ss} \end{bmatrix}^{-1} \times \begin{bmatrix} \sum \left(Q_{xx}t\beta_x + Q_{xy}t\beta_y + Q_{xs}t\beta_{xy} \right) \\ \sum \left(Q_{yx}t\beta_x + Q_{yy}t\beta_y + Q_{ys}t\beta_{xy} \right) \\ 0 \end{bmatrix} \times (\Delta m) \tag{3.11}
$$

Where the superscript "e" denotes the equilibrium strain when the resin saturates with water. These strains are measured in the global system x – y, taking as reference the stress-free dry dimensions. The equilibrium strains are positive since the laminate expands on picking up water. The equilibrium hydric strains expressed in the global system are directly proportional to the amount of water picked up by the resin. The global equilibrium strains rotated to the local 1–2 system in the usual way.

$$
\begin{bmatrix} \varepsilon_1^e \\ \varepsilon_2^e \\ \dfrac{1}{2}\gamma_{12}^e \end{bmatrix} = [T] \times \begin{bmatrix} \varepsilon_x^e \\ \varepsilon_y^e \\ 0 \end{bmatrix}
$$

Where [T] is the rotation matrix defined in Section 3.3.

The residual hydric strains (indicated by the superscript "r") occur when the laminate constrains the embedded plies and prevent their free expansion as they soak up water. The residual strains result from the subtraction of the free expansion of each ply from the equilibrium strain. They can be negative or positive, depending on the coefficient of hydric expansion of each ply.

$$
\begin{bmatrix} \varepsilon_1^r \\ \varepsilon_2^r \\ \gamma_{12}^r \end{bmatrix} = \begin{bmatrix} \varepsilon_1^e - \beta_1 \Delta m \\ \varepsilon_2^e - \beta_2 \Delta m \\ \gamma_{12}^e \end{bmatrix} \tag{3.12}
$$

The residual hydric stresses are not required to compute the load-dependent durability. Their computation, shown below, is not required.

$$
\begin{bmatrix} \sigma_1^r \\ \sigma_2^r \\ \tau_{12}^r \end{bmatrix} = \begin{bmatrix} Q_{11} & Q_{12} & 0 \\ Q_{12} & Q_{22} & 0 \\ 0 & 0 & Q_{ss} \end{bmatrix} \times \begin{bmatrix} \varepsilon_1^r \\ \varepsilon_2^r \\ \gamma_{12}^r \end{bmatrix} \text{kg/cm}
$$

3.5   THE PROTOCOL FOR MECHANICAL STRAINS

The protocol to compute the mechanical strains is like follows:

1. Rotation of the ply matrices [Q] to the global frame
2. Computation of the laminate matrix [A]
3. There are no coefficients of mechanical expansion
4. The mechanical resultants are given.
5. Computation of the global mechanical strains
6. Rotation of the global strains to the local system 1–2.

The computation details are essentially identical to those discussed before and are not repeated here. The laminate matrix [A] is the same. There are no mechanical coefficients of expansion. The mechanical force resultants are given. The mechanical equilibrium strains in the global system x − y are computed in the same way as the equilibrium thermal and hydric strains.

$$
\begin{bmatrix} \varepsilon_x^e \\ \varepsilon_y^e \\ \gamma_{xy}^e \end{bmatrix} = \begin{bmatrix} A_{xx} & A_{xy} & A_{xs} \\ A_{yx} & A_{yy} & A_{ys} \\ A_{sx} & A_{sy} & A_{ss} \end{bmatrix}^{-1} \times \begin{bmatrix} N_x^M \\ N_y^M \\ N_{xy}^M \end{bmatrix}
$$

Since all commercial laminates are balanced

$$
\begin{bmatrix} \varepsilon_x^e \\ \varepsilon_y^e \\ \gamma_{xy}^e \end{bmatrix} = \begin{bmatrix} A_{xx} & A_{xy} & 0 \\ A_{yx} & A_{yy} & 0 \\ 0 & 0 & A_{ss} \end{bmatrix}^{-1} \times \begin{bmatrix} N_x^M \\ N_y^M \\ N_{xy}^M \end{bmatrix} \tag{3.13}
$$

In the absence of external torque, the mechanical shear resultant is zero. The preservation of circularity assures the same equilibrium mechanical strains for all plies. The equilibrium strains rotate from the global to the local system in the usual way.

$$
\begin{bmatrix} \varepsilon_1^e \\ \varepsilon_2^e \\ \dfrac{1}{2}\gamma_{12}^e \end{bmatrix} = \begin{bmatrix} T \end{bmatrix} \times \begin{bmatrix} \varepsilon_x^e \\ \varepsilon_y^e \\ \dfrac{1}{2}\gamma_{xy}^e \end{bmatrix}
$$

The equilibrium strains above are the mechanical strains we are looking for.

Example 3.2

Assuming preservation of circularity, explain the difference between the equilibrium strains and the residual strains.

The preservation of circularity assures the equality of the strains on all embedded plies. The equilibrium strains occur when the mechanical loading, the operating temperature and the water pick up stabilize. The following equations give the equilibrium thermal, hydric and mechanical strains.

$$
\begin{bmatrix} N_x^T \\ N_y^T \\ 0 \end{bmatrix} = \begin{bmatrix} A_{xx} & A_{xy} & 0 \\ A_{yx} & A_{yy} & 0 \\ 0 & 0 & A_{ss} \end{bmatrix} \times \begin{bmatrix} \varepsilon_x^e \\ \varepsilon_y^e \\ 0 \end{bmatrix}_{Thermal} \text{kg/cm} \qquad (3.14)
$$

$$
\begin{bmatrix} N_x^H \\ N_y^H \\ 0 \end{bmatrix} = \begin{bmatrix} A_{xx} & A_{xy} & 0 \\ A_{yx} & A_{yy} & 0 \\ 0 & 0 & A_{ss} \end{bmatrix} \times \begin{bmatrix} \varepsilon_x^e \\ \varepsilon_y^e \\ 0 \end{bmatrix}_{Hydric} \text{kg/cm} \qquad (3.15)
$$

$$
\begin{bmatrix} N_x^M \\ N_y^M \\ N_{xy}^M \end{bmatrix} = \begin{bmatrix} A_{xx} & A_{xy} & 0 \\ A_{yx} & A_{yy} & 0 \\ 0 & 0 & A_{ss} \end{bmatrix} \times \begin{bmatrix} \varepsilon_x^e \\ \varepsilon_y^e \\ \gamma_{xy}^e \end{bmatrix}_{Mechanical} \text{kg/cm} \qquad (3.16)
$$

The above are the laminate equilibrium strains. From the preservation of circularity, they are the same as those of all embedded plies. There are two things to remember about the equilibrium strains:

- They refer to the global laminate system, not the local ply system.
- They are not the strains we are looking for. We are looking for the residual, not the equilibrium, strains.

The residual strains occur on those plies that expand or contract at a different rate than the laminate. The concept of residual strain is applicable to plies, not to laminates. The residual strains depend on the expansion coefficient of each ply and have the same nature as the mechanical strains. They accumulate elastic energy and promote crack growth in exactly the same way as the strains from external mechanical loads.

3.6   TOTAL STRAINS

A simple addition of the mechanical, thermal and hydric strain components gives the total ply strains. This addition is possible, since the three components act in the same ply directions. The ply total strains are

$$
\begin{bmatrix} Total \\ strain \end{bmatrix} = \begin{bmatrix} mechanical \\ strain \end{bmatrix} + \begin{bmatrix} residual \\ thermal\ strain \end{bmatrix} + \begin{bmatrix} residual \\ hydric\ strain \end{bmatrix}
$$

From the protocol we have just described

$$
\begin{bmatrix} \varepsilon_1 \\ \varepsilon_2 \\ \gamma_{12} \end{bmatrix}_{TOTAL} = \begin{bmatrix} \varepsilon_1^e \\ \varepsilon_2^e \\ \gamma_{12}^e \end{bmatrix}_{MEC} + \begin{bmatrix} \varepsilon_1^e - \alpha_1 \left(OT - PT \right) \\ \varepsilon_2^e - \alpha_2 \left(OT - PT \right) \\ \gamma_{12}^e \end{bmatrix}_{TER} + \begin{bmatrix} \varepsilon_1^e - \beta_1 \Delta m \\ \varepsilon_2^e - \beta_2 \Delta m \\ \gamma_{12}^e \end{bmatrix}_{HID}
$$

$$(3.17)$$

The equilibrium strains come from Equations (3.14), (3.15) and (3.16). The commercial product standards usually ignore the residual thermal and hydric strains. The residual strains may have great impact on the analysis of the load-dependent laminate durability.

In conclusion, we have described a simple protocol to compute the mechanical, hydric and thermal strains in circular cylindrical laminates. The computed strains referred to the local ply frame are ready to enter the durability equations. In as much as the UD plies are concerned, the strains in the fiber direction govern the long-term rupture failure of pressure pipes or wind blades. In fact, the strains in the fiber direction of UD plies govern the long-term rupture of any structural laminate. Still on UD plies, the transverse strains govern the long-term infiltration, stiffness and weep failures. The same comments hold for other plies, like those of chopped fibers. We will have more to say about this in the Part II of this book.

3.7   LAMINATE ENGINEERING CONSTANTS

The laminate engineering properties like elastic moduli, coefficients of expansion and others, derive from the constituent plies. The reason we did not develop such properties is our unwavering focus on the strains required to predict the load-dependent durability. However, just as a curiosity, let us develop the expressions to compute the equivalent elastic moduli and coefficients of expansion of balanced circular cylindrical laminates.

Example 3.3

Derive the expressions to compute the coefficients of thermal expansion of circular cylindrical laminates.

We use the step 4 of the thermal protocol. From Equation (3.6) the equilibrium thermal strains are

$$
\begin{bmatrix} \varepsilon_x^e \\ \varepsilon_y^e \\ 0 \end{bmatrix} = \begin{bmatrix} A_{xx} & A_{xy} & 0 \\ A_{yx} & A_{yy} & 0 \\ 0 & 0 & A_{ss} \end{bmatrix}^{-1} \times \begin{bmatrix} \sum\left(Q_{xx}t\alpha_x + Q_{xy}t\alpha_y + Q_{xs}t\alpha_{xy}\right) \\ \sum\left(Q_{yx}t\alpha_x + Q_{yy}t\alpha_y + Q_{ys}t\alpha_{xy}\right) \\ 0 \end{bmatrix} \times (OT - PT)
$$

By definition, the coefficients of thermal expansion are

$$
\begin{bmatrix} \bar{\alpha}_x \\ \bar{\alpha}_y \\ 0 \end{bmatrix} = \begin{bmatrix} A_{xx} & A_{xy} & 0 \\ A_{yx} & A_{yy} & 0 \\ 0 & 0 & A_{ss} \end{bmatrix}^{-1} \times \begin{bmatrix} \sum\left(Q_{xx}t\alpha_x + Q_{xy}t\alpha_y + Q_{xs}t\alpha_{xy}\right) \\ \sum\left(Q_{yx}t\alpha_x + Q_{yy}t\alpha_y + Q_{ys}t\alpha_{xy}\right) \\ 0 \end{bmatrix}
$$

$$
\begin{bmatrix} \bar{\alpha}_x \\ \bar{\alpha}_y \\ 0 \end{bmatrix} = \begin{bmatrix} A_{xx} & A_{xy} & 0 \\ A_{yx} & A_{yy} & 0 \\ 0 & 0 & A_{ss} \end{bmatrix}^{-1} \times \begin{bmatrix} N_x^T \\ N_y^T \\ 0 \end{bmatrix} \times \frac{1}{(OT - PT)} \tag{3.18}
$$

A similar derivation produces the hydric coefficients of expansion

$$
\begin{bmatrix} \bar{\beta}_x \\ \bar{\beta}_y \\ 0 \end{bmatrix} = \begin{bmatrix} A_{xx} & A_{xy} & 0 \\ A_{yx} & A_{yy} & 0 \\ 0 & 0 & A_{ss} \end{bmatrix}^{-1} \times \begin{bmatrix} \sum\left(Q_{xx}t\beta_x + Q_{xy}t\beta_y + Q_{xs}t\beta_{xy}\right) \\ \sum\left(Q_{yx}t\beta_x + Q_{yy}t\beta_y + Q_{ys}t\beta_{xy}\right) \\ 0 \end{bmatrix}
$$

$$
\begin{bmatrix} \bar{\beta}_x \\ \bar{\beta}_y \\ 0 \end{bmatrix} = \begin{bmatrix} A_{xx} & A_{xy} & 0 \\ A_{yx} & A_{yy} & 0 \\ 0 & 0 & A_{ss} \end{bmatrix}^{-1} \times \begin{bmatrix} N_x^H \\ N_y^H \\ 0 \end{bmatrix} \times \frac{1}{\Delta m} \tag{3.19}
$$

Example 3.4

Compute the thermal and hydric coefficients of expansion for ± 70 laminates.

This example requires the laminate matrices $[A]$, $[N^T]$ *and* $[N^H]$ for ± 70 laminates, which are in Table 4.4 of the next chapter. The following matrices came from there.

$$[A]_{\pm70} = \begin{bmatrix} 104000 & 61000 & 0 \\ 61000 & 334000 & 0 \\ 0 & 0 & 66000 \end{bmatrix} \times (t_{UD}) + \begin{bmatrix} 77000 & 23000 & 0 \\ 23000 & 77000 & 0 \\ 0 & 0 & 27000 \end{bmatrix} \times (t_{chop}) \, \text{kg/cm}$$

$$\begin{bmatrix} N_x^T \\ N_y^T \\ N_{xy}^T \end{bmatrix}_{\pm70} = \begin{bmatrix} 5.22 \times t_{UD} + 2.50 \times t_{chop} \\ 8.50 \times t_{UD} + 2.50 \times t_{chop} \\ 0 \end{bmatrix} \times (OT - PT) \, \text{kg/cm}$$

$$\begin{bmatrix} N_x^H \\ N_y^H \\ N_{xy}^H \end{bmatrix}_{\pm70} = \begin{bmatrix} 28000 \times t_{UD} + 25000 \times t_{chop} \\ 44500 \times t_{UD} + 25000 \times t_{chop} \\ 0 \end{bmatrix} \times \Delta m \, \text{kg/cm}$$

The above matrices, and likewise the hydric and thermal coefficients, depend on the ply thicknesses. We illustrate the computations for $t_{UD} = 3.5 \, \text{mm}$ and $t_{chop} = 2.0 \, \text{mm}$.

$$[A]_{\pm70} = \begin{bmatrix} 104000 & 61000 & 0 \\ 61000 & 334000 & 0 \\ 0 & 0 & 66000 \end{bmatrix} \times (0.35) + \begin{bmatrix} 77000 & 23000 & 0 \\ 23000 & 77000 & 0 \\ 0 & 0 & 27000 \end{bmatrix} \times (0.20) \, \text{kg/cm}$$

$$[A]_{\pm70} = \begin{bmatrix} 51800 & 26000 & 0 \\ 26000 & 132000 & 0 \\ 0 & 0 & 28500 \end{bmatrix} \, \text{kg/cm} \tag{3.20}$$

$$\begin{bmatrix} N_x^T \\ N_y^T \\ N_{xy}^T \end{bmatrix}_{\pm70} = \begin{bmatrix} 5.22 \times 0.35 + 2.50 \times 0.20 \\ 8.50 \times 0.35 + 2.50 \times 0.20 \\ 0 \end{bmatrix} \times (OT - PT)$$

$$\begin{bmatrix} N_x^T \\ N_y^T \\ N_{xy}^T \end{bmatrix}_{\pm70} = \begin{bmatrix} 2.33 \\ 3.48 \\ 0 \end{bmatrix} \times (OT - PT)\, \text{kg/cm}$$

$$\begin{bmatrix} N_x^H \\ N_y^H \\ N_{xy}^H \end{bmatrix}_{\pm70} = \begin{bmatrix} 28000 \times 0.35 + 25000 \times 0.20 \\ 44500 \times 0.35 + 25000 \times 0.20 \\ 0 \end{bmatrix} \times \Delta m\, \text{kg/cm}$$

$$\begin{bmatrix} N_x^H \\ N_y^H \\ N_{xy}^H \end{bmatrix}_{\pm70} = \begin{bmatrix} 14800 \\ 20500 \\ 0 \end{bmatrix} \times \Delta m\, \text{kg/cm}$$

Entering the above in Equations (3.18) and (3.19) we have

$$\begin{bmatrix} \bar{\alpha}_x \\ \bar{\alpha}_y \\ 0 \end{bmatrix} = \begin{bmatrix} 51800 & 26000 & 0 \\ 26000 & 132000 & 0 \\ 0 & 0 & 28500 \end{bmatrix}^{-1} \times \begin{bmatrix} 2.33 \\ 3.48 \\ 0 \end{bmatrix}$$

$$\begin{bmatrix} \bar{\alpha}_x \\ \bar{\alpha}_y \\ 0 \end{bmatrix} = \begin{bmatrix} 35.20 \\ 19.42 \\ 0 \end{bmatrix} \times 10^{-6}/\text{C}$$

$$\begin{bmatrix} \bar{\beta}_x \\ \bar{\beta}_y \\ 0 \end{bmatrix} = \begin{bmatrix} 51800 & 26000 & 0 \\ 26000 & 132000 & 0 \\ 0 & 0 & 28500 \end{bmatrix}^{-1} \times \begin{bmatrix} 14800 \\ 20500 \\ 0 \end{bmatrix}$$

$$\begin{bmatrix} \bar{\beta}_x \\ \bar{\beta}_y \\ 0 \end{bmatrix} = \begin{bmatrix} 0.23 \\ 0.11 \\ 0 \end{bmatrix}\, 1/\Delta m$$

The protocol described in this chapter computes the expansion coefficients of any circular cylindrical laminate in a simple and straightforward manner.

Example 3.5

The ply matrices $[Q]_{12}$ of Chapter 1 derive from the elastic properties of the plies. The laminate matrices [A] of this chapter come from the ply matrices $[Q]_{12}$ and their thicknesses. While doing all these computations we said nothing, not a word, about the laminate elastic properties. This example shows how to compute the equivalent, or average, laminate moduli from the stiffness matrix [A]. The protocol is general and applicable to the computation of all elastic moduli of any laminate.

We start with the constitutive equation of balanced circular cylindrical laminates under the general loads N_x, N_y and N_{xy}.

$$\begin{bmatrix} N_x \\ N_y \\ N_{xy} \end{bmatrix} = \begin{bmatrix} A_{xx} & A_{xy} & 0 \\ A_{yx} & A_{yy} & 0 \\ 0 & 0 & A_{ss} \end{bmatrix} \times \begin{bmatrix} \varepsilon_x \\ \varepsilon_y \\ \gamma_{xy} \end{bmatrix} \text{kg/cm}$$

We illustrate the procedure by computing the elastic modulus in the longitudinal laminate direction. To do so, we simulate the test load in the above constitutive equation, i.e., we retain the axial force resultant N_x and make all other forces equal to zero.

$$\begin{bmatrix} N_x \\ 0 \\ 0 \end{bmatrix} = \begin{bmatrix} A_{xx} & A_{xy} & 0 \\ A_{yx} & A_{yy} & 0 \\ 0 & 0 & A_{ss} \end{bmatrix} \times \begin{bmatrix} \varepsilon_x \\ \varepsilon_y \\ 0 \end{bmatrix} \text{kg/cm}$$

Expanding the above, we have

$$N_x = A_{xx}\varepsilon_x + A_{xy}\varepsilon_y$$

$$0 = A_{xx}\varepsilon_x + A_{xy}\varepsilon_y$$

By eliminating ε_y we have

$$N_x = \left(A_{xx} - \frac{A_{xy}^2}{A_{yy}} \right)\varepsilon_x$$

Dividing through by the laminate thickness "t"

$$\sigma_x = \frac{N_x}{t} = \frac{1}{t}\left(A_{xx} - \frac{A_{xy}^2}{A_{yy}} \right)\varepsilon_x$$

By definition of elastic modulus

$$\frac{\sigma_x}{\varepsilon_x} = E_x = \frac{1}{t}\left(A_{xx} - \frac{A_{xy}^2}{A_{yy}}\right)$$

In a like manner we have

$$E_y = \frac{1}{t}\left(A_{yy} - \frac{A_{xy}^2}{A_{xx}}\right)$$

We can easily develop similar equations for the laminate shear modulus and the Poisson ratios. The stiffness matrix [A] allows the computation of all equivalent engineering constants of circular cylindrical laminates.

Example 3.6

Compute the axial and hoop tensile moduli of the laminate described in Example 3.3.

From Example 3.3, Equation (3.20), the laminate stiffness matrix is

$$[A]_{\pm70} = \begin{bmatrix} 51800 & 26000 & 0 \\ 26000 & 132000 & 0 \\ 0 & 0 & 28500 \end{bmatrix} \text{kg/cm}$$

Also from Example 3.3, the laminate thickness is $t = 2.0 + 3.5 = 5.5$ mm. The laminate moduli are

$$E_y = \frac{1}{t}\left(A_{yy} - \frac{A_{xy}^2}{A_{xx}}\right)$$

$$E_y = \frac{1}{0.55}\left(132000 - \frac{26000^2}{51800}\right) = 216000 \text{ kg/cm}^2$$

$$E_x = \frac{1}{t}\left(A_{xx} - \frac{A_{xy}^2}{A_{yy}}\right)$$

$$E_x = \frac{1}{0.55}\left(51800 - \frac{26000^2}{132000}\right) = 84800 \text{ kg/cm}^2$$

APPENDIX 3.1

STRESS-FREE TEMPERATURE

In the first draft of this chapter, I wrote that the resin molecules congeal in a stress-free condition at the peak process temperature, PPT. Any later temperature excursion above the PPT – either in operation or in post-cure – would increase the resin cross-linking with no effect on the stress-free state. For the purpose of residual strain analysis, the post-cure process is just a brief upset, a short-term excursion, into a temperature that happens to be higher than the PPT. If this reasoning is correct, the post-cure temperature is irrelevant in the computation of the residual thermal strains.

I am reluctant to discard this argument. I sometimes feel that my first approach may be correct. If so, the cross-linking attained at the peak process temperature, PPT, is sufficient to congeal the resin in a stress-free condition. Should this feeling be correct, the peak temperature PT to compute the thermal strains would be

$$PT = PPT \quad if\ PPT < HDT$$

$$PT = HDT \quad if\ PPT > HDT$$

As we see, the post-cure treatment has no effect on the stress-free temperature PT, which is equal to the peak process temperature PPT, limited by the resin HDT.

Example 3.7

Let us compute the stress-free temperature PT of the laminate processed as described in the numerical Example 3.1. The HDT of the vinyl ester resin is HDT = 105C. The unassisted peak process temperature is PPT = 70C.

- Suppose the laminate is not post-cured. The stress-free temperature is PT = PPT = 70C.
- Suppose the laminate is post-cured for 1 hour at PCT = 90C. The stress-free temperature is still PT = PPT = 70C.
- Suppose the laminate is post-cured for 1 hour at PCT = 120C. The stress-free temperature is PT = HDT = 105C.

Example 3.8

Compute the stress-free temperature PT of the laminate in Example 3.7, assuming it has been overheated in the cure process and attained a peak process temperature PPT = 120C.

The stress-free temperature is limited by the resin HDT and is computed as PT = HDT = 105C.

Example 3.9

The accurate computation of the residual thermal strains at the ply level requires a good knowledge of the stress-free temperature PT. We have described, without committing ourselves, two equally reasonable approaches to estimate the stress-free temperature. Describe a simple experiment to determine the correct approach.

The experiment considers that warped unbalanced laminates straighten out and regain true flatness at the stress-free temperature. First, we lay up several thin unbalanced cross-ply flat laminates cured at various unassisted PPT. Second, we post-cure some of these laminates. The thin flat unbalanced cross-ply laminates will show substantial warping at room temperature. The experiment continues by heating up the warped laminates until they flatten out. The temperatures that eliminate the warping and flatten out the laminates are the stress-free temperatures.

The above experiment can decide which of the two models presented in this chapter best describes the stress-free temperature.

APPENDIX 3.2

CURE SHRINKAGE AND RESIDUAL STRAINS

The polyester and vinyl ester resins used to make composite laminates display a high volumetric contraction of 6.0% to 8.0% in the cure process. This appendix explains why the resin cure shrinkage, although very high, has not been included in the protocol developed in this chapter.

The residual strains from resin cure shrinkage are relevant at the fiber level, not at the ply level. As discussed in Chapter 10, the characterization process accounts for the residual cure strains at the fiber level which, therefore, are not relevant to the analyst. The high cure shrinkage of the polyester and vinyl ester resins (6.0% to 8.0% in volume) are absorbed as reduction in the laminate thickness. At the stress-free peak temperature PT, the cured and already solid laminate is in the following condition:

- All plies are stress-free.
- The cured, still hot and solid laminate has full contact with the mold surface
- The resin cure shrinkage has been absorbed as thickness reduction.
- The cure shrinkage originates resin-fiber micro-strains, not ply stresses.

The cured, hot and solid laminate shrinks and cools down from the peak temperature PT to the operating temperature OT. The residual thermal strains arise in the plies from this shrinkage.

Example 3.10

Compute the final diameter of a composite cylinder molded in the following conditions:

Mold diameter D = 3000 mm.
Peak cure temperature PT = 85C
Operating temperature OT = 25C
Coefficient of thermal expansion $\alpha_y = 20 \times 10^{-6}$/C

The resin cure shrinkage reduces the laminate thickness and is not relevant. The thermal shrinkage is

$$Thermal\ shrinkage = \alpha_y \left(PT - OT \right) \times D$$

$$Thermal\ shrinkage = 20 \times 10^{-6} \left(85 - 25 \right) \times 3000 = 3.6\ mm$$

The final diameter of the cylinder is 3000 – 3.6 = 2996.4 mm.

4 Standard Matrices of Commercial Laminates

4.1 INTRODUCTION

This chapter describes the commercial laminates used in industrial service and derives their standard matrices. The laminate standard matrices are fundamental in the computation of the total ply strains, which are essential in the analysis of the load-dependent durability. The Part II of this book presents a detailed discussion of this topic. This chapter introduces the fundamental concept of critical ply, whose failure defines the laminate durability. The critical ply depends on the mode of failure. The ply of chopped fibers in the corrosion barrier controls the weeping and the infiltration durability of sanitation and chemical pipes. The ply of UD fibers control the stiffness and the rupture failures of composite equipment in non-chemical services such as wind blades and aircraft parts.

The designer must decide which laminate ply is critical. There is no rule of thumb suggesting the weakest or the strongest ply as critical. For example:

- The ply of chopped fibers is the last to fail in weep and therefore is critical in the weeping failures. However, the first ply to fail in weep – the sand core – may become critical in low class and high stiffness pipes, when its breakdown leads to the immediate failure of the strongest ply of chopped fibers. See example 7.7 in Chapter 7.
- The UD plies loaded in the fiber direction are both the critical and the strongest in terms of laminate rupture.
- The transverse loaded UD plies are both the weakest and the critical in terms of stiffness durability.
- The strongest ply in terms of weeping, the aluminum foil, is never critical, as it breaks down following the failure of the critical chopped fibers (sanitation) or UD fibers (oil pipes)
- The surfacing veil is never the critical ply, as its failure does not compromise the infiltration, weeping, stiffness and rupture durability.

Although not obvious, the choice of the critical ply is easy and does not require much judgement. The failure of the critical ply always defines the laminate durability, as explained in the following section.

4.2   LAMINATE CONSTRUCTION AND THE CRITICAL PLIES

Figure 4.1 shows the three typical laminate constructions used in industrial service. To facilitate the discussion we sometimes refer to such laminates as pipes, in spite of their usage in a variety of other industrial equipment. The reference to pipes facilitates the description of the long-term failure modes with no loss of generality. What follows is a description of the function of each ply in these typical industrial laminates.

- *Liner:* The liner is a thin (0.3 mm) resin-rich inner barrier thought by many to provide water tightness and chemical durability to pipes and tanks. In chemical and sanitation service, the liner consists of a special lightweight veil saturated with resin. In oil pipes, the liner consists of resin alone, with no veil. As discussed later in this book, the resin-rich liner retards, but does not reduce the chemical presence in the corrosion barrier. The liners have a small effect on the durability of the corrosion barriers. Furthermore, they are not critical plies, since they have no effect on the rupture, stiffness, infiltration or weep thresholds.
- *Corrosion barrier:* The industrial laminates used in chemical or sanitation service have at least one inner protective ply of chopped fibers on top of the liner. In non-aggressive sanitation service, this inner ply is the "weep barrier". In chemical service, it is the "corrosion barrier". The corrosion barrier of chopped fibers is the critical ply for infiltration failure in chemical service. The weep barrier is the critical ply for weep failure in sanitation pipes. The oil pipes do not have corrosion/weep barriers. As explained in Chapter 11, the corrosion barrier is also critical to the laminate service life in aggressive chemicals.
- *Structural plies:* The structural wall of industrial laminates consists of several plies of continuous unidirectional UD fibers that provide

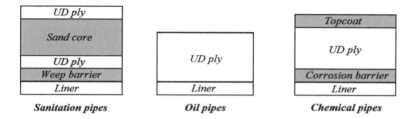

FIGURE 4.1   *Typical laminate constructions for sanitation, chemical and oil service. The corrosion barrier is the critical ply for infiltration failure in chemical service. The weep barrier is the critical ply for weep failure in sanitation pipes. The UD plies are critical for rupture and stiffness failures.*

the high modulus required to accommodate the internal pressure and other loads. The UD plies in the structural wall are critical in all long-term failures involving loss of stiffness and rupture. In oil pipes, which have no barrier of chopped fibers, the UD plies are also critical for weep. The UD plies have no direct contact with aggressive chemicals, and are not critical to infiltration failure.

- **Sand core:** The underground sanitation pipes usually have a core of sand-filled resin to increase their ring stiffness. These core plies are brittle and crack easily on impact, placing stringent demands on the careful handling of sanitation pipeline components. The thickness of the sand core varies according to the application. The sand core is the critical ply for the anomalous failure described in Chapter 20. Sometimes, as explained in the Example 7 of Chapter 7, the sand core is the critical ply for weeping.
- **Topcoat:** The topcoat is a thin (0.3 mm) resin-rich outer ply used for protection against spillage and UV radiation. The topcoat is not critical in any long-term mode of failure.

The foregoing discussion identified the UD and the chopped fibers as the two most important critical plies in industrial service. These two plies determine the durability of most industrial equipment.

4.3   THE STANDARD LAMINATE MATRICES

This chapter is about the computation of the matrices [A], [N^T] and [N^H] of the standard laminates used in commercial service. These laminates consist of a combination of standard plies of known thickness and matrices [Q], [α] and [β]. We start from a general standpoint, which is subsequently simplified and applied to the specific commercial laminates. The standard laminate matrices derived in this chapter are required to compute the total ply strains needed to estimate the load-dependent durability.

The designer feels relaxed and comfortable using the standard laminate mechanical properties, even if aware of their discrepancy from the actual values. As a rule, there are significant differences between the standard and the actual mechanical properties of the laminates. Such discrepancies, however, have no relevance as long as the laminate has the correct number of plies. The source of these differences in properties is the glass loadings. The resin-glass ratios in the actual laminates are never the same as those assumed in the standard construction. Some fabricators may choose to squeeze the laminate to save resin, while others may allow a little more resin to facilitate the fabrication. However, if the number of plies is the same in both laminates, the difference in resin usage is nothing more than a harmless deviation in thickness.

This thickness deviation is taken into account as a fictitious ply of neat resin that is either added to or subtracted from the standard laminate. For all practical purposes, the laminates built with the same plies have equal performance, even if their thickness and resin content differ.

This chapter will start with the protocol to compute the standard matrix and thickness of any laminate. The designer will work with the standard matrix and never use the actual one. However, should the actual matrix $[A]_{actual}$ be required, it can be readily computed by ascribing the difference in thickness between the actual and the standard laminates to a fictitious ply of neat resin that has been either added (thicker laminate) or subtracted (thinner laminate) in the manufacturing process. We assume the fictitious ply of resin spread out uniformly in the actual laminate. The introduction of such fictitious ply of resin allows the computation of the actual laminate matrix $[A]_{actual}$ and other properties, if desired. We will develop several numerical examples to illustrated this.

The embedded UD plies can be laid up to take any orientation with respect to the global laminate frame $x - y$. The ply orientation is the angle α formed between the principal axis 1 and the global axis "x". The ply position is the distance "z" from its center to the laminate mid-surface. The distance "z" and the angle α completely define the position and orientation of the embedded ply. The distance "z" is important in the computation of the flexural strains. The flexural strains, however, complicate the computations without throwing any light on the durability issue. Since our primary concern in this book is the understanding of the mechanics of durability, the distance "z" and the flexural strains are omitted in the discussions that follow.

In this book, the contribution of each individual ply to the laminate come from their properties in the global system $x - y$, with no reference to the distance "z". The next section shows how to rotate the ply properties from the local to the global frame.

4.4   ROTATING THE PLY MATRICES

The computation of the laminate matrix requires rotation of the ply matrices $[Q]$, $[\alpha]$ and $[\beta]$ from their local frames 1–2 to the global frame $x - y$. The laminate matrix allows the computation of the ply strains in the global frame. The computation of the ply strain components requires the back-rotation of the global strains to the local frame. The previous chapter detailed the process of back-rotation. The text that follows is almost a repetition of the procedure outlined in the previous chapter.

The ply matrix $[Q]_{12}$ in the local system 1–2 is

$$[Q]_{12} = \begin{bmatrix} Q_{11} & Q_{12} & 0 \\ Q_{21} & Q_{22} & 0 \\ 0 & 0 & Q_{66} \end{bmatrix} \text{kg/cm}^2$$

When rotated to the global system $x - y$ the above matrix becomes

$$[Q]_{xy} = \begin{bmatrix} Q_{xx} & Q_{xy} & Q_{xs} \\ Q_{yx} & Q_{yy} & Q_{ys} \\ Q_{sx} & Q_{sy} & Q_{ss} \end{bmatrix} \text{kg/cm}^2$$

The equations to compute the elements of the matrix $[Q]_{xy}$ in the global system are the same as those shown in the previous chapter.

$$Q_{xx} = m^4 Q_{11} + n^4 Q_{22} + 2m^2 n^2 Q_{12} + 4m^2 n^2 Q_{66}$$

$$Q_{xy} = Q_{yx} = m^2 n^2 Q_{11} + m^2 n^2 Q_{22} + (m^4 + n^4)Q_{12} - 4m^2 n^2 Q_{66}$$

$$Q_{xs} = Q_{sx} = m^3 n Q_{11} - mn^3 Q_{22} + (mn^3 - m^3 n)Q_{12} + 2(mn^3 - m^3 n)Q_{66} \tag{4.1}$$

$$Q_{yy} = n^4 Q_{11} + m^4 Q_{22} + 2m^2 n^2 Q_{12} + 4m^2 n^2 Q_{66}$$

$$Q_{ys} = Q_{sy} = mn^3 Q_{11} - m^3 n Q_{22} + (m^3 n - mn^3)Q_{12} + 2(m^3 n - mn^3)Q_{66}$$

$$Q_{ss} = m^2 n^2 Q_{11} + m^2 n^2 Q_{22} - 2m^2 n^2 Q_{12} + (m^2 - n^2)^2 Q_{66}$$

Where "m" is the cosine and "n" is the sine of the angle α formed by the local ply axis 1 and the global laminate axis "x". The equations to rotate the hydric and thermal expansion coefficients are also like those of the previous chapter.

$$\alpha_x = m^2 \alpha_1 + n^2 \alpha_2$$

$$\alpha_y = n^2 \alpha_1 + m^2 \alpha_2 \tag{4.2}$$

$$\alpha_{xy} = 2mn(\alpha_2 - \alpha_1)$$

$$\beta_x = m^2 \beta_1 + n^2 \beta_2$$

$$\beta_y = n^2 \beta_1 + m^2 \beta_2 \tag{4.3}$$

$$\beta_{xy} = 2mn(\beta_2 - \beta_1)$$

Where again "m" is the cosine and "n" is the sine of the angle α formed by the local ply axis 1 and the longitudinal global axis "x". This completes the rotation of the ply matrices. The next step is the computation of the laminate matrices.

4.5   COMPUTING THE LAMINATE MATRICES

The general equations to compute the laminate matrices [A], [N^T] and [N^H] were presented in the previous chapter. The laminate stiffness matrix [A] derives from the total contribution of all plies.

$$[A] = \begin{bmatrix} \sum Q_{xx}t & \sum Q_{xy}t & \sum Q_{xs}t \\ \sum Q_{yx}t & \sum Q_{yy}t & \sum Q_{ys}t \\ \sum Q_{sx}t & \sum Q_{sy}t & \sum Q_{ss}t \end{bmatrix} \text{kg/cm} \tag{4.4}$$

In Equation (4.4) "t" is the thickness in cm and [Q] is the matrix in kg/cm^2 of each ply referred to the global frame. The laminate hydric [N^H] and thermal [N^T] force resultants are:

$$\begin{bmatrix} N_x^T \\ N_y^T \\ N_{xy}^T \end{bmatrix} = \begin{bmatrix} \sum \left(Q_{xx}t\alpha_x + Q_{xy}t\alpha_y + Q_{xs}t\alpha_{xy} \right) \\ \sum \left(Q_{yx}t\alpha_x + Q_{yy}t\alpha_y + Q_{ys}t\alpha_{xy} \right) \\ \sum \left(Q_{sx}t\alpha_x + Q_{sy}t\alpha_y + Q_{ss}t\alpha_{xy} \right) \end{bmatrix} \times \left(OT - PT \right) \text{kg/cm} \tag{4.5}$$

$$\begin{bmatrix} N_x^H \\ N_y^H \\ N_{xy}^H \end{bmatrix} = \begin{bmatrix} \sum \left(Q_{xx}t\beta_x + Q_{xy}t\beta_y + Q_{xs}t\beta_{xy} \right) \\ \sum \left(Q_{yx}t\beta_x + Q_{yy}t\beta_y + Q_{ys}t\beta_{xy} \right) \\ \sum \left(Q_{sx}t\beta_x + Q_{sy}t\beta_y + Q_{ss}t\beta_{xy} \right) \end{bmatrix} \times \left(\Delta m \right) \text{kg/cm} \tag{4.6}$$

The thermal and hydric force resultants, as well as the laminate matrix [A], include the contributions from all plies. The thermal resultants depend on the temperature change $\Delta T = OT - PT$. The hydric resultants depend on the amount of water Δm picked up by the resin. In the above equations, "t" is the ply thickness and [Q] is the ply matrix in the global frame. The "alphas" and "betas" are the expansion coefficients of each ply, also in the global system. The thermal resultants take as reference the stress-free temperature PT. The hydric resultants take as reference the stress-free dry condition. The small thickness of composite laminates justifies the same operating temperature OT and water pick up Δm in all plies.

We are now ready to compute the matrices of standard commercial laminates. From these matrices, we compute the ply total strains and the load-dependent durability.

4.6   MATRICES OF ANGLE-PLY ± 55 LAMINATES

The circular cylindrical ± 55 laminates find use in pressure vessels and oil pipes. They consist exclusively of ± 55 UD plies, with no weep/corrosion barrier of chopped fibers. The absence of the barrier of chopped fibers is acceptable in applications involving benign environments, such as those found in the oil and gas industry. We proceed to compute the matrices of ± 55 standard laminates.

From Chapter 1 the local stiffness matrix $[Q]_{12}$ of standard ± 55 UD plies is

$$
\begin{bmatrix}
Q_{11} & Q_{12} & 0 \\
Q_{12} & Q_{22} & 0 \\
0 & 0 & Q_{66}
\end{bmatrix}
=
\begin{bmatrix}
400000 & 30000 & 0 \\
30000 & 100000 & 0 \\
0 & 0 & 35000
\end{bmatrix}
\text{kg/cm}^2
$$

The elements of the ply matrix $[Q]_{xy}$ in the global frame are computed as indicated in Equation (4.1) for $\alpha = 55$ degrees.

$$
\begin{bmatrix}
Q_{xx} & Q_{xy} & Q_{xs} \\
Q_{xy} & Q_{yy} & Q_{ys} \\
Q_{xs} & Q_{ys} & Q_{ss}
\end{bmatrix}_{\pm 55}
=
\begin{bmatrix}
132470 & 96220 & \pm 46370 \\
96220 & 235070 & \pm 94580 \\
\pm 46370 & \pm 94580 & 101220
\end{bmatrix}
\text{kg/cm}^2
\qquad (4.7)
$$

The above is the rotated UD ply matrix for $\alpha \pm 55$ degrees. The ± sign identifies the UD ply orientations +55 or –55 degrees. Equation (4.4) computes the laminate matrix [A].

$$
[A]_{\pm 55} = \left(\frac{t}{2}\right) \times
\begin{bmatrix}
Q_{xx} & Q_{xy} & Q_{xs} \\
Q_{yx} & Q_{yy} & Q_{ys} \\
Q_{sx} & Q_{sy} & Q_{ss}
\end{bmatrix}_{+55}
+
\left(\frac{t}{2}\right) \times
\begin{bmatrix}
Q_{xx} & Q_{xy} & Q_{xs} \\
Q_{yx} & Q_{yy} & Q_{ys} \\
Q_{sx} & Q_{sy} & Q_{ss}
\end{bmatrix}_{-55}
$$

Entering the values from Equation (4.7) and taking into account that all commercial laminates are balanced, we obtain

$$[A]_{\pm55} = \begin{bmatrix} 132470 & 96220 & 0 \\ 96220 & 235070 & 0 \\ 0 & 0 & 101220 \end{bmatrix} \times (t)\text{kg/cm} \qquad (4.8)$$

In Equation (4.8) "t" is the total laminate thickness in cm, obtained by adding the thicknesses of all individual plies. We hope the reader will not get confused that we have used the same letter "t" to designate both the ply and the laminate thickness.

The coefficients of hydric expansion of the individual ± 55 plies in the global frame come from Equation (4.3).

$$\beta_x = m^2\beta_1 + n^2\beta_2 = 0.33 \times 0.04 + 0.67 \times 0.24 = 0.18$$

$$\beta_y = n^2\beta_1 + m^2\beta_2 = 0.67 \times 0.04 + 0.33 \times 0.24 = 0.11 \qquad (4.9)$$

$$\beta_{xy} = 2mn \times (\beta_2 - \beta_1) = \pm0.94 \times (0.24 - 0.04) = \pm0.19$$

The hydric force resultants come from entering the appropriate values of (4.7) and (4.9) in Equation (4.6).

$$\begin{bmatrix} N_x^H \\ N_y^H \\ N_{xy}^H \end{bmatrix}_{\pm55} = \begin{bmatrix} 132470 \times 0.18 + 96220 \times 0.11 + 46370 \times 0.19 \\ 96220 \times 0.18 + 235070 \times 0.11 + 94580 \times 0.19 \\ 0 \end{bmatrix} \times (t) \times (\Delta m)$$

$$\begin{bmatrix} N_x^H \\ N_y^H \\ N_{xy}^H \end{bmatrix}_{\pm55} = \begin{bmatrix} 43240 \\ 61150 \\ 0 \end{bmatrix} \times (t) \times (\Delta m)\text{kg/cm} \qquad (4.10)$$

Equation (4.10) computes the hydric force resultants of ± 55 laminates. These resultants are directly proportional to the laminate thickness "t" and the water pick up "Δm".

The steps involved in the derivation of the thermal resultants are similar to the above. The coefficients of thermal expansion of the UD plies in the global system come from (4.3) with m = cos (±55) = 0.573 and n = sin (±55) = ± 0.819.

$$\alpha_x = (0.573)^2 \times 7.5 \times 10^{-6} + (0.819)^2 \times 45.0 \times 10^{-6} = 32. \times 10^{-6}$$

$$\alpha_y = (0.819)^2 \times 7.5 \times 10^{-6} + (0.573)^2 \times 45.0 \times 10^{-6} = 19.9 \times 10^{-6}$$

$$\alpha_{xy} = \pm2 \times 0.573 \times 0.819 \times (45.0 - 7.5) \times 10^{-6} = \pm35.2 \times 10^{-6}$$

The above enter Equation (4.5) to give

$$
\begin{bmatrix} N_x^T \\ N_y^T \\ N_{xy}^T \end{bmatrix}_{\pm55} = \begin{bmatrix} 7.86 \\ 11.14 \\ 0 \end{bmatrix} \times (t) \times (OT - PT)\text{kg/cm} \tag{4.11}
$$

In the above, t is the laminate thickness in cm. The temperatures enter the equation in degrees Celcius. The thermal resultants are usually negative since in most applications PT > OT. The thermal resultants are proportional to the laminate thickness "t" and to the temperature drop $\Delta T = PT - OT$.

Table 4.1 shows all the matrices of angle-ply ± 55 laminates.

TABLE 4.1
Matrices of standard angle-ply ± 55 laminates.

Angle-ply ± 55 laminate matrices in the global frame

$$
[A]_{\pm55} = \begin{bmatrix} 132470 & 96220 & 0 \\ 96220 & 235070 & 0 \\ 0 & 0 & 101220 \end{bmatrix} \times (t)\,\text{kg/cm}
$$

$$
\begin{bmatrix} \alpha_x \\ \alpha_y \\ \alpha_{xy} \end{bmatrix}_{\pm55} = \begin{bmatrix} 32.6 \\ 19.9 \\ \pm35.2 \end{bmatrix} \times 10^{-6}\,/°C
$$

These are ply properties, not laminate properties. For the laminate properties, see Examples 1 and 2 in the next chapter

$$
\begin{bmatrix} \beta_x \\ \beta_y \\ \beta_{xy} \end{bmatrix}_{\pm55} = \begin{bmatrix} 0.18 \\ 0.11 \\ \pm0.19 \end{bmatrix}
$$

These are ply properties, not laminate properties. For the laminate properties, see Examples 1 and 2 in the next chapter

$$
\begin{bmatrix} N_x^T \\ N_y^T \\ N_{xy}^T \end{bmatrix}_{\pm55} = \begin{bmatrix} 7.86 \\ 11.14 \\ 0 \end{bmatrix} \times (t) \times (OT - PT)\,\text{kg/cm}
$$

$$
\begin{bmatrix} N_x^H \\ N_y^H \\ N_{xy}^H \end{bmatrix}_{\pm55} = \begin{bmatrix} 43240 \\ 61150 \\ 0 \end{bmatrix} \times (t) \times (\Delta m)\,\text{kg/cm}
$$

4.7 MATRICES OF ANGLE-PLY ± 70 LAMINATES

The ± 70 angle-ply laminates find use in vertical storage tanks and underground sanitation pipes. The water and urban sewage of sanitation pipes are benign environments that do no harm to the corrosion/weep barrier. The preserved weep barrier is included in the computations of the laminate matrices. The computation of the ± 70 standard laminate matrices follows the same protocol outlined in the previous section. First, we rotate the UD ply matrices [Q] from the local into the global system. The chopped ply matrices are isotropic and do not require rotation.

To rotate the ± 70 UD stiffness matrix [Q] to the global system we take $m = \cos(\pm70) = 0.34$ and $n = \sin(\pm70) = \pm 0.94$ in Equation (4.1).

$$\begin{bmatrix} Q_{xx} & Q_{xy} & Q_{xs} \\ Q_{xy} & Q_{yy} & Q_{ys} \\ Q_{xs} & Q_{ys} & Q_{ss} \end{bmatrix}_{\pm70} = \begin{bmatrix} 104000 & 61000 & \pm11300 \\ 61000 & 334000 & \pm85140 \\ \pm11300 & \pm85140 & 66000 \end{bmatrix} kg/cm^2$$

The matrix [Q] of the isotropic chopped ply is invariant

$$\begin{bmatrix} Q_{xx} & Q_{xy} & Q_{xs} \\ Q_{yx} & Q_{yy} & Q_{ys} \\ Q_{sx} & Q_{sy} & Q_{ss} \end{bmatrix}_{chop} = \begin{bmatrix} Q_{11} & Q_{12} & 0 \\ Q_{12} & Q_{22} & 0 \\ 0 & 0 & Q_{66} \end{bmatrix}_{chop} = \begin{bmatrix} 77000 & 23000 & 0 \\ 23000 & 77000 & 0 \\ 0 & 0 & 27000 \end{bmatrix} kg/cm^2$$

The ± 70 standard laminate matrix [A] comes from Equation (4.4)

$$[A]_{\pm70} = \left(\frac{t_{UD}}{2}\right) \times \begin{bmatrix} Q_{xx} & Q_{xy} & Q_{xs} \\ Q_{yx} & Q_{yy} & Q_{ys} \\ Q_{sx} & Q_{sy} & Q_{ss} \end{bmatrix}_{+70} + \left(\frac{t_{UD}}{2}\right) \times \begin{bmatrix} Q_{xx} & Q_{xy} & Q_{xs} \\ Q_{yx} & Q_{yy} & Q_{ys} \\ Q_{sx} & Q_{sy} & Q_{ss} \end{bmatrix}_{-70}$$

$$+ \left(t_{chop}\right) \times \begin{bmatrix} Q_{xx} & Q_{xy} & Q_{xs} \\ Q_{yx} & Q_{yy} & Q_{ys} \\ Q_{sx} & Q_{sy} & Q_{ss} \end{bmatrix}_{chop}$$

$$[A]_{\pm70} = \begin{bmatrix} 104000 & 61000 & 0 \\ 61000 & 334000 & 0 \\ 0 & 0 & 66000 \end{bmatrix} \times (t_{UD}) + \begin{bmatrix} 77000 & 23000 & 0 \\ 23000 & 77000 & 0 \\ 0 & 0 & 27000 \end{bmatrix}$$

$$\times (t_{chop}) \, kg/cm$$

In the above, t_{UD} and t_{chop} are the total thicknesses of the UD and the chopped fiber plies respectively.

The thermal and hydric coefficients of the UD ply in the global system are.

$$\alpha_x = (\cos 70)^2 (7.5 \times 10^{-6}) + (sen70)^2 (45.0 \times 10^{-6}) = 40.61 \times 10^{-6}$$

$$\alpha_y = (sen70)^2 (7.5 \times 10^{-6}) + (\cos 70)^2 (45.0 \times 10^{-6}) = 11.89 \times 10^{-6}$$

$$\alpha_{xy} = 2 \times (\cos 70)(sen70)(45.0 - 7.5) \times 10^{-6} = \pm 24.10 \times 10^{-6}$$

$$\beta_x = (\cos 70)^2 (0.04) + (sen70)^2 (0.24) = 0.22$$

$$\beta_y = (sen70)^2 (0.04) + (\cos 70)^2 (0.24) = 0.06$$

$$\beta_{xy} = 2(\cos 70)(sen70)(0.24 - 0.04) = \pm 0.13$$

$$\begin{bmatrix} \beta_x \\ \beta_y \\ \beta_{xy} \end{bmatrix}_{\pm70} = \begin{bmatrix} 0.22 \\ 0.06 \\ \pm 0.13 \end{bmatrix} \quad \begin{bmatrix} \beta_x \\ \beta_y \\ \beta_{xy} \end{bmatrix}_{chop} = \begin{bmatrix} \beta_1 \\ \beta_2 \\ \beta_3 \end{bmatrix}_{chop} = \begin{bmatrix} 0.25 \\ 0.25 \\ 0 \end{bmatrix}$$

$$\begin{bmatrix} \alpha_x \\ \alpha_y \\ \alpha_{xy} \end{bmatrix}_{\pm70} = \begin{bmatrix} 40.61 \\ 11.89 \\ \pm 24.10 \end{bmatrix} \times 10^{-6}/^\circ C \quad \begin{bmatrix} \alpha_x \\ \alpha_y \\ \alpha_{xy} \end{bmatrix}_{chop} = \begin{bmatrix} \alpha_1 \\ \alpha_2 \\ \alpha_{12} \end{bmatrix}_{chop}$$

$$= \begin{bmatrix} 25.0 \\ 25.0 \\ 0 \end{bmatrix} \times 10^{-6}/^\circ C$$

The thermal resultants from the general Equation (4.5) are

$$
\begin{bmatrix} N_x^T \\ N_y^T \\ N_{xy}^T \end{bmatrix}_{\pm 70} = \begin{bmatrix} Q_{xx} & Q_{xy} & Q_{xs} \\ Q_{yx} & Q_{yy} & Q_{ys} \\ Q_{sx} & Q_{sy} & Q_{ss} \end{bmatrix}_{chop} \begin{bmatrix} \alpha_x \\ \alpha_y \\ \alpha_{xy} \end{bmatrix}_{chop} (t_{chop}) \times (OT - PT)
$$

$$
+ \begin{bmatrix} Q_{xx} & Q_{xy} & Q_{xs} \\ Q_{yx} & Q_{yy} & Q_{ys} \\ Q_{sx} & Q_{sy} & Q_{ss} \end{bmatrix}_{+70} \begin{bmatrix} \alpha_x \\ \alpha_y \\ \alpha_{xy} \end{bmatrix}_{+70} \left(\frac{t_{UD}}{2} \right) \times (OT - PT)
$$

$$
+ \begin{bmatrix} Q_{xx} & Q_{xy} & Q_{xs} \\ Q_{yx} & Q_{yy} & Q_{ys} \\ Q_{sx} & Q_{sy} & Q_{ss} \end{bmatrix}_{-70} \begin{bmatrix} \alpha_x \\ \alpha_y \\ \alpha_{xy} \end{bmatrix}_{-70} \left(\frac{t_{UD}}{2} \right) \times (OT - PT)
$$

Entering the appropriate numerical values in the above, we obtain.

$$
\begin{bmatrix} N_x^T \\ N_y^T \\ N_{xy}^T \end{bmatrix}_{\pm 70} = \begin{bmatrix} 5.22 \\ 8.50 \\ 0 \end{bmatrix} \times (t_{UD}) \times (OT - PT) + \begin{bmatrix} 2.50 \\ 2.50 \\ 0 \end{bmatrix} \times (t_{chop}) \times (OT - PT)
$$

$$
\begin{bmatrix} N_x^T \\ N_y^T \\ N_{xy}^T \end{bmatrix}_{\pm 70} = \begin{bmatrix} 5.22 \times t_{UD} + 2.50 \times t_{chop} \\ 8.50 \times t_{UD} + 2.50 \times t_{chop} \\ 0 \end{bmatrix} \times (OT - PT) \text{kg/cm} \tag{4.12}
$$

The derivation of the hydric resultants is similar

$$
\begin{bmatrix} N_x^H \\ N_y^H \\ N_{xy}^H \end{bmatrix} = \begin{bmatrix} \sum \left(Q_{xx} t \beta_x + Q_{xy} t \beta_y + Q_{xs} t \beta_{xy} \right) \\ \sum \left(Q_{yx} t \beta_x + Q_{yy} t \beta_y + Q_{ys} t \beta_{xy} \right) \\ \sum \left(Q_{sx} t \beta_x + Q_{sy} t \beta_y + Q_{ss} t \beta_{xy} \right) \end{bmatrix} \times (\Delta m)
$$

$$
\begin{bmatrix} N_x^H \\ N_y^H \\ N_{xy}^H \end{bmatrix}_{\pm 70} = \begin{bmatrix} 28000 \\ 44500 \\ 0 \end{bmatrix} \times (t_{UD}) \times (\Delta m) + \begin{bmatrix} 25000 \\ 25000 \\ 0 \end{bmatrix} \times (t_{chop}) \times (\Delta m)
$$

$$
\begin{bmatrix} N_x^H \\ N_y^H \\ N_{xy}^H \end{bmatrix}_{\pm 70} = \begin{bmatrix} 28000 \times t_{UD} + 25000 \times t_{chop} \\ 44500 \times t_{UD} + 25000 \times t_{chop} \\ 0 \end{bmatrix} \times \Delta m \text{kg/cm} \tag{4.13}
$$

Table 4.2 summarizes all of the above. The expressions in Table 4.2 compute the strains in circular cylindrical $\pm$ 70 laminates with preserved – non-destroyed – corrosion barriers.

4.8 MATRICES OF HOOP-CHOP LAMINATES

The hoop-chop laminates combine UD, chopped and sand plies. The UD plies are laid up with the fibers in the hoop direction, at an angle $\alpha = 90$

TABLE 4.2
Standard matrices of ± 70 angle-ply laminates with corrosion barrier.

Angle-ply ± 70 laminate matrices in the global system.

$$
[A]_{\pm 70} = \begin{bmatrix} 104000 & 61000 & 0 \\ 61000 & 334000 & 0 \\ 0 & 0 & 66000 \end{bmatrix} \times (t_{UD}) + \begin{bmatrix} 77000 & 23000 & 0 \\ 23000 & 77000 & 0 \\ 0 & 0 & 27000 \end{bmatrix} \times (t_{chop}) \text{ kg/cm}
$$

$$
\begin{bmatrix} \alpha_x \\ \alpha_y \\ \alpha_{xy} \end{bmatrix}_{\pm 70} = \begin{bmatrix} 40.61 \\ 11.89 \\ \pm 24.10 \end{bmatrix} \times 10^{-6}/{}^{\circ}C \qquad \begin{bmatrix} \alpha_x \\ \alpha_y \\ \alpha_{xy} \end{bmatrix}_{chop} = \begin{bmatrix} \alpha_1 \\ \alpha_2 \\ \alpha_{12} \end{bmatrix}_{chop} = \begin{bmatrix} 25.0 \\ 25.0 \\ 0 \end{bmatrix} \times 10^{-6}/{}^{\circ}C
$$

These are ply properties, not laminate properties. For the laminate properties, see Examples 1 and 2 in the next chapter.

$$
\begin{bmatrix} \beta_x \\ \beta_y \\ \beta_{xy} \end{bmatrix}_{\pm 70} = \begin{bmatrix} 0.22 \\ 0.06 \\ \pm 0.13 \end{bmatrix} \qquad \begin{bmatrix} \beta_x \\ \beta_y \\ \beta_{xy} \end{bmatrix}_{chop} = \begin{bmatrix} \beta_1 \\ \beta_2 \\ \beta_3 \end{bmatrix}_{chop} = \begin{bmatrix} 0.25 \\ 0.25 \\ 0 \end{bmatrix}
$$

These are ply properties, not laminate properties. For the laminate properties, see Examples 1 and 2 in the next chapter.

$$
\begin{bmatrix} N_x^T \\ N_y^T \\ N_{xy}^T \end{bmatrix}_{\pm 70} = \begin{bmatrix} 5.22 \times t_{UD} + 2.50 \times t_{chop} \\ 8.50 \times t_{UD} + 2.50 \times t_{chop} \\ 0 \end{bmatrix} \times (OT - PT) \text{kg/cm}
$$

$$
\begin{bmatrix} N_x^H \\ N_y^H \\ N_{xy}^H \end{bmatrix}_{\pm 70} = \begin{bmatrix} 28000 \times t_{UD} + 25000 \times t_{chop} \\ 44500 \times t_{UD} + 25000 \times t_{chop} \\ 0 \end{bmatrix} \times \Delta m \text{kg/cm}
$$

---

degrees to the longitudinal axis. From Equation (1), and taking into account that $m = \cos(90) = 0$ and $n = \sin(90) = 1$, the $[Q]_{xy}$ matrix for the UD ply in the global system is

$$
\begin{bmatrix} Q_{xx} & Q_{xy} & Q_{xs} \\ Q_{xy} & Q_{yy} & Q_{ys} \\ Q_{xs} & Q_{ys} & Q_{ss} \end{bmatrix}_{UD} = \begin{bmatrix} 100000 & 30000 & 0 \\ 30000 & 400000 & 0 \\ 0 & 0 & 35000 \end{bmatrix} \text{kg/cm}^2
$$

The plies of chopped fibers and sand are isotropic and need no rotation.

$$
\begin{bmatrix} Q_{xx} & Q_{xy} & Q_{xs} \\ Q_{yx} & Q_{yy} & Q_{ys} \\ Q_{sx} & Q_{sy} & Q_{ss} \end{bmatrix}_{chop} = \begin{bmatrix} Q_{11} & Q_{12} & 0 \\ Q_{12} & Q_{22} & 0 \\ 0 & 0 & Q_{66} \end{bmatrix}_{chop}
$$

$$
= \begin{bmatrix} 77000 & 23000 & 0 \\ 23000 & 77000 & 0 \\ 0 & 0 & 27000 \end{bmatrix} \text{kg/cm}^2
$$

$$
\begin{bmatrix} Q_{xx} & Q_{xy} & Q_{xs} \\ Q_{xy} & Q_{yy} & Q_{ys} \\ Q_{xs} & Q_{ys} & Q_{ss} \end{bmatrix}_{SAND} = \begin{bmatrix} Q_{11} & Q_{12} & 0 \\ Q_{21} & Q_{22} & 0 \\ 0 & 0 & Q_{66} \end{bmatrix}_{SAND}
$$

$$
= \begin{bmatrix} 66000 & 20000 & 0 \\ 20000 & 66000 & 0 \\ 0 & 0 & 23000 \end{bmatrix} \text{kg/cm}^2
$$

The coefficients of thermal expansion of the UD plies in the global system are

$$\alpha_x = (\cos 90)^2 (7.5 \times 10^{-6}) + (sen90)^2 (45.0 \times 10^{-6}) = 45.00 \times 10^{-6}$$

$$\alpha_y = (sen90)^2 (7.5 \times 10^{-6}) + (\cos 90)^2 (45.0 \times 10^{-6}) = 7.50 \times 10^{-6}$$

$$\alpha_{xy} = 0$$

In a like manner, the global hydric expansion coefficients of UD plies are

$$\beta_x = (\cos 90)^2 (0.04) + (sen 90)^2 (0.24) = 0.24$$

$$\beta_y = (sen 90)^2 (0.04) + (\cos 90)^2 (0.24) = 0.04$$

$$\beta_{xy} = 0$$

The plies of sand and chopped fibers are isotropic and have the same properties in all reference frames.

$$\begin{bmatrix} \alpha_x \\ \alpha_y \\ \alpha_{xy} \end{bmatrix}_{UD} = \begin{bmatrix} \alpha_2 \\ \alpha_1 \\ \alpha_{12} \end{bmatrix}_{UD} = \begin{bmatrix} 45.0 \\ 7.5 \\ 0 \end{bmatrix} \times 10^{-6}/{}^{\circ}C \qquad \begin{bmatrix} \beta_x \\ \beta_y \\ \beta_{xy} \end{bmatrix}_{UD} = \begin{bmatrix} \beta_2 \\ \beta_1 \\ \beta_{12} \end{bmatrix}_{UD} = \begin{bmatrix} 0.24 \\ 0.04 \\ 0.0 \end{bmatrix}$$

$$\begin{bmatrix} \alpha_x \\ \alpha_y \\ \alpha_{xy} \end{bmatrix}_{chop} = \begin{bmatrix} \alpha_1 \\ \alpha_2 \\ \alpha_{12} \end{bmatrix}_{chop} = \begin{bmatrix} 25.0 \\ 25.0 \\ 0.0 \end{bmatrix} \times 10^{-6}/{}^{\circ}C \qquad \begin{bmatrix} \alpha_x \\ \alpha_y \\ \alpha_{xy} \end{bmatrix}_{sand}$$

$$= \begin{bmatrix} \alpha_1 \\ \alpha_2 \\ \alpha_{12} \end{bmatrix}_{sand} = \begin{bmatrix} 13.0 \\ 13.0 \\ 0.0 \end{bmatrix} \times 10^{-6}/{}^{\circ}C$$

$$\begin{bmatrix} \beta_x \\ \beta_y \\ \beta_{xy} \end{bmatrix}_{chop} = \begin{bmatrix} \beta_1 \\ \beta_2 \\ \beta_{12} \end{bmatrix}_{chop} = \begin{bmatrix} 0.25 \\ 0.25 \\ 0.0 \end{bmatrix} \qquad \begin{bmatrix} \beta_x \\ \beta_y \\ \beta_{xy} \end{bmatrix}_{sand} = \begin{bmatrix} \beta_1 \\ \beta_2 \\ \beta_{12} \end{bmatrix}_{sand} = \begin{bmatrix} 0.28 \\ 0.28 \\ 0 \end{bmatrix}$$

The stiffness matrix [A] of hoop-chop laminates is

$$[A] = (t_{UD}) \times \begin{bmatrix} Q_{22} & Q_{21} & 0 \\ Q_{12} & Q_{11} & 0 \\ 0 & 0 & Q_{66} \end{bmatrix}_{UD} + (t_{chop}) \times \begin{bmatrix} Q_{11} & Q_{12} & 0 \\ Q_{21} & Q_{22} & 0 \\ 0 & 0 & Q_{66} \end{bmatrix}_{chop}$$

$$+ (t_{sand}) \times \begin{bmatrix} Q_{11} & Q_{12} & 0 \\ Q_{21} & Q_{22} & 0 \\ 0 & 0 & Q_{66} \end{bmatrix}_{sand}$$

Entering the numerical values just developed in these equations, we obtain

$$[A] = (t_{UD}) \times \begin{bmatrix} 100000 & 30000 & 0 \\ 30000 & 400000 & 0 \\ 0 & 0 & 35000 \end{bmatrix} + (t_{chop}) \times \begin{bmatrix} 77000 & 23000 & 0 \\ 23000 & 77000 & 0 \\ 0 & 0 & 27000 \end{bmatrix}$$

$$+ (t_{sand}) \times \begin{bmatrix} 66000 & 20000 & 0 \\ 20000 & 66000 & 0 \\ 0 & 0 & 23000 \end{bmatrix} kg/cm$$

The thermal and hydric resultants are

$$\begin{bmatrix} N_x^H \\ N_y^H \\ N_{xy}^H \end{bmatrix} = \begin{bmatrix} 25200 \times t_{UD} + 25000 \times t_{chop} + 24080 \times t_{sand} \\ 23200 \times t_{UD} + 25000 \times t_{chop} + 24080 \times t_{sand} \\ 0 \end{bmatrix} \times (\Delta m) kg/cm \quad (4.14)$$

$$\begin{bmatrix} N_x^T \\ N_y^T \\ N_{xy}^T \end{bmatrix} = \begin{bmatrix} 4.73 \times t_{UD} + 2.50 \times t_{chop} + 1.12 \times t_{sand} \\ 4.35 \times t_{UD} + 2.50 \times t_{chop} + 1.12 \times t_{sand} \\ 0 \end{bmatrix} \times (OT - PT) kg/cm \quad (4.15)$$

The computed results are in Table 4.3.

TABLE 4.3

Matrices of standard hoop-chop laminates with sand core.

Hoop-chop matrices in the global system.

$$[A]_{HC} = \begin{bmatrix} 100000 & 30000 & 0 \\ 30000 & 400000 & 0 \\ 0 & 0 & 35000 \end{bmatrix} \times (t_{UD}) + \begin{bmatrix} 77000 & 23000 & 0 \\ 23000 & 77000 & 0 \\ 0 & 0 & 27000 \end{bmatrix} \times (t_{chop})$$

$$+ \begin{bmatrix} 66000 & 20000 & 0 \\ 20000 & 66000 & 0 \\ 0 & 0 & 23000 \end{bmatrix} \times (t_{sand})$$

$$\begin{bmatrix} \alpha_x \\ \alpha_y \\ \alpha_{xy} \end{bmatrix}_{UD} = \begin{bmatrix} \alpha_2 \\ \alpha_1 \\ \alpha_{12} \end{bmatrix}_{UD} = \begin{bmatrix} 45.0 \\ 7.5 \\ 0 \end{bmatrix} \times 10^{-6}/°C$$

$$\begin{bmatrix} \alpha_x \\ \alpha_y \\ \alpha_{xy} \end{bmatrix}_{chop} = \begin{bmatrix} \alpha_1 \\ \alpha_2 \\ \alpha_{12} \end{bmatrix}_{chop} = \begin{bmatrix} 25.0 \\ 25.0 \\ 0 \end{bmatrix} \times 10^{-6}/°C$$

(Continued)

TABLE 4.3 (*Continued*)
Matrices of standard hoop-chop laminates with sand core.

$$
\begin{bmatrix} \alpha_x \\ \alpha_y \\ \alpha_{xy} \end{bmatrix}_{sand} = \begin{bmatrix} \alpha_1 \\ \alpha_2 \\ \alpha_{12} \end{bmatrix}_{sand} = \begin{bmatrix} 13.0 \\ 13.0 \\ 0 \end{bmatrix} \times 10^{-6}/^\circ C
$$

These are ply properties, not laminate properties. For the laminate properties, see the Examples 1 and 2 in the next chapter.

$$
\begin{bmatrix} \beta_x \\ \beta_y \\ \beta_{xy} \end{bmatrix}_{UD} = \begin{bmatrix} \beta_2 \\ \beta_1 \\ \beta_{12} \end{bmatrix}_{UD} = \begin{bmatrix} 0.24 \\ 0.04 \\ 0.0 \end{bmatrix} \qquad \begin{bmatrix} \beta_x \\ \beta_y \\ \beta_{xy} \end{bmatrix}_{sand} = \begin{bmatrix} \beta_1 \\ \beta_2 \\ \beta_{12} \end{bmatrix}_{sand} = \begin{bmatrix} 0.28 \\ 0.28 \\ 0 \end{bmatrix}
$$

$$
\begin{bmatrix} \beta_x \\ \beta_y \\ \beta_{xy} \end{bmatrix}_{chop} = \begin{bmatrix} \beta_1 \\ \beta_2 \\ \beta_{12} \end{bmatrix}_{chop} = \begin{bmatrix} 0.25 \\ 0.25 \\ 0.0 \end{bmatrix}
$$

These are ply properties, not laminate properties. For the laminate properties, see the Examples 1 and 2 in the next chapter.

$$
\begin{bmatrix} N_x^T \\ N_y^T \\ N_{xy}^T \end{bmatrix} = \begin{bmatrix} 4.73 \times t_{UD} + 2.50 \times t_{chop} + 1.12 \times t_{sand} \\ 4.35 \times t_{UD} + 2.50 \times t_{chop} + 1.12 \times t_{sand} \\ 0 \end{bmatrix} \times (OT - PT)\,\text{kg/cm}
$$

$$
\begin{bmatrix} N_x^H \\ N_y^H \\ N_{xy}^H \end{bmatrix} = \begin{bmatrix} 25200 \times t_{UD} + 25000 \times t_{chop} + 24080 \times t_{sand} \\ 23200 \times t_{UD} + 25000 \times t_{chop} + 24080 \times t_{sand} \\ 0 \end{bmatrix} \times (\Delta m)\,\text{kg/cm}
$$

This completes our task. We have developed the standard matrices [A], [N^T] and [N^H] for the three most important commercial laminates. These matrices enter in the computation of the thermal, hydric and mechanical strains of embedded plies.

Example 4.1

Derive the matrices of ± 55 laminates with corrosion barriers.

Benign chemicals, like water, oil and urban sewage, preserve the corrosion barrier. The preserved barrier contributes to the laminate matrices. The stiffness matrix [A] of ± 55 laminates with preserved barriers is

$$
[A]_{\pm55} = \begin{bmatrix} 132470 & 96220 & 0 \\ 96220 & 235070 & 0 \\ 0 & 0 & 101220 \end{bmatrix} \times (t_{UD}) + \begin{bmatrix} 77000 & 23000 & 0 \\ 23000 & 77000 & 0 \\ 0 & 0 & 27000 \end{bmatrix}
$$

$$
\times \left(t_{chop} \right) \text{kg/cm}
$$

The hydric and thermal resultants should include the contribution of the preserved corrosion barrier

$$
\begin{bmatrix} N_x^H \\ N_y^H \\ N_{xy}^H \end{bmatrix}_{\pm55} = \begin{bmatrix} 43240 \\ 61150 \\ 0 \end{bmatrix} \times (t_{UD}) \times (\Delta m) + \begin{bmatrix} 25000 \\ 25000 \\ 0 \end{bmatrix} \times \left(t_{chop} \right) \times \Delta m \text{kg/cm}
$$

$$
\begin{bmatrix} N_x^T \\ N_y^T \\ N_{xy}^T \end{bmatrix}_{\pm55} = \begin{bmatrix} 7.86 \\ 11.14 \\ 0 \end{bmatrix} \times (t_{UD}) \times (OT - PT) + \begin{bmatrix} 2.50 \\ 2.50 \\ 0 \end{bmatrix} \times \left(t_{chop} \right) \times \left(OT - PT \right) \text{kg/cm}
$$

The laminate matrices of any layup come from a simple addition of the partial matrices of each ply.

Example 4.2

Compute the matrices of ± 70 laminates operating in corrosive environments.

The aggressive environments destroy the corrosion barriers. The desired matrices are obtained by simply setting $t_{chop} = 0$ in the equations listed in Table 4.2. For the stiffness matrix [A] we have

$$
[A]_{\pm70} = \begin{bmatrix} 104000 & 61000 & 0 \\ 61000 & 334000 & 0 \\ 0 & 0 & 66000 \end{bmatrix} \times (t_{UD}) + \begin{bmatrix} 77000 & 23000 & 0 \\ 23000 & 77000 & 0 \\ 0 & 0 & 27000 \end{bmatrix} \times \left(t_{chop} \right)
$$

Setting $t_{chop} = 0$ we have

$$
[A]_{\pm70} = \begin{bmatrix} 104000 & 61000 & 0 \\ 61000 & 334000 & 0 \\ 0 & 0 & 66000 \end{bmatrix} \times (t_{UD})
$$

The derivation of the thermal and hydric resultants is similar

$$
\begin{bmatrix} N_x^T \\ N_y^T \\ N_{xy}^T \end{bmatrix}_{\pm 70} = \begin{bmatrix} 5.22 \\ 8.50 \\ 0 \end{bmatrix} \times (t_{UD}) \times (OT - PT) + \begin{bmatrix} 2.50 \\ 2.50 \\ 0 \end{bmatrix} \times (t_{chop}) \times (OT - PT)
$$

$$
\begin{bmatrix} N_x^T \\ N_y^T \\ N_{xy}^T \end{bmatrix}_{\pm 70} = \begin{bmatrix} 5.22 \times t_{UD} \\ 8.50 \times t_{UD} \\ 0 \end{bmatrix} \times (OT - PT)
$$

$$
\begin{bmatrix} N_x^H \\ N_y^H \\ N_{xy}^H \end{bmatrix}_{\pm 70} = \begin{bmatrix} 28000 \\ 44500 \\ 0 \end{bmatrix} \times (t_{UD}) \times (\Delta m) + \begin{bmatrix} 25000 \\ 25000 \\ 0 \end{bmatrix} \times (t_{chop}) \times (\Delta m)
$$

$$
\begin{bmatrix} N_x^H \\ N_y^H \\ N_{xy}^H \end{bmatrix}_{\pm 70} = \begin{bmatrix} 28000 \times t_{UD} \\ 44500 \times t_{UD} \\ 0 \end{bmatrix} \times \Delta m
$$

Example 4.3

This example illustrates the computation of the actual matrix [A]$_{actual}$ of real laminates. The standard laminate matrix [A] and thickness t come from the standard plies. The actual laminate thickness t$_{actual}$ comes from measurement on the finished part. Let us illustrate the computation of the actual matrix [A]$_{actual}$ of the real laminate.

The real laminate is a combination of a standard laminate of matrix [A] and thickness t, with a fictitious ply of neat resin of thickness t$_{actual}$ - t. The actual stiffness matrix [A]$_{actual}$ is

$$
[A]_{actual} = [A] + [Q]_{resin} \times (t_{actual} - t)
$$

Where [Q]$_{resin}$ is the standard resin matrix listed in Chapter 1. As we see, the actual matrix is obtained by adding the standard matrix [A] and a fictitious ply of neat resin. In most cases, the difference between the actual and standard laminate thickness is small. This, coupled with the low resin stiffness, results in the near equality of the actual and standard laminate matrices.

Example 4.4

This numerical example shows how to obtain the standard ply stiffness matrices $[Q]_{12}$ from lab tests on actual plies.

The actual ply properties obtained in lab tests are never the same as those of the standard ply. The standard ply is a mathematical fiction, impossible to replicate in the real world. The lab tests provide actual values for the engineering properties, thickness and glass weight in kg/m² of the tested ply. Of the preceding quantities, only the glass weight is the same in both the actual and in the standard plies. The following equation shows how the standard ply matrix $[Q]_{12}$ is obtained from actual lab test results.

$$
[Q]_{12} = \begin{bmatrix} \dfrac{E_1}{1-v_{12}v_{21}} & \dfrac{v_{21}E_1}{1-v_{12}v_{21}} & 0 \\ \dfrac{v_{12}E_2}{1-v_{12}v_{21}} & \dfrac{E_2}{1-v_{12}v_{21}} & 0 \\ 0 & 0 & G_{12} \end{bmatrix} \times \dfrac{t_{actual}}{t} - \begin{bmatrix} 33000 & 10000 & 0 \\ 10000 & 33000 & 0 \\ 0 & 0 & 11500 \end{bmatrix} \times \dfrac{t_{actual}-t}{t}
$$

The only unknown in the above equation is the standard ply thickness "t", which is readily computed from the measured glass weight and the formulas listed in Table 1.3 of Chapter 1. The quantities in the above equation are all known and the standard ply matrix $[Q]_{12}$ can be readily computed. See the next example.

Example 4.5

A lab characterization of an UD ply yielded the following data.

Modulus in the fiber direction	$E_1 = 331\,000$ kg/cm²
Transverse modulus	$E_2 = 85\,000$ kg/cm²
Major Poisson ratio	$v_{12} = 0.31$
Minor Poisson ratio	$v_{21} = 0.08$
Shear modulus	$G_{12} = 31\,000$ kg/cm²
Actual ply thickness	3.0 mm
Glass weight	3400 g/m²

Let us compute the standard ply stiffness matrix [Q] from the above lab results. First, we compute the standard ply thickness from the measured glass weight. For that, we use the appropriate equation listed in Table 1.3 of Chapter 1.

$$
\left[\dfrac{kg}{m^2}\right]_{UD} = 1.35 \times t_{UD}
$$

$$3.4 = 1.35 \times t_{UD}$$

$$t_{UD} = 2.5 \ mm$$

This is the standard ply thickness, corresponding to the measured 3400 g/m² of glass. The actual ply thickness is 3.0 mm, meaning an excess 3.0 – 2.5 = 0.5 mm of resin. Next, we enter the measured lab data in the equation of Example 4.4

$$
[Q]_{12} = \begin{bmatrix} \dfrac{331000}{1-0.08\times0.31} & \dfrac{0.08\times331000}{1-0.08\times0.31} & 0 \\ \dfrac{0.31\times85000}{1-0.08\times0.31} & \dfrac{85000}{1-0.08\times0.31} & 0 \\ 0 & 0 & 31000 \end{bmatrix} \times \dfrac{3.0}{2.5}
$$

$$
- \begin{bmatrix} 33000 & 10000 & 0 \\ 10000 & 33000 & 0 \\ 0 & 0 & 11500 \end{bmatrix} \times \dfrac{3.0-2.5}{2.5}
$$

$$
[Q]_{12} = \begin{bmatrix} 400700 & 30500 & 0 \\ 30500 & 98000 & 0 \\ 0 & 0 & 34900 \end{bmatrix} \ kg/cm^2
$$

The measured stiffness matrix is nearly equal to the proposed standard listed in Table 1.6.

This numerical example brings to mind an interesting comment by Ever Barbero in his book "Finite Element Analysis of Composite Materials". See reference 30.

"The elastic properties of unidirectional plies can be computed using micromechanics or with experimental data. In the analysis of most composite structures, it is usual to avoid the micromechanics approach and to obtain experimentally the properties of the UD plies. However, the experimental approach is not ideal, because a change of constituents or volume fraction of reinforcement invalidates the material data and requires a new experimental program for the new material. It is better to calculate the elastic properties of the ply using micromechanics formulas. Unfortunately, the micromechanics formulas are not accurate to predict strength, so experimental work cannot be ruled out completely."

The above statement ignores the concepts of standard plies and the mathematical corrections to the experimental data. As shown in Example 4.5, the experimental lab test results serve well to derive the standard ply data.

5 Total Strains in ± 55 Angle-Ply Laminates

5.1 INTRODUCTION

This chapter illustrates the computation of the numerical strain values of plies embedded in standard oil pipes under internal pressure. The computations will combine the protocol developed in Chapter 3 with the numerical matrices derived in Chapter 4. The strains computed in this chapter will serve as inputs in a few numerical examples discussed in the Part II of this book.

The pipe consists of several ± 55 UD plies and has no corrosion barrier. The operating conditions are like follows:

- The internal pressure is P
- The laminate thickness is t.
- The diameter is D
- The pipeline operates non-anchored and aboveground.
- The operating temperature is OT.
- The water absorbed by the resin matrix is $\Delta m = 1.0\%$.

5.2 THERMAL STRAINS

The thermal resultants $\left[N^T \right]$ and the stiffness matrix [A] of ± 55 laminates are listed in Table 4.3 of Chapter 4.

$$[A]_{\pm55} = \begin{bmatrix} 132,470 & 96,220 & 0 \\ 96,220 & 235,070 & 0 \\ 0 & 0 & 101,220 \end{bmatrix} \times (t)\,\mathrm{kg/cm}$$

$$\begin{bmatrix} N_x^T \\ N_y^T \\ N_{xy}^T \end{bmatrix}_{\pm55} = \begin{bmatrix} 7.86 \\ 11.14 \\ 0 \end{bmatrix} \times (t) \times (OT - PT)\,\mathrm{kg/cm}$$

From Chapter 3, the equilibrium thermal strains are

$$
\begin{bmatrix} N_x^T \\ N_y^T \\ 0 \end{bmatrix} = \begin{bmatrix} A_{xx} & A_{xy} & 0 \\ A_{yx} & A_{yy} & 0 \\ 0 & 0 & A_{ss} \end{bmatrix} \times \begin{bmatrix} \varepsilon_x^e \\ \varepsilon_y^e \\ \gamma_{xy}^e \end{bmatrix} \tag{5.1}
$$

Combining the above and cancelling out the laminate thickness t we have

$$
\begin{bmatrix} 7.86 \\ 11.14 \\ 0 \end{bmatrix} \times (OT - PT) = \begin{bmatrix} 132,470 & 96,220 & 0 \\ 96,220 & 235,070 & 0 \\ 0 & 0 & 101,220 \end{bmatrix} \times \begin{bmatrix} \varepsilon_x^e \\ \varepsilon_y^e \\ \gamma_{xy}^e \end{bmatrix}
$$

The global equilibrium thermal strains are

$$
\begin{bmatrix} \varepsilon_x^e \\ \varepsilon_y^e \\ \gamma_{xy}^e \end{bmatrix} = \begin{bmatrix} 0.3545 \\ 0.3288 \\ 0 \end{bmatrix} \times 10^{-4} (OT - PT) \tag{5.2}
$$

The embedded plies move together and take the above equilibrium strains as the laminate cools from PT to OT. The preservation of circularity ensures that all plies have the same global equilibrium strain. The analysis of durability requires the strains in the local ply frame. The rotation from the global system is done by multiplication with the matrix [T] below, where "m" and "n" are the cosine and the sine of the angle $\alpha = \pm 55$ degrees respectively.

$$
[T] = \begin{bmatrix} m^2 & n^2 & 2mn \\ n^2 & m^2 & -2mn \\ -mn & mn & m^2 - n^2 \end{bmatrix}
$$

For $\alpha = \pm 55$ degrees, we have

$$
[T] = \begin{bmatrix} 0.33 & 0.67 & \pm 0.94 \\ 0.67 & 0.33 & \mp 0.94 \\ \mp 0.47 & \pm 0.47 & -0.34 \end{bmatrix}
$$

The rotated strain components are

$$\begin{bmatrix} \varepsilon_1^e \\ \varepsilon_2^e \\ \frac{1}{2}\gamma_{12}^e \end{bmatrix} = \begin{bmatrix} 0.33 & 0.67 & \pm 0.94 \\ 0.67 & 0.33 & \mp 0.94 \\ \mp 0.47 & \pm 0.47 & -0.34 \end{bmatrix} \times \begin{bmatrix} \varepsilon_x^e \\ \varepsilon_y^e \\ 0 \end{bmatrix}$$

$$\begin{bmatrix} \varepsilon_1^e \\ \varepsilon_2^e \\ \gamma_{12}^e \end{bmatrix} = \begin{bmatrix} 0.3373 \\ 0.3460 \\ \pm 0.0241 \end{bmatrix} \times 10^{-4}\left(OT - PT\right)$$

The above are the thermal equilibrium strains in the local ply frame. The equilibrium strains are generally negative since OT is usually less than PT. The differences between the laminate equilibrium strain and the free ply contractions give the residual thermal strains.

$$\begin{bmatrix} \varepsilon_1^r \\ \varepsilon_2^r \\ \gamma_{12}^r \end{bmatrix} = \begin{bmatrix} \varepsilon_1^e - \alpha_1\left(OT - PT\right) \\ \varepsilon_2^e - \alpha_2\left(OT - PT\right) \\ \gamma_{12}^e \end{bmatrix}$$

$$\begin{bmatrix} \varepsilon_1^r \\ \varepsilon_2^r \\ \gamma_{12}^r \end{bmatrix} = \begin{bmatrix} 0.3373 \times 10^{-4}\left(OT - PT\right) - 7.5 \times 10^{-6}\left(OT - PT\right) \\ 0.3460 \times 10^{-4}\left(OT - PT\right) - 45.0 \times 10^{-6}\left(OT - PT\right) \\ \pm 0.0241 \times 10^{-4} \times \left(OT - PT\right) \end{bmatrix}$$

$$\begin{bmatrix} \varepsilon_1^r \\ \varepsilon_2^r \\ \gamma_{12}^r \end{bmatrix} = \begin{bmatrix} +0.2623 \\ -0.1040 \\ \pm 0.0241 \end{bmatrix} \times 10^{-4} \times \left(OT - PT\right) \tag{5.3}$$

Equation (5.3) computes the residual thermal strains on all ± 55 plies. In general OT<PT and the thermal strains are negative – compressive – in the fiber direction and positive – tensile – in the transverse direction. The compressive strains in the fiber direction increase the rupture durability, while the tensile strains in the transverse direction are detrimental to the stiffness and weep failures.

From Equation (5.3), the ply transverse strain is

$$\epsilon_2^r = -0.1040 \times 10^{-4} \times \left(OT - PT \right)$$

Assuming OT < PT, Equation (5.3) indicates a tensile transverse residual thermal strain, which is detrimental to stiffness and weep. However, from the same Equation (5.3), this strain decreases as the operating temperature OT increases. In other words, increments in the operating temperature OT decrease the transverse tensile strain ϵ_2 and favor the stiffness/weep performance. This amazing conclusion contradicts the widely accepted wisdom that higher operating temperatures are detrimental to weep performance. In fact, higher operating temperatures alleviate the UD ply transverse tensile strain and actually improves the infiltration and weep performance of the pipe. We will have more to say about this topic in the Part II of this book.

The residual thermal strains in ± 55 plies depend exclusively on the temperature drop (PT − OT) and are the same for all standard laminates, regardless of wall thickness and cylinder diameter. This conclusion holds also for the hydric strains discussed in the next section.

5.3   HYDRIC STRAINS

The computation of the residual hydric strains on ± 55 plies follows the same protocol to compute the residual thermal strains. The computation is straightforward and, for the most part, is as described in Chapter 4. Skipping the details and going directly to the results, the residual hydric strains on the local ply frame of ± 55 laminates are.

$$\begin{bmatrix} \epsilon_1^r \\ \epsilon_2^r \\ \gamma_{12}^r \end{bmatrix} = \begin{bmatrix} 1475 \\ -446 \\ \mp 216 \end{bmatrix} \times 10^{-4} \left(\Delta m \right) \qquad (5.4)$$

The residual hydric strains are tensile in the fiber direction and compressive in the transverse direction. The transverse compression is a desirable situation for weep performance, as explained in Chapter 13.

The reader may wonder why the residual thermal and hydric strains have opposite signs. The reason is simple and obvious. In the thermal case, the laminates shrink from a high stress-free temperature PT whereas in the hydric case they expand from a stress-free dry condition.

5.4   MECHANICAL STRAINS

The non-anchored pipeline is pressurized to develop a 2:1 loading, meaning the hoop force resultant is twice as large as the longitudinal force resultant. This loading condition is present in pressure vessels and in non-anchored above ground pipelines. The mechanical force resultants on pressurized circular cylinders under 2:1 loadings are

$$
\begin{bmatrix} N_x^M \\ N_y^M \\ N_{xy}^M \end{bmatrix} = \begin{bmatrix} 0.25 \\ 0.50 \\ 0 \end{bmatrix} \times (PD)
$$

Where P is the internal pressure and D is the pipe diameter. The equilibrium mechanical strains result from the applied mechanical force resultants and the laminate matrix [A]. For ± 55 laminates we have

$$
\begin{bmatrix} 132470 & 96220 & 0 \\ 96220 & 235070 & 0 \\ 0 & 0 & 101220 \end{bmatrix} \times \begin{bmatrix} \varepsilon_x^e \\ \varepsilon_y^e \\ \gamma_{xy}^e \end{bmatrix} = \begin{bmatrix} 0.25 \\ 0.50 \\ 0 \end{bmatrix} \times \left(\frac{PD}{t} \right)
$$

$$
\begin{bmatrix} \varepsilon_x^e \\ \varepsilon_y^e \\ \gamma_{xy}^e \end{bmatrix} = \begin{bmatrix} 0.0049 \\ 0.0193 \\ 0 \end{bmatrix} \times 10^{-4} \times \left(\frac{PD}{t} \right)
$$

The mechanical strains for 2:1 loadings are tensile and – unlike the residual thermal and hydric strains – depend on the cylinder thickness and diameter. Following the protocol of Chapter 3, the mechanical strain components in the ply frame are.

$$
\begin{bmatrix} \varepsilon_1^e \\ \varepsilon_2^e \\ \frac{1}{2}\gamma_{12}^e \end{bmatrix} = \begin{bmatrix} 0.33 & 0.67 & \pm 0.94 \\ 0.67 & 0.33 & \mp 0.94 \\ \mp 0.47 & \pm 0.47 & -0.34 \end{bmatrix} \times \begin{bmatrix} 0.0049 \\ 0.0193 \\ 0 \end{bmatrix} \times 10^{-4} \times \left(\frac{PD}{t} \right)
$$

$$
\begin{bmatrix} \varepsilon_1^e \\ \varepsilon_2^e \\ \gamma_{12}^e \end{bmatrix}_{\pm 55} = \begin{bmatrix} +0.015 \\ +0.010 \\ \pm 0.007 \end{bmatrix} \times 10^{-4} \times \left(\frac{PD}{t} \right)
$$

These are the local mechanical strains on ± 55 plies subjected to 2:1 pressure loadings. We proceed now to compute the total ply strains.

5.5   TOTAL STRAINS

The total strains on the ply frames 1–2, come by adding the mechanical, thermal and hydric components

$$
\begin{bmatrix} Total \\ strain \end{bmatrix} = \begin{bmatrix} mechanical \\ strain \end{bmatrix} + \begin{bmatrix} residual \\ thermal\ strain \end{bmatrix} + \begin{bmatrix} residual \\ hydric\ strain \end{bmatrix}
$$

$$
\begin{bmatrix} \epsilon_1^e \\ \epsilon_2^e \\ \gamma_{12}^e \end{bmatrix} = \begin{bmatrix} 0.015 \\ 0.010 \\ \pm 0.007 \end{bmatrix} \times 10^{-4} \times \left(\frac{PD}{t} \right) + \begin{bmatrix} 0.262 \\ -0.104 \\ \pm 0.024 \end{bmatrix} \times 10^{-4} \times \left(OT - PT \right)
$$

$$
+ \begin{bmatrix} 1475 \\ -446 \\ \mp 216 \end{bmatrix} \times 10^{-4} \left(\Delta m \right)
$$

(5.5)

Equation (5.5) computes the total strains on the UD plies of standard ± 55 laminates subjected to 2:1 loadings.

Example 5.1

Compute the mechanical stresses on circular cylindrical standard ± 55 angle-ply laminates subjected to a 2:1 loading.

The mechanical stresses come from the constitutive equation

$$
\begin{bmatrix} \sigma_1 \\ \sigma_2 \\ \tau_{12} \end{bmatrix}_{\pm 55} = \begin{bmatrix} 400000 & 30000 & 0 \\ 30000 & 100000 & 0 \\ 0 & 0 & 35000 \end{bmatrix} \times \begin{bmatrix} 0.015 \\ 0.010 \\ \pm 0,007 \end{bmatrix} \times 10^{-4} \times \left(\frac{PD}{t} \right)
$$

$$
\begin{bmatrix} \sigma_1 \\ \sigma_2 \\ \tau_{12} \end{bmatrix}_{\pm 55} = \begin{bmatrix} 0.630 \\ 0.145 \\ \pm 0.025 \end{bmatrix} \times \left(\frac{PD}{t} \right)
$$

Assuming $P = 10$ bar, $D = 500$ mm and $t = 5.0$ mm, we have

$$\begin{bmatrix} \sigma_1 \\ \sigma_2 \\ \tau_{12} \end{bmatrix}_{\pm55} = \begin{bmatrix} 0.630 \\ 0.145 \\ \pm0.025 \end{bmatrix} \times \left(\frac{10 \times 500}{5} \right)$$

$$\begin{bmatrix} \sigma_1 \\ \sigma_2 \\ \tau_{12} \end{bmatrix}_{\pm55} = \begin{bmatrix} 630 \\ 145 \\ \pm25 \end{bmatrix} kg/cm^2$$

The load-dependent durability analysis of composites requires the ply strains, not the stresses. The reader is asked to compare the relation 630/145 = 4.5 between the fiber and transverse stresses with the same relation 0.015/0.010 = 1.5 for the strains. This enormous difference results from the strong Poisson coupling.

Example 5.2

Compute the weep minimum thickness of a ± 55 degrees circular cylindrical oil pipe made of vinyl ester resin. The operating conditions are

- The HDT of the vinyl ester resin matrix is HDT = 105C
- The peak post-cure temperature is PCT = 100C.
- The pipes carry water at room temperature OT = RT = 25C
- The operating pressure is P = 10 bar under a 2:1 loading.
- The pipe diameter is D = 500 mm
- The water pick up of vinyl ester resins is Δm = 1.0%
- The allowable total strain in the direction transverse to the fibers is ε_2 = 0.15%.

The design is for weeping and the pipe thickness is determined to keep the transverse strain less than the allowable ε_2 = 0.15%.

From Equation (5.5), the total strain component in the transverse ply direction is

$$0.010 \times 10^{-4} \times \frac{PD}{t} - 0.104 \times 10^{-4} \left(OT - PT \right) - 446 \times 10^{-4} \times \Delta m = \epsilon_2 \qquad (5.6)$$

Entering the problem data in the above

$$0.010 \times 10^{-4} \times \frac{10 \times 500}{t} - 0.104 \times 10^{-4} \left(25 - 100 \right) - 446 \times 10^{-4} \times 0.01 = 0.0015$$

$$\frac{0.005}{t} + 0.0008 - 0.0004 = 0.0015$$

Solving for the laminate thickness t we have

$t = 4.5\,\text{mm}$

The computed thickness – 4.5 mm – meets the stated design condition of 0.15% strain in the transverse ply direction. The question is..... How long would this pipeline last before weeping? The unified equation discussed in Chapter 16 gives the answer to this question.

Example 5.3

This numerical example compares the residual transverse strains of UD plies embedded in ± 55 standard laminates at different operating conditions.

The transverse strain is computed from Equation (5.6), repeated below to facilitate the presentation.

$$\epsilon_2 = 0.010 \times 10^{-4} \times \frac{PD}{t} + 0.104 \times 10^{-4}\left(PT - OT\right) - 446 \times 10^{-4} \times \Delta m$$

The hydric strain component varies from zero in dry air ($\Delta m = 0$) to a compressive – 0.05% in saturated conditions ($\Delta m = 1.0\%$). Assuming a typical allowable transverse strain of 0.15%, the water saturation alone improves the pipe rating in 0.05/0.15 = 33%. This is a surprising result. Most people would expect otherwise.

The thermal strain component decreases with increments in the operating temperature, meaning better pipe performance at higher OT. As an example, consider the pipes post-cured at PCT = 100C and operating at two different temperatures, say OT = 25C and OT = 80C. From Equation (5.6) the residual thermal strains are + 0.08% at 25C and + 0.02% at 80C, indicating a substantially improved performance at higher temperatures. This unexpected conclusion runs against the conventional wisdom, which degrades the pipe performance at higher operating temperatures. In Chapter 10, we discuss and reject the classical arguments in support of this false pipe degradation.

Example 5.4

Let us estimate the rupture durability of the pipe described in Example 5.2.

The total strain in the fiber direction controls the long-term rupture failure.

$$\epsilon_1 = 0.015 \times 10^{-4}\left(\frac{PD}{t}\right) + 1475 \times 10^{-4}\,\Delta m - 0.262 \times 10^{-4}\left(PT - OT\right)$$

$$\epsilon_1 = 0.015 \times 10^{-4}\left(\frac{10 \times 500}{4.5}\right) + 1475 \times 10^{-4} \times 0.01 - 0.262 \times 10^{-4}\left(100 - 25\right)$$

$$\epsilon_1 = 0.0017 + 0.0015 - 0.0020$$

$$\epsilon_1 = 0.0012$$

The compressive thermal strain (– 0.20%) in the fiber direction improves the long-term rupture performance. The tensile hydric strain (+ 0.15%) reduces the long-term performance. On balance, the inclusion of the residual strains reduces the mechanical strain 0.17% to 0.12%. This indicates the importance of the thermal and hydric residual strains. The residual strains play an important role in the determination of the load-dependent durability.

The question is… How long will it take before the glass fibers fail? This is a question for the Part II of this book.

Example 5.5

It is required to compute the minimum thickness of the pipe described in Example 5.2, assuming the operating temperature increased from OT = 25C to OT = 80C.

The mode of failure is weep and the allowable transverse strain is $\epsilon_2 = 0.15\%$. The total strain in the transverse direction of the UD plies is, as before

$$\epsilon_2 = 0.010 \times 10^{-4} \left(\frac{PD}{t} \right) - 446 \times 10^{-4} \Delta m + 0.104 \times 10^{-4} \left(PT - OT \right)$$

Entering the given operating data in the above, we have

$$\epsilon_2 = 0.010 \times 10^{-4} \left(\frac{10 \times 500}{t} \right) - 446 \times 10^{-4} \times 0.01 + 0.104 \times 10^{-4} \left(100 - 80 \right)$$

$$0.0015 = \left(\frac{0.0050}{t} \right) - 0.0004 + 0.0002$$

From which we obtain the minimum pipe thickness.

$$t = 3.0 \, \text{mm}$$

The required pipe thickness decreases from 4.5 mm at OT = 25C to 3.0 mm at OT = 80C. The commercial pipe standards and specifications say otherwise and demand larger pipe thicknesses at higher operating temperatures.

There are two arguments supporting the idea that higher operating temperatures improve the weep performance of oil pipes. The first argument, developed in this example, is the drop in the transverse residual thermal strain with increments in the operating temperature. The second argument – detailed in Chapter 10 – is the increased resin plasticization and the reduced free-ply residual strains at higher temperatures. These arguments taken together form a compelling platform suggesting improved weep performance at higher temperatures.

On the other hand, the strain in the fiber direction increases with the operating temperature. This, as discussed in the Part II of the book, is detrimental to long-term fiber rupture. However, since the long-term weep failure is far more important than the long-term rupture, the argument in favor of improved performance of oil pipes at higher operating temperatures still stands.

6 Total Strains in ± 70 Angle-Ply Laminates

6.1 INTRODUCTION

The corrosion barrier of ± 70 cylindrical laminates in chemical service may or may not be included in the structural design, depending on the severity of the chemical attack. In benign chemicals, those that do not attack the resin, the corrosion barrier is preserved and included in the design. In aggressive services, those that attack the resin, the destroyed corrosion barrier is not included. However, regardless of inclusion or exclusion, destruction or preservation, the corrosion barrier remains as the critical ply governing the laminate design in the transportation or storage of liquids. In mild service, where chemical infiltration is not a problem, the weep barrier controls the allowable strain. In aggressive service, where chemical infiltration is a problem, the corrosion barrier also controls the allowable strain. As we see, even if excluded as a structural contributor, the corrosion barrier remains as the critical ply controlling the laminate design in all cases involving liquid service.

The exclusion of the corrosion barrier facilitates the structural analysis and the design equations are similar to those developed in the previous chapter for ± 55 oil pipes. The preservation of the corrosion barrier complicates the design equations, forcing the choice of the laminate construction by trial and error. In this chapter, we assume applications in highly aggressive chemical service and ignore the corrosion barrier. The next chapter discusses the applications in mild services that preserve the corrosion barrier. The following computations combine the protocol developed in Chapter 3 with the numerical matrices derived in Chapter 4.

Our discussions assume circular cylindrical laminates operating in the following conditions:

- The corrosion barrier is ignored in the design
- The internal pressure is P
- The structural laminate thickness is t_{UD}
- The cylinder diameter is D.
- The operating temperature is OT.
- The water absorbed by the resin is $\Delta m = 1.0\%$.

6.2 THERMAL STRAINS

The thermal resultants $\left[N^T\right]$ and the stiffness matrix [A] of standard ± 70 laminates are obtained by setting $t_{chop} = 0$ in the appropriate matrices of Table 4.3, Chapter 4.

$$
\left[A\right]_{\pm 70} = \begin{bmatrix} 104000 & 61000 & 0 \\ 61000 & 334000 & 0 \\ 0 & 0 & 66000 \end{bmatrix} \times (t_{UD}) \text{ kg/cm}
$$

$$
\begin{bmatrix} N^T_x \\ N^T_y \\ N^T_{xy} \end{bmatrix}_{\pm 70} = \begin{bmatrix} 5,22 \times t_{UD} \\ 8,50 \times t_{UD} \\ 0 \end{bmatrix} \times (OT - PT) \text{ kg/cm}
$$

The reader will note the exclusion of the corrosion barrier from the laminate matrix. However, even when excluded, the corrosion barrier remains as the critical ply in control of the weep and infiltration failures. From Chapter 3, the equilibrium thermal strains are

$$
\begin{bmatrix} N^T_x \\ N^T_y \\ 0 \end{bmatrix} = \begin{bmatrix} A_{xx} & A_{xy} & 0 \\ A_{yx} & A_{yy} & 0 \\ 0 & 0 & A_{ss} \end{bmatrix} \times \begin{bmatrix} \varepsilon^e_x \\ \varepsilon^e_y \\ \gamma^e_{xy} \end{bmatrix} \tag{6.1}
$$

Combining the above equations and cancelling out the laminate thickness t_{UD}, we have

$$
\begin{bmatrix} 5.22 \\ 8.50 \\ 0 \end{bmatrix} \times (OT - PT) = \begin{bmatrix} 104000 & 61000 & 0 \\ 61000 & 334000 & 0 \\ 0 & 0 & 66000 \end{bmatrix} \times \begin{bmatrix} \varepsilon^e_x \\ \varepsilon^e_y \\ \gamma^e_{xy} \end{bmatrix}
$$

From this equation, the global thermal equilibrium strains components are

$$
\begin{bmatrix} \varepsilon^e_x \\ \varepsilon^e_y \\ \gamma^e_{xy} \end{bmatrix} = \begin{bmatrix} 0.395 \\ 0.182 \\ 0 \end{bmatrix} \times 10^{-4} (OT - PT) \tag{6.2}
$$

The embedded plies move together and take the above equilibrium strains as they cool down from the stress-free temperature PT to the operating temperature OT. The preservation of circularity ensures the same value of the equilibrium global strain to all plies. The global strains are rotated to the local systems 1–2 of the UD plies through the matrix [T] of Chapter 3, where "m" and "n" are the cosine and sine of the angle $\alpha = \pm 70^\circ$.

$$[T] = \begin{bmatrix} m^2 & n^2 & 2mn \\ n^2 & m^2 & -2mn \\ -mn & mn & m^2 - n^2 \end{bmatrix}$$

For $\alpha = \pm 70$ degrees

$$[T] = \begin{bmatrix} 0.12 & 0.88 & \pm 0.64 \\ 0.88 & 0.12 & \mp 0.64 \\ \mp 0.32 & \pm 0.32 & -0.77 \end{bmatrix}$$

$$\begin{bmatrix} \epsilon_1^e \\ \epsilon_2^e \\ \frac{1}{2}\gamma_{12}^e \end{bmatrix}_{UD} = \begin{bmatrix} 0.12 & 0.88 & \pm 0.64 \\ 0.88 & 0.12 & \mp 0.64 \\ \mp 0.32 & \pm 0.32 & -0.77 \end{bmatrix} \times \begin{bmatrix} \epsilon_x^e \\ \epsilon_y^e \\ 0 \end{bmatrix}$$

$$\begin{bmatrix} \epsilon_1^e \\ \epsilon_2^e \\ \gamma_{12}^e \end{bmatrix}_{UD} = \begin{bmatrix} 0.208 \\ 0.369 \\ \pm 0.068 \end{bmatrix} \times 10^{-4} \left(OT - PT \right)$$

The above are the local thermal equilibrium strain components of the ± 70 UD plies. The equilibrium strain components are generally negative since OT is usually less than PT. The maximum equilibrium strains of the ignored isotropic corrosion barrier are equal to those of the laminate in the global system. As we know, the isotropic plies do not require strain rotation. The residual thermal strain components come from a subtraction of the free ply contractions from the equilibrium strains.

$$\begin{bmatrix} \varepsilon_1^r \\ \varepsilon_2^r \\ \gamma_{12}^r \end{bmatrix} = \begin{bmatrix} \varepsilon_1^e - \alpha_1(OT - PT) \\ \varepsilon_2^e - \alpha_2(OT - PT) \\ \gamma_{12}^e \end{bmatrix}$$

For the UD plies we have

$$\begin{bmatrix} \varepsilon_1^r \\ \varepsilon_2^r \\ \gamma_{12}^r \end{bmatrix}_{\pm 70} = \begin{bmatrix} 0.208\times10^{-4}(OT-PT)-7.5\times10^{-6}(OT-PT) \\ 0.369\times10^{-4}(OT-PT)-45.0\times10^{-6}(OT-PT) \\ \pm0.068\times10^{-4}\times(OT-PT) \end{bmatrix}$$

$$\begin{bmatrix} \varepsilon_1^r \\ \varepsilon_2^r \\ \gamma_{12}^r \end{bmatrix}_{\pm 70} = \begin{bmatrix} +0.133 \\ -0.081 \\ \pm0.068 \end{bmatrix}\times10^{-4}\times(OT-PT) \tag{6.3}$$

Equation (6.3) computes the residual thermal strain components on the UD plies.

To compute the residual thermal strain components on the corrosion barrier, we subtract the free ply contractions from the global equilibrium strain components.

$$\begin{bmatrix} \varepsilon_1^r \\ \varepsilon_2^r \\ \gamma_{12}^r \end{bmatrix}_{chop} = \begin{bmatrix} 0.395\times10^{-4}(OT-PT)-25\times10^{-6}(OT-PT) \\ 0.182\times10^{-4}(OT-PT)-25\times10^{-6}(OT-PT) \\ 0 \end{bmatrix}$$

$$\begin{bmatrix} \epsilon_x^r \\ \epsilon_y^r \\ \gamma_{xy}^r \end{bmatrix}_{chop} = \begin{bmatrix} \epsilon_1^r \\ \epsilon_2^r \\ \gamma_{12}^r \end{bmatrix}_{chop} = \begin{bmatrix} 0.145 \\ -0.068 \\ 0 \end{bmatrix}\times10^{-4}\left(OT-PT\right)$$

The residual thermal strain components on both the UD and the chopped plies depend only on the temperature drop (PT – OT), regardless of the cylinder thickness and diameter. This conclusion holds for the hydric strains as well.

6.3   HYDRIC STRAINS

The computation of the residual hydric strain components follows the same protocol as for the thermal strains. Skipping the details and going directly to the final numbers, the residual hydric strain components in the local ply frames are.

$$\begin{bmatrix} \epsilon_1^r \\ \epsilon_2^r \\ \gamma_{12}^r \end{bmatrix}_{UD} = \begin{bmatrix} -60.0 \\ 34.0 \\ \mp1300 \end{bmatrix}\times10^{-4}\times\left(\Delta m\right) \tag{6.4}$$

$$\begin{bmatrix} \varepsilon_1^r \\ \varepsilon_2^r \\ \gamma_{12}^r \end{bmatrix}_{chop} = \begin{bmatrix} -36.0 \\ -1560.0 \\ 0 \end{bmatrix} \times 10^{-4} \times (\Delta m) \qquad (6.4A)$$

The residual hydric strains on the corrosion barrier are both compressive. This is a highly desirable condition, since compressive strains favor the infiltration and weep performances. The full importance of this statement will be apparent in Chapter 8.

6.4   MECHANICAL STRAINS ON VERTICAL STORAGE TANKS

The hoop force on cylinders of vertical storage tanks is $N_y = PD/2$, where P is the hydrostatic pressure and D is the tank diameter. The longitudinal forces N_x in such tanks are usually equal to zero. Furthermore, the torque N_{xy} is also equal to zero. The mechanical force resultants acting on the cylindrical wall of vertical storage tanks are

$$\begin{bmatrix} N_x^M \\ N_y^M \\ N_{xy}^M \end{bmatrix}_{\pm70} = \begin{bmatrix} 0 \\ 0.50 \\ 0 \end{bmatrix} \times (PD)$$

As discussed in Chapter 3, we compute the equilibrium mechanical strains, of any laminate from the force resultants and the laminate matrix [A]. Assuming standard ± 70 angle-ply laminates without the corrosion barrier, we have

$$\begin{bmatrix} 104000 & 61000 & 0 \\ 61000 & 334000 & 0 \\ 0 & 0 & 66000 \end{bmatrix} \times \begin{bmatrix} \varepsilon_x^e \\ \varepsilon_y^e \\ \gamma_{xy}^e \end{bmatrix} = \begin{bmatrix} 0 \\ 0.50 \\ 0 \end{bmatrix} \times \left(\frac{PD}{t_{UD}} \right)$$

$$\begin{bmatrix} \varepsilon_x^e \\ \varepsilon_y^e \\ \gamma_{xy}^e \end{bmatrix}_{\pm70} = \begin{bmatrix} -0.0098 \\ 0.0168 \\ 0 \end{bmatrix} \times 10^{-4} \times \left(\frac{PD}{t_{UD}} \right)$$

Following the protocol of Chapter 3, the global equilibrium mechanical strains rotated to the local UD ply system are

$$
\begin{bmatrix} \epsilon_1^e \\ \epsilon_2^e \\ \frac{1}{2}\gamma_{12}^e \end{bmatrix}_{UD} = \begin{bmatrix} 0.12 & 0.88 & \pm 0.64 \\ 0.88 & 0.12 & \mp 0.64 \\ \mp 0.64 & \pm 0.32 & -0.77 \end{bmatrix} \times \begin{bmatrix} -0.0098 \\ 0.0168 \\ 0 \end{bmatrix} \times 10^{-4} \times \left(\frac{PD}{t_{UD}} \right)
$$

$$
\begin{bmatrix} \varepsilon_1^e \\ \varepsilon_2^e \\ \gamma_{12}^e \end{bmatrix}_{\pm 70} = \begin{bmatrix} 0.014 \\ -0.007 \\ \pm 0.009 \end{bmatrix} \times 10^{-4} \times \left(\frac{PD}{t_{UD}} \right)
$$

The corrosion barrier is isotropic and requires no strain rotation. The equilibrium strain components on the corrosion barrier are

$$
\begin{bmatrix} \varepsilon_x^e \\ \varepsilon_y^e \\ \gamma_{xy}^e \end{bmatrix}_{chop} = \begin{bmatrix} -0.0098 \\ 0.0168 \\ 0 \end{bmatrix} \times 10^{-4} \times \left(\frac{PD}{t_{UD}} \right)
$$

These are the local mechanical strain components on the UD and chopped plies embedded in standard $\pm$ 70 vertical storage tanks under hydrostatic loading. The corrosion barrier, we repeat, were excluded from the computations. We proceed now to compute the mechanical strains on $\pm$ 70 above ground pipes.

6.5   MECHANICAL STRAINS IN NON-ANCHORED ABOVE GROUND PIPES

The $\pm$ 70 above ground pipes usually operate fully anchored to eliminate the axial component of the pressure force. The reader is referred Example 6.5 for a full numerical discussion of this situation. In this section, we discuss the case of free, non-anchored, above ground $\pm$ 70 pipes. The mechanical force resultants from the internal pressure P are

$$
\begin{bmatrix} N_x^M \\ N_y^M \\ N_{xy}^M \end{bmatrix}_{\pm 70} = \begin{bmatrix} 0.25 \\ 0.50 \\ 0 \end{bmatrix} \times (PD)
$$

The above force resultants are applicable to non-anchored pipes subjected to the internal pressure P that produces a 2:1 loading on the pipe wall. To compute the equilibrium mechanical strains we make use of the force resultants and the laminate matrix [A], as shown in the protocol developed in Chapter 3. For standard ± 70 angle-ply laminates without the corrosion barrier, we have

$$
\begin{bmatrix}
104000 & 61000 & 0 \\
61000 & 334000 & 0 \\
0 & 0 & 66000
\end{bmatrix}
\times
\begin{bmatrix}
\varepsilon_x^e \\
\varepsilon_y^e \\
\gamma_{xy}^e
\end{bmatrix}
=
\begin{bmatrix}
0.25 \\
0.50 \\
0
\end{bmatrix}
\times \left(\frac{PD}{t_{UD}} \right)
$$

From this equation, the equilibrium strain components in the global system are

$$
\begin{bmatrix}
\varepsilon_x^e \\
\varepsilon_y^e \\
\gamma_{xy}^e
\end{bmatrix}_{\pm 70}
=
\begin{bmatrix}
0.0171 \\
0.0118 \\
0
\end{bmatrix}
\times 10^{-4} \times \left(\frac{PD}{t_{UD}} \right)
$$

Following the protocol of Chapter 3, the global equilibrium strains rotated to the local UD ply system are

$$
\begin{bmatrix}
\epsilon_1^e \\
\epsilon_2^e \\
\tfrac{1}{2}\gamma_{12}^e
\end{bmatrix}_{UD}
=
\begin{bmatrix}
0.12 & 0.88 & \pm 0.64 \\
0.88 & 0.12 & \mp 0.64 \\
\mp 0.32 & \pm 0.32 & -0.77
\end{bmatrix}
\times
\begin{bmatrix}
-0.0171 \\
0.0118 \\
0
\end{bmatrix}
\times 10^{-4} \times \left(\frac{PD}{t_{UD}} \right)
$$

$$
\begin{bmatrix}
\epsilon_1^e \\
\epsilon_2^e \\
\gamma_{12}^e
\end{bmatrix}_{UD}
=
\begin{bmatrix}
0.010 \\
0.016 \\
\mp 0.002
\end{bmatrix}
\times 10^{-4} \times \left(\frac{PD}{t_{UD}} \right)
$$

The corrosion barrier is isotropic and requires no strain rotation. The principal strains, in the hoop and axial directions of the corrosion barrier are equal to those on the laminate

$$
\begin{bmatrix}
\varepsilon_x^e \\
\varepsilon_y^e \\
\gamma_{xy}^e
\end{bmatrix}_{chop}
=
\begin{bmatrix}
0.0171 \\
0.0118 \\
0
\end{bmatrix}
\times 10^{-4} \times \left(\frac{PD}{t_{UD}} \right)
$$

These are the local mechanical strains on the UD and chopped plies embedded on free, non-anchored, standard ± 70 pipes subjected to a pressure P. We proceed now to compute the total strains.

6.6   TOTAL STRAINS

The total strains on vertical ± 70 storage tanks and on non-anchored above ground ± 70 pipes are obtained by adding the mechanical, thermal and hydric components computed in the preceding sections. The following equation applies.

$$
\begin{bmatrix} Total \\ strain \end{bmatrix} = \begin{bmatrix} mechanical \\ strain \end{bmatrix} + \begin{bmatrix} residual \\ thermal\ strain \end{bmatrix} + \begin{bmatrix} residual \\ hydric\ strain \end{bmatrix}
$$

6.6.1   VERTICAL TANKS

For vertical tanks, the total strains are

$$
\begin{bmatrix} \epsilon_1 \\ \epsilon_2 \\ \epsilon_3 \end{bmatrix}_{UD} = \begin{bmatrix} 0.014 \\ -0.007 \\ \mp 0.009 \end{bmatrix} \times 10^{-4} \times \left(\frac{PD}{t_{UD}} \right) + \begin{bmatrix} 0.133 \\ -0.081 \\ \pm 0.068 \end{bmatrix} \times 10^{-4} \times \left(OT - PT \right)
$$

$$
+ \begin{bmatrix} -60.0 \\ 34.0 \\ \mp 1300 \end{bmatrix} \times 10^{-4} \times \Delta m
$$

(6.5)

$$
\begin{bmatrix} \epsilon_x \\ \epsilon_y \\ \gamma_{xy} \end{bmatrix}_{chop} = \begin{bmatrix} -0.010 \\ 0.017 \\ 0 \end{bmatrix} \times 10^{-4} \frac{PD}{t_{UD}} + \begin{bmatrix} 0.145 \\ -0.068 \\ 0 \end{bmatrix} \times 10^{-4} \left(OT - PT \right)
$$

$$
+ \begin{bmatrix} -36.0 \\ -1560 \\ 0 \end{bmatrix} \times 10^{-4} \Delta m
$$

(6.6)

Equation (6.5) computes the local total strain components on the UD plies embedded in standard ± 70 laminates of vertical storage tanks subjected to hydrostatic loadings. Equation (6.6) computes the total strains on the corrosion barrier of the same vertical storage tanks.

The total hoop strain component on the corrosion barrier controls the tank design.

$$\epsilon_y = \left(0.017 \frac{PD}{t_{UD}} + 0.068 \left(PT - OT \right) - 1560 \Delta m \right) \times 10^{-4} \qquad (6.6A)$$

The above equation indicates that higher operating temperatures OT reduce the hoop strain on the corrosion barrier and favor the infiltration performance. We will return to this later in a numerical example.

6.6.2 ABOVE GROUND NON-ANCHORED PIPES

For above ground non-anchored pipes, the total strains are

$$\begin{bmatrix} \epsilon_1 \\ \epsilon_2 \\ \epsilon_3 \end{bmatrix}_{UD} = \begin{bmatrix} 0.010 \\ 0.016 \\ \mp 0.002 \end{bmatrix} \times 10^{-4} \times \left(\frac{PD}{t_{UD}} \right) + \begin{bmatrix} 0.133 \\ -0.081 \\ \pm 0.068 \end{bmatrix} \times 10^{-4} \times \left(OT - PT \right)$$

$$+ \begin{bmatrix} -60.0 \\ 34.0 \\ \mp 1300 \end{bmatrix} \times 10^{-4} \times \Delta m \qquad (6.7)$$

$$\begin{bmatrix} \epsilon_x \\ \epsilon_y \\ \gamma_{xy} \end{bmatrix}_{chop} = \begin{bmatrix} 0.017 \\ 0.012 \\ 0 \end{bmatrix} \times 10^{-4} \frac{PD}{t_{UD}} + \begin{bmatrix} 0.145 \\ -0.068 \\ 0 \end{bmatrix} \times 10^{-4} \left(OT - PT \right)$$

$$+ \begin{bmatrix} -36.0 \\ -1560 \\ 0 \end{bmatrix} \times 10^{-4} \Delta m \qquad (6.8)$$

Equation (6.7) computes the total strain components on the UD plies of standard ± 70 laminates of above ground non-anchored pipelines under internal pressure. Equation (6.8) computes the total strains on the corrosion barrier of such pipes. The total strain on the corrosion barrier controls the pipe design.

Equation (6.8) indicates that higher operating temperatures OT reduce the hoop strain and increase the axial strain on the corrosion barrier. The overall effect is easy to compute, and allow the designer to decide whether increments in the operating temperature are favorable or detrimental to the infiltration or weep performance of the pipeline.

Example 6.1

Compute the minimum structural thickness of a standard ± 70 vertical cylindrical storage tank. The following information applies.

- The HDT of the vinyl ester resin is HDT = 105C
- The peak post-cure temperature is PCT = 100C.
- The tank stores 20% HCl at room temperature OT = RT = 25C
- The operating hydrostatic pressure is P = 1.0 bar.
- The tank diameter is D = 5000 mm
- The water pick up is Δm = 1.0%
- The allowable total strain in the corrosion barrier is 0.25%.

The corrosion barrier of chopped fibers is the critical ply governing the design of chemical storage tanks. The failure criterion in such cases is infiltration. See Chapter 8. From Equation (6.6A), the maximum total strain on the corrosion barrier is in the hoop direction y. The design criterion is infiltration and the wall thickness is determined to keep the strain in the hoop direction of the corrosion barrier less than 0.25%. From Equation (6.6A), we have

$$\epsilon_y = 0.017 \times 10^{-4} \frac{PD}{t_{UD}} + 0.068 \times 10^{-4} \left(PT - OT \right) - 1560 \times 10^{-4} \Delta m$$

Entering the problem data in the above

$$0.017 \times 10^{-4} \times \frac{1.0 \times 5000}{t_{UD}} + 0.068 \times 10^{-4} \left(100 - 25 \right) - 1560 \times 10^{-4} \times 0.01 < 0.0025$$

$$\frac{0.009}{t_{UD}} + 0.0005 - 0.0016 < 0.0025$$

Solving for the laminate thickness t_{UD} we have

$$t_{UD} > 2.6 \ mm$$

The minimum computed thickness – $t_{UD} = 2.6$ mm – meets the stated design criterion. The question is..... How thick should the corrosion barrier be for a desired service life of, say, 25 years? The answer to this question is in Chapter 11.

Example 6.2

This numerical example compares the residual strains on the corrosion barrier of standard ± 70 vertical storage tanks at different operating conditions.

To facilitate the presentation, we repeat the equation to compute the total hoop strain in the corrosion barrier of vertical tanks, Equation (6.6A).

$$\epsilon_y = 0.017 \times 10^{-4} \frac{PD}{t_{UD}} + 0.068 \times 10^{-4} \left(PT - OT \right) - 1560 \times 10^{-4} \Delta m$$

The residual hoop hydric strain component varies from zero in air ($\Delta m = 0$) to a compressive − 0.16% in saturated conditions ($\Delta m = 1.0\%$). Assuming a typical allowable infiltration strain of 0.25%, the water saturation may significantly improve the tank rating. This is a surprising result.

The residual hoop thermal strain component decreases with increments in the operating temperature, meaning better infiltration performance at higher OT. This result is also unexpected.

The preceding analysis indicates improved infiltration performance of tanks operating at higher temperature and saturated in water.

Example 6.3

Let us estimate the rupture durability of the vertical storage tank described in Example 6.1. This requires knowledge of the total strain in the fiber direction. From Equation (6.5)

$$\epsilon_1 = 0.014 \times 10^{-4} \left(\frac{PD}{t_{UD}} \right) - 60 \times 10^{-4} \Delta m - 0.133 \times 10^{-4} \left(PT - OT \right)$$

$$\epsilon_1 = 0.014 \times 10^{-4} \left(\frac{1 \times 5000}{2.6} \right) - 60 \times 10^{-4} \times 0.01 - 0.133 \times 10^{-4} \left(100 - 25 \right)$$

$$\epsilon_1 = 0.0027 - 0.0001 - 0.0010$$

$$\epsilon_1 = 0.0016$$

The compressive thermal and hydric residual strains reduce the mechanical strain on the fiber from 0.27% to 0.16%, thereby substantially improving the tank performance. The residual strains are important in design. They play a significant role in the computation of the total strains and the load-dependent durability.

The question is…. How long will it take until the glass fibers fail in this case? The answer to this question is in the Part II of this book.

Example 6.4

Consider the vertical storage tank of Example 6.1. What would the required structural thickness be if the operating temperature increased from OT = 25C to OT = 80C? The mode of failure is infiltration and the allowable hoop strain is 0.25%. The maximum total strain on the corrosion barrier is in the hoop direction

$$\epsilon_y = 0.017 \times 10^{-4} \frac{PD}{t_{UD}} + 0.068 \times 10^{-4} \left(PT - OT \right) - 1560 \times 10^{-4} \Delta m$$

Entering the given operating data in the above, we have

$$0.0025 = 0.017 \times 10^{-4} \times \frac{1 \times 5000}{t_{UD}} + 0.068 \times 10^{-4} \left(100 - 80 \right) - 1560 \times 10^{-4} \times 0.01$$

From which we obtain

$$t_{UD} = 2.2 \text{ mm}$$

The required structural wall thickness decreases from 2.6 mm at OT = 25C to 2.2 mm at OT = 80C. This is because the residual thermal hoop strain on the corrosion barrier decreases with increments in the operating temperature.

The reader will also note that the residual hydric hoop strain in the corrosion barrier is always compressive, indicating improved infiltration performance of tanks operating saturated in water and at higher temperatures. This discussion leaves no doubt about the improved infiltration performance of vertical storage tanks operating at higher temperatures. However, it says not a word about the durability of the corrosion barrier. For information on the service life of the corrosion barrier, see Chapter 11.

Example 6.5

Compute the axial anchor force Nx and the total strains on an above ground, anchored standard ± 70 vinyl ester pipeline operating in the following conditions:

- The HDT of the vinyl ester resin is HDT = 105C
- The peak post-cure temperature is PCT = 100C.
- The pipeline carries a 20% solution of HCl at room temperature OT = RT = 25C
- The operating pressure is P = 15.0 bar.
- There is no external torque
- The pipe diameter is D = 500 mm
- The structural thickness is t_{UD} = 10.0 mm
- The water pick up is Δm = 1.0%
- The above ground pipeline operates fully anchored.

The total strains are

$$
\begin{bmatrix} Total \\ strain \end{bmatrix} = \begin{bmatrix} mechanical \\ strain \end{bmatrix} + \begin{bmatrix} residual \\ thermal\ strain \end{bmatrix} + \begin{bmatrix} residual \\ hydric\ strain \end{bmatrix}
$$

For the UD plies we have

$$
\begin{bmatrix} \epsilon_1 \\ \epsilon_2 \\ \gamma_{12} \end{bmatrix}_{UD} = \begin{bmatrix} \epsilon_1 \\ \epsilon_2 \\ \gamma_{12} \end{bmatrix}_{MECH} + \begin{bmatrix} 0.133 \\ -0.081 \\ \pm 0.068 \end{bmatrix} \times 10^{-4} \times (OT - PT)
$$

$$
+ \begin{bmatrix} -60.0 \\ 34.0 \\ \mp 1300 \end{bmatrix} \times 10^{-4} \times \Delta m
$$

(6.9)

For the chopped plies we have

$$
\begin{bmatrix} \epsilon_x \\ \epsilon_y \\ \gamma_{xy} \end{bmatrix}_{chop} = \begin{bmatrix} \epsilon_x \\ \epsilon_y \\ 0 \end{bmatrix}_{MECH} + \begin{bmatrix} 0.145 \\ -0.068 \\ 0 \end{bmatrix} \times 10^{-4} (OT - PT)
$$

$$
+ \begin{bmatrix} -36.0 \\ -1560 \\ 0 \end{bmatrix} \times 10^{-4} \Delta m
$$

(6.10)

Let us compute the mechanical strains in this case. The mechanical strains come from the general Equation (3.15) of Chapter 3.

$$
\begin{bmatrix} \varepsilon_x^e \\ \varepsilon_y^e \\ \gamma_{xy} \end{bmatrix} = \begin{bmatrix} A_{xx} & A_{xy} & 0 \\ A_{yx} & A_{yy} & 0 \\ 0 & 0 & A_{ss} \end{bmatrix}^{-1} \times \begin{bmatrix} N_x^M \\ N_y^M \\ N_{xy}^M \end{bmatrix}
$$

In fully anchored pipelines, the axial strain $\varepsilon_x = 0$. The external torque is $N_{xy} = 0$. The internal pressure P develops a 2:1 loading condition. The axial anchor load Nx is unknown. Entering this in the above equation

$$
\begin{bmatrix} 0 \\ \varepsilon_y^e \\ 0 \end{bmatrix} = \begin{bmatrix} A_{xx} & A_{xy} & 0 \\ A_{yx} & A_{yy} & 0 \\ 0 & 0 & A_{ss} \end{bmatrix}^{-1} \times \begin{bmatrix} PD/4 + N_x \\ PD/2 \\ 0 \end{bmatrix}
$$

$$
\begin{bmatrix} 0 \\ \varepsilon_y^e \\ 0 \end{bmatrix} = \begin{bmatrix} 104000 & 61000 & 0 \\ 61000 & 334000 & 0 \\ 0 & 0 & 66000 \end{bmatrix}^{-1} \times \begin{bmatrix} PD/4t_{UD} + N_x/t_{UD} \\ PD/2t_{UD} \\ 0 \end{bmatrix}
$$

Entering $P = 15.0$ bar, $D = 500\,mm$ and $t_{UD} = 10.0\,mm$ in the above we obtain $Nx = -118.6$ kg/cm and $\varepsilon_y^e = 0.00112$. The mechanical strains in the local frame of the UD plies are

$$
\begin{bmatrix} \varepsilon_1^e \\ \varepsilon_2^e \\ \tfrac{1}{2}\gamma_{12}^e \end{bmatrix} = \begin{bmatrix} m^2 & n^2 & 2mn \\ n^2 & m^2 & -2mn \\ -mn & mn & m^2-n^2 \end{bmatrix} \times \begin{bmatrix} \varepsilon_x^e \\ \varepsilon_y^e \\ 0 \end{bmatrix}
$$

$$
\begin{bmatrix} \varepsilon_1^e \\ \varepsilon_2^e \\ \tfrac{1}{2}\gamma_{12}^e \end{bmatrix}_{UD} = \begin{bmatrix} 0.12 & 0.88 & \pm 0.64 \\ 0.88 & 0.12 & \mp 0.64 \\ \mp 0.32 & \pm 0.32 & -0.77 \end{bmatrix} \times \begin{bmatrix} 0 \\ 0.00112 \\ 0 \end{bmatrix}
$$

$$
\begin{bmatrix} \varepsilon_1^e \\ \varepsilon_2^e \\ \gamma_{12}^e \end{bmatrix}_{MECH} = \begin{bmatrix} 0.001 \\ 0.000 \\ \pm 0.001 \end{bmatrix}
$$

From Equation (6.9) the total strains in the local frame of the UD plies are

$$
\begin{bmatrix} \epsilon_1 \\ \epsilon_2 \\ \gamma_{12} \end{bmatrix}_{UD} = \begin{bmatrix} \epsilon_1 \\ \epsilon_2 \\ \gamma_{12} \end{bmatrix}_{MECH} + \begin{bmatrix} 0.133 \\ -0.081 \\ \pm 0.068 \end{bmatrix} \times 10^{-4} \times (OT-PT)
$$

$$
+ \begin{bmatrix} -60.0 \\ 34.0 \\ \mp 1300 \end{bmatrix} \times 10^{-4} \times \Delta m
$$

$$
\begin{bmatrix} \epsilon_1 \\ \epsilon_2 \\ \gamma_{12} \end{bmatrix}_{UD} = \begin{bmatrix} 0.001 \\ 0.000 \\ \pm 0.001 \end{bmatrix}_{MECH} + \begin{bmatrix} 0.133 \\ -0.081 \\ \pm 0.068 \end{bmatrix} \times 10^{-4} \times (25-100)
$$

$$
+ \begin{bmatrix} -60.0 \\ 34.0 \\ \mp 1300 \end{bmatrix} \times 10^{-4} \times 0.01
$$

$$\begin{bmatrix} \epsilon_1 \\ \epsilon_2 \\ \gamma_{12} \end{bmatrix}_{UD} = \begin{bmatrix} 0.000 \\ 0.001 \\ \mp 0.001 \end{bmatrix}$$

From Equation (6.10) the total strains on the corrosion barrier of chopped fibers are

$$\begin{bmatrix} \epsilon_x \\ \epsilon_y \\ \gamma_{xy} \end{bmatrix}_{chop} = \begin{bmatrix} \epsilon_x \\ \epsilon_y \\ 0 \end{bmatrix}_{MECH} + \begin{bmatrix} 0.145 \\ -0.068 \\ 0 \end{bmatrix} \times 10^{-4} \left(OT - PT \right) + \begin{bmatrix} -36.0 \\ -1560 \\ 0 \end{bmatrix} \times 10^{-4} \Delta m$$

$$\begin{bmatrix} \epsilon_x \\ \epsilon_y \\ \gamma_{xy} \end{bmatrix}_{chop} = \begin{bmatrix} 0.000 \\ 0.001 \\ 0 \end{bmatrix}_{MECH} + \begin{bmatrix} 0.145 \\ -0.068 \\ 0 \end{bmatrix} \times 10^{-4} \left(25 - 100 \right) + \begin{bmatrix} -36.0 \\ -1560 \\ 0 \end{bmatrix} \times 10^{-4} \times 0.01$$

$$\begin{bmatrix} \epsilon_x \\ \epsilon_y \\ \gamma_{xy} \end{bmatrix}_{chop} = \begin{bmatrix} 0.001 \\ 0.000 \\ 0.000 \end{bmatrix}$$

In addition to the above, we should also take into consideration the mechanical residual strain from the axial force Nx = − 118.6 kg/cm.

$$\epsilon_x^r = \frac{N_x}{t_{UD} \times \bar{E}_x}$$

Where $\bar{E}_x$ is the laminate axial modulus

$$\bar{E}_x = 104000 - \frac{61000^2}{334000} = 92859 \text{ kg/cm}^2$$

$$\epsilon_x^r = \frac{-118.6}{1.0 \times 92859} = -0.0013$$

Introducing this correction, the actual total strains on the corrosion barrier become

$$\begin{bmatrix} \epsilon_x \\ \epsilon_y \\ \gamma_{xy} \end{bmatrix}_{chop} = \begin{bmatrix} 0.000 \\ 0.000 \\ 0.000 \end{bmatrix}$$

The total strains on the corrosion barrier of this specific pipeline are negligibly small, close to zero.

Example 6.6

Determine the minimum operating temperatures for ±70 vertical storage tanks.

The following analysis ignores the resin behavior at extremely low temperatures, and may not reflect the actual laminate response. The total hoop strain on the corrosion barrier controls the tank design. The mode of failure is infiltration and the allowable strain is 0.20%. The total hoop strain on the corrosion barrier should never exceed 0.20%.

From Equation (6.6A) we have

$$\epsilon_y = \left(0.017\frac{PD}{t_{UD}} + 0.068\left(PT - OT\right) - 1560\Delta m\right)\times 10^{-4} \tag{6.6A}$$

The minimum operating temperature OT is determined so that

$$\left(0.017\frac{PD}{t_{UD}} + 0.068\left(PT - OT\right) - 1560\Delta m\right)\times 10^{-4} < 0.002$$

Developing the above, we have

$$OT > 0.25\times\frac{PD}{t_{UD}} + PT - 22900\times\Delta m - 294$$

Assuming the stress free temperature PT = 100C and Δm = 0.001, we have

$$OT > 0.25\times\frac{PD}{t_{UD}} + 100 - 22900\times 0.001 - 294$$

$$OT > 0.25\times\frac{PD}{t_{UD}} - 217$$

The minimum operating temperature for empty tanks is

$$OT > -217C$$

Example 6.7

What would the minimum operating temperature be, assuming an allowable strain of 0.10%?

The computation protocol is the same

$$\left(0.017\frac{PD}{t_{UD}} + 0.068\left(PT - OT\right) - 1560\Delta m\right) \times 10^{-4} < 0.001$$

Developing the above, we have

$$OT > 0.25 \times \frac{PD}{t_{UD}} + PT - 22900 \times \Delta m - 147$$

Assuming the stress free temperature PT = 100C and Δm = 0.001, we have

$$OT > 0.25 \times \frac{PD}{t_{UD}} + 100 - 22900 \times 0.001 - 147$$

$$OT > 0.25 \times \frac{PD}{t_{UD}} - 70$$

The minimum operating temperature for empty tanks is

$$OT > -70C$$

7 Total Strains in Hoop-Chop Sanitation Pipes

7.1 INTRODUCTION

This chapter computes the total strains of plies embedded in hoop-chop laminates used in sanitation pipes. The computation combines the protocol developed in Chapter 3 with the numerical matrices of Chapter 4. The computed total strains enter directly in the unified equation (see Chapter 16) to estimate the laminate load-dependent durability.

The assumed design conditions are

- The corrosion barrier is fully preserved and included in the design
- The internal pressure is P
- The laminate construction is determined by trial and error.
- The pipeline is anchored, implying zero axial strain.
- The operating temperature is OT = 25C.
- The absorbed water is $\Delta m = 1.0\%$.
- The design criterion is weeping, with the allowable hoop strain $\varepsilon_y = 0.40\%$.

The structural contribution from three distinct plies – UD, sand core and chopped fibers – complicates the design equations, which become too lengthy and unwieldy. To facilitate the mathematical manipulation the ply strains are computed numerically from arbitrarily chosen laminate constructions and the final design is established by trial and error. The computed numerical total strain components cannot exceed the allowable values. The entire process is repeated for a different laminate until a satisfactory solution is found. The analysis performed by trial and error uses simple numerical matrices to avoid the otherwise complex equations.

7.2 THERMAL STRAINS IN HOOP-CHOP LAMINATES

The thermal resultants $\left[N^T \right]$ and the stiffness matrix [A] of hoop-chop laminates are in Table 4.3 of Chapter 4.

$$[A]_{HC} = \begin{bmatrix} 100000 & 30000 & 0 \\ 30000 & 400000 & 0 \\ 0 & 0 & 35000 \end{bmatrix} \times (t_{UD}) + \begin{bmatrix} 77000 & 23000 & 0 \\ 23000 & 77000 & 0 \\ 0 & 0 & 27000 \end{bmatrix}$$

$$\times (t_{chop}) + \begin{bmatrix} 66000 & 20000 & 0 \\ 20000 & 66000 & 0 \\ 0 & 0 & 23000 \end{bmatrix} \times (t_{sand}) \, \text{kg/cm}$$

$$\begin{bmatrix} N_x^T \\ N_y^T \\ N_{xy}^T \end{bmatrix} = \begin{bmatrix} 4.73 \times t_{UD} + 2.50 \times t_{chop} + 1.12 \times t_{sand} \\ 4.35 \times t_{UD} + 2.50 \times t_{chop} + 1.12 \times t_{sand} \\ 0 \end{bmatrix} \times (OT - PT) \, \text{kg/cm}$$

The equilibrium thermal strains come from Equation (3.1) of Chapter 3.

$$\begin{bmatrix} N_x^T \\ N_y^T \\ 0 \end{bmatrix} = \begin{bmatrix} A_{xx} & A_{xy} & 0 \\ A_{yx} & A_{yy} & 0 \\ 0 & 0 & A_{ss} \end{bmatrix} \times \begin{bmatrix} \varepsilon_x^e \\ \varepsilon_y^e \\ \gamma_{xy}^e \end{bmatrix} \qquad (7.1)$$

As we can see, the laminate thicknesses do not cancel out in this case, making the computations too lengthy and awkward. To facilitate the algebraic manipulations, a tentative laminate construction is arbitrarily chosen to compute a numerical value for the residual thermal strain components. The residual hydric and mechanical strains are computed for the same chosen laminate construction. The total strains are compared with the allowable value and the entire procedure is repeated until a satisfactory solution is found. Usually three trials suffice to arrive at the proper design.

7.3 HYDRIC STRAINS IN HOOP-CHOP LAMINATES

The hydric equilibrium strains are computed from an equation similar to (7.1), substituting the hydric resultants for the thermal resultants.

$$\begin{bmatrix} N_x^H \\ N_y^H \\ N_{xy}^H \end{bmatrix} = \begin{bmatrix} 25200 \times t_{UD} + 25000 \times t_{chop} + 24080 \times t_{sand} \\ 23200 \times t_{UD} + 25000 \times t_{chop} + 24080 \times t_{sand} \\ 0 \end{bmatrix} \times (\Delta m)$$

As mentioned, the hydric equilibrium strains are computed by assigning arbitrary thicknesses to all three plies.

7.4   MECHANICAL STRAINS ON HOOP-CHOP SANITATION PIPES

The hoop-chop sanitation pipes usually work underground with the soil friction absorbing all longitudinal movement. The internal pressure is P. The hoop force resultant is $N_y = PD/2$. The longitudinal force from the soil friction is Nx. The axial mechanical strain is $\varepsilon_x = 0$. The torque is also zero. Such a loading condition is typical of underground pipes. The equilibrium mechanical strains are computed from the mechanical force resultants and the laminate matrix [A]. For underground hoop-chop pipes we have

$$
\left[A \right]_{HC} \times \begin{bmatrix} 0 \\ \varepsilon_y^e \\ 0 \end{bmatrix} = \begin{bmatrix} N_x \\ \dfrac{PD}{2} \\ 0 \end{bmatrix}
\tag{7.2}
$$

$$
\left(\begin{bmatrix} 100 & 30 & 0 \\ 30 & 400 & 0 \\ 0 & 0 & 35 \end{bmatrix} \times t_{UD} + \begin{bmatrix} 77 & 23 & 0 \\ 23 & 77 & 0 \\ 0 & 0 & 27 \end{bmatrix} \times t_{chop} + \begin{bmatrix} 66 & 20 & 0 \\ 20 & 66 & 0 \\ 0 & 0 & 23 \end{bmatrix} \times t_{sand} \right)
$$

$$
\times 10^3 \times \begin{bmatrix} 0 \\ \epsilon_y^e \\ 0 \end{bmatrix} = \begin{bmatrix} N_x \\ PD/2 \\ 0 \end{bmatrix}
$$

The compute the mechanical hoop strain and the soil friction force, we arbitrarily assign the thicknesses of the three plies, as we did in the computation of the thermal and hydric residual strains.

7.5   TOTAL STRAINS

The total strains are the sum result of the mechanical, thermal and hydric components

$$
\begin{bmatrix} Total \\ strain \end{bmatrix} = \begin{bmatrix} mechanical \\ strain \end{bmatrix} + \begin{bmatrix} residual \\ thermal\ strain \end{bmatrix} + \begin{bmatrix} residual \\ hydric\ strain \end{bmatrix}
$$

Example 7.1

Determine an acceptable laminate construction for a hoop-chop pipe laminate operating in the following conditions. The design is for weep failure.

- The HDT of the polyester resin matrix is HDT = 100C.
- The peak post-cure temperature is PCT = 100C.
- The pipe carries water at room temperature OT = RT = 25C.
- The operating pressure is P = 10 bar.
- The pipe diameter is D = 500 mm.
- The water pick up is Δm = 1.0%.
- The weep threshold is 0.80% and the allowable strain is 0.80/2 = 0.40%.

The design criterion is weeping. The critical ply is the weep barrier of chopped fibers, not the UD plies as in the case of the oil pipes discussed in Chapter 5. The laminate construction is determined by trial and error until a solution is found that keeps the strain in the weep barrier less than 0.80/2 = 0.40%.

Let us tentatively set t_{chop} = 4.0 mm, t_{UD} = 3.0 mm and t_{sand} = 5.0 mm to compute the strains. If necessary, we choose another set of tentative thicknesses until we have adequate values for the strains. The arbitrarily set ply thicknesses produce the following laminate stiffness matrix

$$[A]_{HC} = \begin{bmatrix} 100000 & 30000 & 0 \\ 30000 & 400000 & 0 \\ 0 & 0 & 35000 \end{bmatrix} \times (t_{UD}) + \begin{bmatrix} 77000 & 23000 & 0 \\ 23000 & 77000 & 0 \\ 0 & 0 & 27000 \end{bmatrix}$$

$$\times (t_{chop}) + \begin{bmatrix} 66000 & 20000 & 0 \\ 20000 & 66000 & 0 \\ 0 & 0 & 23000 \end{bmatrix} \times (t_{sand})$$

$$[A]_{HC} = \begin{bmatrix} 100000 & 30000 & 0 \\ 30000 & 400000 & 0 \\ 0 & 0 & 35000 \end{bmatrix} \times 0.3 + \begin{bmatrix} 77000 & 23000 & 0 \\ 23000 & 77000 & 0 \\ 0 & 0 & 27000 \end{bmatrix}$$

$$\times 0.4 + \begin{bmatrix} 66000 & 20000 & 0 \\ 20000 & 66000 & 0 \\ 0 & 0 & 23000 \end{bmatrix} \times 0.5$$

$$[A]_{HC} = \begin{bmatrix} 93800 & 28200 & 0 \\ 28200 & 183800 & 0 \\ 0 & 0 & 32800 \end{bmatrix} \text{kg/cm}$$

The thermal resultants are

$$\begin{bmatrix} N_x^T \\ N_y^T \\ N_{xy}^T \end{bmatrix} = \begin{bmatrix} 4.73 \times t_{UD} + 2.50 \times t_{chop} + 1.12 \times t_{sand} \\ 4.35 \times t_{UD} + 2.50 \times t_{chop} + 1.12 \times t_{sand} \\ 0 \end{bmatrix} \times (OT - PT) \, \text{kg/cm}$$

The stress-free temperature is PT = PCT = 100C. The operating temperature is OT = 25C. The arbitrarily chosen thicknesses are known. Entering these in the above equation we obtain.

$$\begin{bmatrix} N_x^T \\ N_y^T \\ N_{xy}^T \end{bmatrix}_{HC} = - \begin{bmatrix} 223.4 \\ 214.9 \\ 0 \end{bmatrix} \, \text{kg/cm}$$

From Equation (7.1) the equilibrium thermal strains are

$$\begin{bmatrix} N_x^T \\ N_y^T \\ 0 \end{bmatrix} = \begin{bmatrix} A_{xx} & A_{xy} & 0 \\ A_{yx} & A_{yy} & 0 \\ 0 & 0 & A_{ss} \end{bmatrix} \times \begin{bmatrix} \varepsilon_x^e \\ \varepsilon_y^e \\ \gamma_{xy}^e \end{bmatrix}$$

$$- \begin{bmatrix} 223.4 \\ 214.9 \\ 0 \end{bmatrix} = \begin{bmatrix} 93800 & 28200 & 0 \\ 28200 & 183800 & 0 \\ 0 & 0 & 32800 \end{bmatrix} \times \begin{bmatrix} \varepsilon_x^e \\ \varepsilon_y^e \\ \gamma_{xy}^e \end{bmatrix}$$

$$\begin{bmatrix} \varepsilon_x^e \\ \varepsilon_y^e \\ \gamma_{xy}^e \end{bmatrix}_{ther} = \begin{bmatrix} -0.00084 \\ -0.00213 \\ 0 \end{bmatrix}$$

The above are the equilibrium thermal strain components in the global frame. For hoop-chop laminates, the global frame coincides with the local frames of all plies. Strain rotations to the local frames are, therefore, not required in this case. The residual thermal strains on all plies in the local frames are

$$\begin{bmatrix} \varepsilon_x \\ \varepsilon_y \\ \gamma_{xy} \end{bmatrix}_{UD}^{thermal} = \begin{bmatrix} -0.00084 - 45.0 \times 10^{-6} (25 - 100) \\ -0.00213 - 7.5 \times 10^{-6} (25 - 100) \\ 0 \end{bmatrix} = \begin{bmatrix} 0.00254 \\ -0.00157 \\ 0 \end{bmatrix}$$

$$
\begin{bmatrix} \epsilon_x \\ \epsilon_y \\ \gamma_{xy} \end{bmatrix}_{chop}^{thermal} = \begin{bmatrix} -0.00084 - 25.0 \times 10^{-6} \left(25 - 100\right) \\ -0.00213 - 25.0 \times 10^{-6} \left(25 - 100\right) \\ 0 \end{bmatrix} = \begin{bmatrix} 0.00104 \\ -0.00026 \\ 0 \end{bmatrix}
$$

$$
\begin{bmatrix} \epsilon_x \\ \epsilon_y \\ \gamma_{xy} \end{bmatrix}_{sand}^{thermal} = \begin{bmatrix} -0.00084 - 13.0 \times 10^{-6} \left(25 - 100\right) \\ -0.00213 - 13.0 \times 10^{-6} \left(25 - 100\right) \\ 0 \end{bmatrix} = \begin{bmatrix} 0.00014 \\ -0.00116 \\ 0 \end{bmatrix}
$$

The derivation of the residual hydric strains in the global frame follows the same protocol, entering the hydric resultants in place of the thermal resultants. From Table 4.3, Chapter 4, we have

$$
\begin{bmatrix} N_x^H \\ N_y^H \\ N_{xy}^H \end{bmatrix} = \begin{bmatrix} 25200 \times t_{UD} + 25000 \times t_{chop} + 24080 \times t_{sand} \\ 23200 \times t_{UD} + 25000 \times t_{chop} + 24080 \times t_{sand} \\ 0 \end{bmatrix} \times (\Delta m)
$$

$$
\begin{bmatrix} N_x^H \\ N_y^H \\ N_{xy}^H \end{bmatrix} = \begin{bmatrix} 25200 \times 0.3 + 25000 \times 0.4 + 24080 \times 0.5 \\ 23200 \times 0.3 + 25000 \times 0.4 + 24080 \times 0.5 \\ 0 \end{bmatrix} \times (0.01)
$$

$$
\begin{bmatrix} N_x^H \\ N_y^H \\ N_{xy}^H \end{bmatrix} = \begin{bmatrix} 296 \\ 290 \\ 0 \end{bmatrix} \text{kg/cm}
$$

Skipping the computation details, the residual hydric strains are

$$
\begin{bmatrix} \epsilon_x \\ \epsilon_y \\ \gamma_{xy} \end{bmatrix}_{UD}^{hydric} = \begin{bmatrix} 0.00115 - 0.24 \times 0.01 \\ 0.00350 - 0.04 \times 0.01 \\ 0 \end{bmatrix} = \begin{bmatrix} -0.00125 \\ 0.00310 \\ 0 \end{bmatrix}
$$

$$
\begin{bmatrix} \epsilon_x \\ \epsilon_y \\ \gamma_{xy} \end{bmatrix}_{chop}^{hydric} = \begin{bmatrix} 0.00115 - 0.25 \times 0.01 \\ 0.00350 - 0.25 \times 0.01 \\ 0 \end{bmatrix} = \begin{bmatrix} -0.00135 \\ 0.00100 \\ 0 \end{bmatrix}
$$

$$
\begin{bmatrix} \epsilon_x \\ \epsilon_y \\ \gamma_{xy} \end{bmatrix}_{sand}^{hydric} = \begin{bmatrix} 0.00115 - 0.28 \times 0.01 \\ 0.00350 - 0.28 \times 0.01 \\ 0 \end{bmatrix} = \begin{bmatrix} -0.00165 \\ 0.00070 \\ 0 \end{bmatrix}
$$

The mechanical strains in the global system are

$$
[A]_{HC} \times \begin{bmatrix} 0 \\ \varepsilon_y^e \\ 0 \end{bmatrix} = \begin{bmatrix} N_x \\ \dfrac{PD}{2} \\ 0 \end{bmatrix}
$$

$$
\begin{bmatrix} 93800 & 28200 & 0 \\ 28200 & 183800 & 0 \\ 0 & 0 & 32800 \end{bmatrix} \times \begin{bmatrix} 0 \\ \varepsilon_y^e \\ 0 \end{bmatrix} = \begin{bmatrix} N_x \\ \dfrac{10 \times 50}{2} \\ 0 \end{bmatrix} \tag{7.3}
$$

From which we have $\epsilon_y = 0.00136$ and $N_x = 38.4 \ kg / cm$.

The mechanical strains in the local systems of all plies are equal to those in the global system, since in hoop-chop laminates all systems coincide. The total strains in each ply are

$$
\begin{bmatrix} \epsilon_x \\ \epsilon_y \\ \gamma_{xy} \end{bmatrix}_{UD} = \begin{bmatrix} 0 \\ 0.00136 \\ 0 \end{bmatrix} + \begin{bmatrix} 0.00254 \\ -0.00157 \\ 0 \end{bmatrix} + \begin{bmatrix} -0.00125 \\ 0.00310 \\ 0 \end{bmatrix} = \begin{bmatrix} 0.13\% \\ 0.29\% \\ 0 \end{bmatrix}
$$

$$
\begin{bmatrix} \epsilon_x \\ \epsilon_y \\ \gamma_{xy} \end{bmatrix}_{chop} = \begin{bmatrix} 0 \\ 0.00136 \\ 0 \end{bmatrix} + \begin{bmatrix} 0.00104 \\ -0.00026 \\ 0 \end{bmatrix} + \begin{bmatrix} -0.00135 \\ 0.00100 \\ 0 \end{bmatrix} = \begin{bmatrix} -0.03\% \\ 0.21\% \\ 0 \end{bmatrix} \tag{7.4}
$$

$$
\begin{bmatrix} \epsilon_1 \\ \epsilon_2 \\ \gamma_{12} \end{bmatrix}_{sand} = \begin{bmatrix} 0 \\ 0.00136 \\ 0 \end{bmatrix} + \begin{bmatrix} 0.00014 \\ -0.00116 \\ 0 \end{bmatrix} + \begin{bmatrix} -0.00165 \\ 0.00070 \\ 0 \end{bmatrix} = \begin{bmatrix} -0.15\% \\ 0.10\% \\ 0 \end{bmatrix}
$$

The computed maximum total strain in the weep barrier of chopped fibers is 0.21%, in the hoop direction. This is less than the allowable 0.40%, meaning the chosen laminate is too conservative and could be cut down a bit, perhaps by eliminating the sand core, or reducing the UD ply. A satisfactory solution is reached in two or three trials.

Example 7.2

Let us estimate the weep pressure rating of the hoop-chop pipe with the arbitrary ply thicknesses of Example 7.1.

We begin by computing the hoop mechanical strain component required to produce a total strain of 0.40% in the weep barrier of the pipe. The total hoop strain in the weep barrier is computed from Equation (7.4).

$$\epsilon_y - 0.00026 + 0.00100 = 0.00400$$

The required mechanical hoop strain component is

$$\epsilon_y = 0.00326(0.326\%)$$

The hoop component of the mechanical strain connects to the operating pressure by the expression below, taken from Equation (7.3).

$$183800 \times \epsilon_y = \frac{P \times 50}{2}$$

Solving the above for $\epsilon_y = 0.00326$ gives the weep pressure rating P = 24 bar.

Example 7.3

Let us compute the rupture durability of the pipe described in Example 7.1 operating at rated pressure.

The total strain in the fiber direction at the rated pressure P = 24 bar is

$$\epsilon_y = \epsilon_1 = 0.00326 - 0.00157 + 0.00310 = 0.00479$$

The thermal strain component is compressive in the fiber direction, which improves the durability. The hydric strain component is tensile and detrimental. On balance the total strain in the fiber direction increases from a pure mechanical component of 0.326% to a total value 0.479% when the residual strains are accounted for. The residual hydric and thermal strains cannot be ignored.

The question is…. How long will it take until the glass fibers fail? The answer to this question is in the Part II of this book.

The underground pipes in sewage service usually work at high burial depths and require large ring stiffness to resist the soil weight. Also, these pipes usually operate in gravity service subjected to zero internal pressure. Such operating conditions require sewer pipes with a thick sand core, to provide

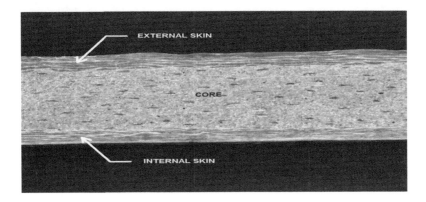

FIGURE 7.1 *The cross-section of a typical sewer pipe wall shows a large sand core compared to a thin inner and outer laminates. The sand core provides the required ring stiffness.*

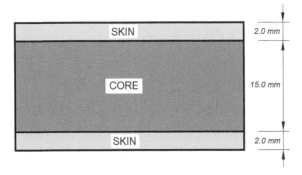

FIGURE 7.2 *Typical wall construction of a gravity sewer pipe showing the large sand core compared to the overall thickness.*

the required ring stiffness, and a thin structural thickness, to handle the low pressure of gravity service. Figures 7.1 and 7.2 show the cross-section of a typical laminate used in gravity sewer service.

Example 7.4

Compute the standard laminate matrices for the pipe construction shown in Figure 7.2. The large sand core compared to the overall laminate thickness indicates a gravity sewer pipe operating at low pressure and high burial depth. Also note the absence of UD plies. The computations are performed in accordance with the protocol described in Chapter 3. The ply properties are tabulated in Chapter 1. The liner thickness is ignored.

The stiffness matrix [A] is

$$[A] = \begin{bmatrix} 77000 \times 0.4 + 66000 \times 1.5 & 23000 \times 0.4 + 20000 \times 1.5 & 0 \\ 23000 \times 0.4 + 20000 \times 1.5 & 77000 \times 0.4 + 66000 \times 1.5 & 0 \\ 0 & 0 & 27000 \times 0.4 + 23000 \times 1.5 \end{bmatrix}$$

$$[A] = \begin{bmatrix} 129800 & 39200 & 0 \\ 39200 & 129800 & 0 \\ 0 & 0 & 45300 \end{bmatrix} \text{kg/cm}$$

The components of the thermal resultant are

$$[N^T] = \begin{bmatrix} (77000 + 23000) \times 0.4 \times 25 \times 10^{-6} + (66000 + 20000) \times 1.5 \times 13 \times 10^{-6} \\ same \\ 0 \end{bmatrix} \times (OT - PT)$$

$$[N^T] = \begin{bmatrix} 2.68 \\ 2.68 \\ 0 \end{bmatrix} \times (OT - PT) \text{ kg/cm}$$

The components of the hydric resultant are

$$[N^H] = \begin{bmatrix} (77000 + 23000) \times 0.4 \times 0.25 + (66000 + 20000) \times 1.5 \times 0.28 \\ same \\ 0 \end{bmatrix} \times \Delta m$$

$$[N^H] = \begin{bmatrix} 46120 \\ 46120 \\ 0 \end{bmatrix} \times \Delta m \text{ kg/cm}$$

Example 7.5

Compute the stiffness matrix [A] of the pipe depicted in Figure 7.2 in the absence of the sand core.

The computation is performed by entering $t_{sand} = 0$ in the matrix [A] of Example 7.4.

$$[A] = \begin{bmatrix} 77000 \times 0.4 & 23000 \times 0.4 & 0 \\ 23000 \times 0.4 & 77000 \times 0.4 & 0 \\ 0 & 0 & 27000 \times 0.4 \end{bmatrix}$$

$$
[A] = \begin{bmatrix} 30800 & 9200 & 0 \\ 9200 & 30800 & 0 \\ 0 & 0 & 10800 \end{bmatrix} \text{ kg/cm}
$$

A comparison of the above matrix with the one computed in the previous example shows the significant contribution of the sand core to the stiffness of sewer pipes designed for low pressure and high depths of burial.

Example 7.6

Compute the total strains on the pipe depicted in Figure 7.2 for PT = 100C, OT = 25C and Δm = 0.01. Assume a healthy sand core.

First, we compute the equilibrium thermal and hydric strains from the equations developed in Chapter 3. The thermal equilibrium strains are

$$
\begin{bmatrix} \varepsilon_x^e \\ \varepsilon_y^e \\ 0 \end{bmatrix} = \begin{bmatrix} A_{xx} & A_{xy} & 0 \\ A_{yx} & A_{yy} & 0 \\ 0 & 0 & A_{ss} \end{bmatrix}^{-1} \times \begin{bmatrix} N_x^T \\ N_y^T \\ 0 \end{bmatrix}
$$

$$
\begin{bmatrix} \varepsilon_x^e \\ \varepsilon_y^e \\ 0 \end{bmatrix} = \begin{bmatrix} 129800 & 39200 & 0 \\ 39200 & 129800 & 0 \\ 0 & 0 & 45300 \end{bmatrix}^{-1} \times \begin{bmatrix} 2.68 \\ 2.68 \\ 0 \end{bmatrix} \times (25 - 100)
$$

$$
\begin{bmatrix} \varepsilon_x^e \\ \varepsilon_y^e \\ 0 \end{bmatrix}_{thermal} = \begin{bmatrix} -1.19 \times 10^{-3} \\ -1.19 \times 10^{-3} \\ 0 \end{bmatrix}
$$

The hydric equilibrium strains are

$$
\begin{bmatrix} \varepsilon_x^e \\ \varepsilon_y^e \\ 0 \end{bmatrix} = \begin{bmatrix} 129800 & 39200 & 0 \\ 39200 & 129800 & 0 \\ 0 & 0 & 45300 \end{bmatrix}^{-1} \times \begin{bmatrix} 46120 \\ 46120 \\ 0 \end{bmatrix} \times 0.01
$$

$$
\begin{bmatrix} \varepsilon_x^e \\ \varepsilon_y^e \\ 0 \end{bmatrix}_{hydric} = \begin{bmatrix} 2.73 \times 10^{-3} \\ 2.73 \times 10^{-3} \\ 0 \end{bmatrix}
$$

The residual thermal strains in the global system come from the general formula in Chapter 3.

$$
\begin{bmatrix} \varepsilon_x^r \\ \varepsilon_y^r \\ \gamma_{xy}^r \end{bmatrix} = \begin{bmatrix} \varepsilon_x^e - \alpha_1(OT - PT) \\ \varepsilon_y^e - \alpha_2(OT - PT) \\ \gamma_{12}^e \end{bmatrix}
$$

For the ply of chopped fibers we have

$$
\begin{bmatrix} \varepsilon_x^r \\ \varepsilon_y^r \\ \gamma_{xy}^r \end{bmatrix}_{chop} = \begin{bmatrix} -1.19 \times 10^{-3} - 25 \times 10^{-6}(25 - 100) \\ same \\ 0 \end{bmatrix}
$$

$$
\begin{bmatrix} \varepsilon_x^r \\ \varepsilon_y^r \\ \gamma_{xy}^r \end{bmatrix}_{chop} = \begin{bmatrix} 6.85 \times 10^{-4} \\ same \\ 0 \end{bmatrix}
$$

For the core ply we have

$$
\begin{bmatrix} \varepsilon_x^r \\ \varepsilon_y^r \\ \gamma_{xy}^r \end{bmatrix}_{core} = \begin{bmatrix} -1.19 \times 10^{-3} - 13 \times 10^{-6}(25 - 100) \\ same \\ 0 \end{bmatrix}
$$

$$
\begin{bmatrix} \varepsilon_x^r \\ \varepsilon_y^r \\ \gamma_{xy}^r \end{bmatrix}_{core} = \begin{bmatrix} -2.15 \times 10^{-4} \\ same \\ 0 \end{bmatrix}
$$

The residual hydric strains in the global system are computed from the general equation

$$
\begin{bmatrix} \varepsilon_x^r \\ \varepsilon_y^r \\ \gamma_{xy}^r \end{bmatrix} = \begin{bmatrix} \varepsilon_x^e - \beta_1 \Delta m \\ \varepsilon_y^e - \beta_2 \Delta m \\ \gamma_{12}^e \end{bmatrix}
$$

For the ply of chopped fibers we have

$$\begin{bmatrix} \varepsilon_x^r \\ \varepsilon_y^r \\ \gamma_{xy}^r \end{bmatrix}_{chop} = \begin{bmatrix} 2.73 \times 10^{-3} - 0.25 \times 0.01 \\ same \\ 0 \end{bmatrix}$$

$$\begin{bmatrix} \varepsilon_x^r \\ \varepsilon_y^r \\ \gamma_{xy}^r \end{bmatrix}_{chop} = \begin{bmatrix} 2.28 \times 10^{-4} \\ same \\ 0 \end{bmatrix}$$

For the sand core ply we have

$$\begin{bmatrix} \varepsilon_x^r \\ \varepsilon_y^r \\ \gamma_{xy}^r \end{bmatrix}_{core} = \begin{bmatrix} 2.73 \times 10^{-3} - 0.28 \times 0.01 \\ same \\ 0 \end{bmatrix}$$

$$\begin{bmatrix} \varepsilon_x^r \\ \varepsilon_y^r \\ \gamma_{xy}^r \end{bmatrix}_{core} = \begin{bmatrix} -7.22 \times 10^{-5} \\ same \\ 0 \end{bmatrix}$$

The mechanical strains are

$$\begin{bmatrix} \varepsilon_x^e \\ \varepsilon_y^e \\ \gamma_{xy} \end{bmatrix} = \begin{bmatrix} A_{xx} & A_{xy} & 0 \\ A_{yx} & A_{yy} & 0 \\ 0 & 0 & A_{ss} \end{bmatrix}^{-1} \times \begin{bmatrix} N_x^M \\ N_y^M \\ N_{xy}^M \end{bmatrix}$$

The soil friction prevents all pipe expansions and contractions in the axial direction. Therefore, taking $\varepsilon_x = 0$ we have

$$\begin{bmatrix} 0 \\ \varepsilon_y^e \\ 0 \end{bmatrix} = \begin{bmatrix} 129800 & 39200 & 0 \\ 39200 & 129800 & 0 \\ 0 & 0 & 45300 \end{bmatrix}^{-1} \times \begin{bmatrix} N_x^M \\ P \times \Phi/2 \\ 0 \end{bmatrix}$$

Solving the above we obtain

$$\epsilon_y = \frac{P \times \varnothing}{2 \times 129800} = 3.85 \times 10^{-6} P \times \varnothing$$

$$N_x = 0.151 P \times \varnothing$$

Finally, the total strains are

$$\begin{bmatrix} \varepsilon_x \\ \varepsilon_y \\ \gamma_{xy} \end{bmatrix}_{total}^{chop} = \begin{bmatrix} 0 \\ 3.85 \times 10^{-6} P\Phi \\ 0 \end{bmatrix}_{MEC} + \begin{bmatrix} 6.85 \times 10^{-4} \\ same \\ 0 \end{bmatrix}_{TER} + \begin{bmatrix} 2.28 \times 10^{-4} \\ same \\ 0 \end{bmatrix}_{HYD}$$

$$\begin{bmatrix} \varepsilon_x \\ \varepsilon_y \\ \gamma_{xy} \end{bmatrix}_{total}^{core} = \begin{bmatrix} 0 \\ 3.85 \times 10^{-6} P\Phi \\ 0 \end{bmatrix}_{MEC} + \begin{bmatrix} -2.15 \times 10^{-4} \\ same \\ 0 \end{bmatrix}_{TER} + \begin{bmatrix} -7.22 \times 10^{-5} \\ same \\ 0 \end{bmatrix}_{HYD}$$

Example 7.7

Compute the pressure rating of the pipe in Figures 7.1 and 7.2.

The pressure rating of sanitation pipes is usually determined by limiting the maximum strain in the weep barrier to ½ the weep threshold. Assuming a typical weep threshold of 0.80%, the design strain would be

$$design\ strain = \frac{0.80\%}{2} = 0.40\%$$

Such a large design strain is perfectly acceptable for high-pressure sanitation pipes in which the sand core plays a minor part in the overall stiffness. This is not the case in low-pressure sanitation pipes. From Example 7.5, we learned that heavy cracking of the sand core causes a severe loss of pipe stiffness in low-pressure sanitation pipes. This is usually true for all low-pressure gravity pipes in which the matrix [A] is highly dependent on the sand core. In these cases it is safer to design the pipes for core rupture, instead of weeping of the chopped fibers. The rationale is that once the core ruptures and the pipe matrix [A] drops, the weep barrier by itself may not control the failure.

Going back to Example 7.5, we see that the hoop stiffness drops from 30800+99000 = 129800 kg/cm in healthy pipes to 30800 kg/cm in over-strained pipes with cracked cores. Such a low hoop stiffness value would produce strains above the weep threshold of the chopped plies. From this new perspective, the design strain in the case of extremely low-pressure sanitation pipes should be ½ the rupture strain of the sand core. Assuming the rupture strength (in Chapter 8 we call it the rupture threshold) of the sand core to be 0.30%, the design strain should be

$$design\ strain = \frac{0.30\%}{2} = 0.15\%$$

Taking the above as the design strain the pipe's pressure rating is

$$3.85 \times 10^{-6}\ P \times \varnothing - 2.15 \times 10^{-4} - 7.22 \times 10^{-5} = 0.0015$$

$$P = \frac{464.2}{D}$$

Assuming a diameter D = 1000 mm we have

$$P = \frac{464.2}{100} = 4.6\ bar$$

The failure of the weak sand core precipitates the collapse of the strong chopped plies. The critical ply in this case is the sand core, not the chopped fibers.

APPENDIX 7.1

RESIDUAL MECHANICAL STRAINS

We end this chapter discussing a detail overlooked in Chapter 3. The reader will remember that in developing the protocol to compute the mechanical strains – see Chapter 3 – we emphatically denied the existence of residual mechanical strains. This simplification was necessary to facilitate the presentation that would otherwise become too complex. In reality, however, the laminates constrained by reactive external forces develop residual mechanical strains. Since this is an exception, rather than a rule, we have decided to exclude the residual mechanical strains from the protocol and compute them separately every time they arise, as in the presence of external support reactions. The residual mechanical strains are then added to the strains computed by the protocol. This is better explained in a numerical example.

Example 7.8

Suppose a pressurized underground pipeline. The soil friction prevents the axial movement of the pipes and the mechanical axial strain is $\varepsilon_x = 0$. This strain condition is used to compute the reactive axial force Nx, which takes the exact value required to make $\varepsilon_x = 0$, as discussed in the preceding numerical examples of this chapter. There is no doubt that the axial mechanical strain is zero in this case. However, we must not forget the presence of the residual mechanical axial strain, arising from the reactive force Nx and preventing the free Poisson contraction of the pipe subjected to the internal pressure P.

Let us compute the residual mechanical strains for the underground pipeline discussed in Example 7.1. The equilibrium mechanical strains and the reactive force Nx come from Equation (7.3)

$$
\begin{bmatrix} 93800 & 28200 & 0 \\ 28200 & 183800 & 0 \\ 0 & 0 & 32800 \end{bmatrix} \times \begin{bmatrix} 0 \\ \varepsilon_y^e \\ 0 \end{bmatrix} = \begin{bmatrix} N_x \\ \dfrac{10 \times 50}{2} \\ 0 \end{bmatrix} \tag{7.3}
$$

From which we have $\epsilon_y = 0.136\%$ and $N_x = 38.4\ kg/cm$. The equilibrium axial strain is $\varepsilon_x = 0$.

The free mechanical strains that would occur in the absence of the reactive force Nx are

$$
\begin{bmatrix} 93800 & 28200 & 0 \\ 28200 & 183800 & 0 \\ 0 & 0 & 32800 \end{bmatrix} \times \begin{bmatrix} \varepsilon_x^e \\ \varepsilon_y^e \\ 0 \end{bmatrix} = \begin{bmatrix} 0 \\ \dfrac{10 \times 50}{2} \\ 0 \end{bmatrix}
$$

Solving the above we have

$\epsilon_y = 0.143\%$

$\epsilon_x = -0.043\%$

By definition, the residual mechanical strains are computed by subtracting the free mechanical strains from the equilibrium strains.

$$
\begin{bmatrix} \epsilon_y^r \\ \epsilon_x^r \end{bmatrix} = \begin{bmatrix} 0.136 - 0.143 \\ 0 - (-0.043) \end{bmatrix} = \begin{bmatrix} -0.007\% \\ +0.043\% \end{bmatrix}
$$

Again by definition, the total mechanical strains are obtained by adding the free and the residual strains components.

$$\begin{bmatrix} \epsilon_y^{total} \\ \epsilon_x^{total} \end{bmatrix} = \begin{bmatrix} 0.143 - 0.007 \\ -0.043 + 0.043 \end{bmatrix} = \begin{bmatrix} +0.136\% \\ zero \end{bmatrix}$$

As we see, the total mechanical strains remain unchanged at ε_y = 0.136% and ε_x = 0. The residual axial strain, however, is ε_x = +0.043%. This residual strain comes from the reactive force Nx counteracting the Poisson contraction that would otherwise occur. This example illustrates a situation when the equilibrium axial strain is zero, and yet the residual axial strain is not zero.

As discussed in the Part II of this book, the higher hoop strain – not the axial strain – controls the load-dependent weep and strain-corrosion durability of underground sanitation pipes. The controlling hoop strains remain unchanged in this process, and the small axial residual strains are irrelevant. The simplified protocol developed in Chapter 3 is, therefore, valid in all practical cases.

Part II

Computation of Durability

The Part II of this book is about the durability of composite equipment in industrial service. We start with a description of the eight modes of long-term failure, followed by the quantitative models to predict the equipment durability. The Part II of the book presents quantitative solutions to all eight modes of long-term laminate failure. To clarify the concepts, we introduce many numerical examples along the discussions.

The estimation of the load-dependent durability of composite equipment requires knowledge of the total ply strain components as computed in the Part I of this book.

8 The Eight Modes of Long-Term Failure

8.1 INTRODUCTION

The modes of deterioration leading to long-term laminate damage and failure fall into two broad categories: the load-dependent and the load-independent. This book assumes that all modes of deterioration, regardless of their category, are independent and do not interact. This assumption allows the study of the load-dependent durability cases independent of the chemical in contact with the laminate. Conversely, the applied loads have no effect on the chemical durability. The computation of the load-independent durability is amazingly direct and simple. By contrast, the computations involved in the load-dependent durability cases are relatively complex and may require some effort on the part of the analyst. These topics are the subject of the subsequent chapters.

This book assumes that the degradation of a single ply, the critical ply, determines the fitness for use and durability of the entire laminate. There is a critical ply for every mode of failure. As examples, the chopped plies control the leakage failure of pressurized pipes in liquid service, while the unidirectional (UD) plies control the stiffness and rupture failures in dry applications such as aircraft parts and wind blades. In all cases, a specified level of accumulated damages defines the failure. No laminate lasts longer than its critical ply.

The estimation of the durability of any laminate requires knowledge of the following:

- The mode of failure and its degradation mechanism
- The critical ply
- The definition of a failure criterion.

In all cases, the laminate fails when the critical ply accumulates a damage level defined by a predetermined criterion. Every mode of failure has its own mode of deterioration and failure criterion. For example:

- In applications where long-term deflection under load is critical, as in wind blades, the mode of failure is loss of stiffness, the degradation mechanism is cyclic fatigue, and the failure criterion is the critical level of accumulated damages in the UD ply.

- In services involving fluid containment or transmission, the long-term mode of failure is weeping, the degradation mechanism is cyclic fatigue, and the failure criterion is the critical accumulated crack densities in the UD or chopped fiber plies.
- In other cases, the mode of failure could be an objectionable fading of the initial color caused by UV radiation and the failure criterion would be the deviation from an agreed upon color standard.

The computation of the long-term laminate durability requires a clear understanding of the modes of failure, the identification of the critical plies and the definition of the failure criteria. There is a trend in academic circles to consider the laminate gross structural deterioration as the leading indicator of failure. This approach, useful in the design for long-term rupture, is of little use in industrial service, where the prevailing design criteria derive from other modes of failure, such as infiltration, weeping, and loss of stiffness. The study of laminate durability based on the ultimate strength, just prior to rupture, is unrealistic in most industrial service.

Our approach to the analysis of durability focuses on industrial equipment and the length of time until they reach certain critical damage states defined by the failure criteria. This chapter introduces the eight modes of laminate long-term failure in industrial service, identifies the critical plies, and discusses the failure criteria in each case. The subsequent chapters provide additional details.

8.2 THE EIGHT MODES OF LONG-TERM FAILURE

The eight modes of long-term failure defining the durability of industrial laminates are:

- The **chemical failure,** defined as the chemical penetration, attack, and deterioration of the corrosion barrier. The applied loads have no effect on the chemical durability of laminates operating below the infiltration threshold. This topic is addressed in Chapter 10.
- The **infiltration failure,** characterized by crack densities large enough to allow the fast ingress of aggressive chemicals in the corrosion barrier. The applied loadings define the infiltration life, regardless of the aggressive chemicals. This topic is addressed in Chapter 13.
- The **weep failure,** characterized by crack densities large enough to allow the leakage of fluids. The applied loads govern the weep failure, regardless of the chemical involved (see Chapter 13).

- The **stiffness failure,** characterized by accumulated damages large enough to cause unacceptable loss of laminate stiffness. The damage (crack densities + delamination) leading to stiffness failure come from the applied loadings, regardless of the aggressive chemical (see Chapter 13).
- The **fiber rupture failure,** caused by accumulated fiber damages from strain-corrosion or cyclic fatigue. This topic is in Chapter 12.
- The **laminate strain-corrosion failure,** characterized by the growth of a large crack cutting across the fibers and perpendicular to the laminate. The laminate strain-corrosion comes from the combined action of chemical attack and bending loads. This topic is discussed in Chapter 14.
- The **abrasion failure,** defined as the loss of laminate thickness from abrasive action. The abrasion life results from friction and laminate wear (see Chapter 15).
- The **anomalous failure,** characterized by the growth of large, non-osmotic, water blisters in the laminate. The anomalous deterioration develops in large pre-existing delamination cracks usually present in the sand core of sanitation pipes. The discussion of this topic is in Chapter 20.

The eight above listed modes of failure exhaust all known causes of long-term disability of industrial laminates. The failure criteria, separating the acceptable from the non-acceptable laminates, depend on the failure mode. As examples, the failure criterion for chemical durability is the full penetration of the corrosion barrier by the contained aggressive fluid. The failure criteria for the load-dependent modes of failure—infiltration, weeping, stiffness, and rupture—are the equality of the long-term ply strengths and the applied loads. Table 8.1 summarizes the eight modes of long-term failure with their critical plies and the failure criteria for static loads.

The three load-independent modes of long-term failure—anomalous, abrasion, and chemical—result from clear and simple deterioration mechanisms. This facilitates their analysis. The protocols to estimate the load-independent lifetimes presented in this book are essentially complete and require little more than experimental work to develop the necessary computation parameters. The subsequent chapters will clarify this.

By contrast, the four load-dependent modes of long-term failure— infiltration, weeping, stiffness, and rupture—are not so clear and simple and require some intellectual effort on the part of the designer. The study of the load-dependent durability provides the central pieces of this book. The load-dependent failure modes are so important and appear so often in

TABLE 8.1

The eight modes of long-term failure and their critical plies.

Failure mode	Failure criterion	Critical ply	Comments
Chemical attack	Full penetration of the corrosion barrier	Corrosion barrier of chopped fibers	Corrosion barriers consisting of chopped fibers and premium resins.
Abrasion	Full wear of the abrasion barrier	Abrasion barrier	Abrasion barrier consisting of tough resins filled with hard ceramic materials.
Anomalous	Zero anomalous blisters	Sand core	Anomalous blisters triggered by core delamination.
Strain-corrosion	Strain = S_b = Ss Strain = Sc = Tw	Weep barrier of chopped fibers	The two listed failure criteria are valid for sanitation or mild chemical service.
Infiltration	Strain = Ti = infiltration threshold	Corrosion barrier of chopped fibers	Corrosion barrier consisting of chopped fibers and premium resins.
Weeping	Strain = Tw = weep threshold	Weep barrier in sanitation pipes UD plies in oil pipes	Applicable to UD and chopped fibers in oil or sanitation service
Stiffness	Strain = Ts = stiffness threshold	Transverse loaded UD plies	Applicable to UD plies. Not relevant for chopped plies.
Rupture	Strain = Ss = long-term fiber static strength	UD plies loaded in the fiber direction	Applicable to UD plies. Not relevant for chopped plies.

the discussion of laminate durability that the ply short-term strengths have received special names—failure thresholds—and symbols (see Table 8.2). The ply failure thresholds are fundamental in the analysis of long-term load-dependent durability.

Of the eight long-term modes of failure, the laminate strain-corrosion is the most elusive and difficult to quantify. This is because of the enormous variety of chemicals found in industrial applications, together with the difficulty to estimate the laminate bending strains. The best protective strategy against strain-corrosion is to reinforce the laminate areas subjected to bending strains in the presence of chemicals. This is an empirical process, not yet quantified. The regression equations to predict the durability of laminates in strain-corrosion service are too expensive to develop. The exception to this statement is the widely spread application involving underground sanitation pipes that operate in stable and well-known environments. The underground sanitation pipes are the only composite application with available strain-corrosion regression equations. In Chapter 14, we present a full description

TABLE 8.2

The four long-term load-dependent modes of failure and their corresponding short-term strengths. The term "rupture strain" is the same as "rupture threshold."

Failure mode	Short-term strengths	Symbol
Infiltration	Infiltration threshold	Ti
Weeping	Weep threshold	Tw
Loss of stiffness	Stiffness threshold	Ts
Rupture	Rupture threshold	Tr

of the general process of laminate strain-corrosion, together with our proposed solution to the specific case of sanitation pipes.

Table 8.2 lists the four thresholds—short-term ply strengths—required in the analysis of long-term load-dependent failures. Figure 8.1 shows these thresholds in relation to the operating strain. The failure criterion for load-dependent failures, as explained, is the equality of the short-term ply strengths and the operating strain. In this book, we use the maximum-strain failure criterion and never allow the operating strain to exceed the failure thresholds. There is much evidence to support the maximum-strain failure as the criterion of choice for composites.

The rest of this chapter is dedicated to the study of the critical plies and their load-dependent short-term strengths, also known as failure thresholds.

Example 8.1

Solvent attack was not included in the list of the eight modes of long-term failure. Explain why.

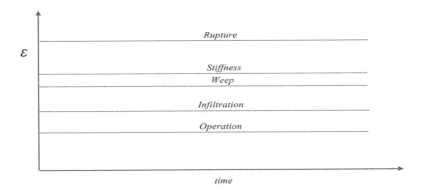

FIGURE 8.1 *The static operating strain plotted in relation to the four load-dependent short-term strengths.*

The solvent diffusion in composite laminates is fast and leads to failure in short immersion times. The solvent failures, when they happen, are short-term processes. The impermeable pipes discussed in Chapter 20 provide a complete solution to the short-term solvent failure problem.

Example 8.2

The eight modes of long-term failure listed in the introduction does not include creep. Why is that so?

The industrial laminates operate below the failure thresholds, subjected to very low strains. Any industrial equipment strained above the failure threshold is accounted as failed. The high strains leading to creep are certainly above the failure thresholds and for that reason are not significant. The creep design based on such high strains produces equipment operating in a forbidden, failed zone.

8.3 CRITICAL PLIES

As explained in Chapter 4, the critical ply is directly responsible for the laminate failure. The failure of the critical ply terminates the laminate capability to remain in service. The concept of critical ply is universal in the short-term analysis of composites. What we do in this book is merely extend this concept to the long-term. The critical ply depends on the mode of failure and is easy to identify. As an example, the sand core is the only ply developing anomalous failure (see Chapter 20) and therefore is the critical ply in this mode of failure. In some situations, the identification of the critical ply requires some elaboration. Such is the case in Example 7.7, where the sand core is the critical weep ply, instead of the "usual" ply of chopped fibers. The examples abound in this regard. Take the case of the cross-ply laminate $(0, 90)_S$, with two outer "0" plies and two inner "90" plies. The critical ply for this laminate could be either the "90" or the "0" ply, depending on the mode of failure, stiffness, or rupture. The designers as a rule have no problem identifying the critical ply. This situation is not a concern.

The following is a summary of the eight modes of long-term failure and their "usual" critical plies:

- For the three load-independent modes of failure

  - Chemical → the critical ply is the corrosion barrier of chopped fibers.
  - Abrasion → the ceramic filled abrasion barrier is obviously the critical.
  - Anomalous → the sand core is the only ply developing anomalous failure.

- For the four load-dependent modes of failure

  - Infiltration → the corrosion barrier of chopped fibers.
  - Weep → the corrosion barrier in sanitation pipes, or the UD ply in oil pipes
  - Stiffness → the transverse loaded UD plies
  - Rupture → the UD plies loaded in the fiber direction are critical. The laminate may infiltrate, weep, lose stiffness, etc. The corrosion barrier and the liner may degenerate. All of the above may occur, but no laminate will rupture before the UD plies.

- For the strain-corrosion failure

  - The critical ply is the one nearest to the aggressive chemical, usually the corrosion barrier of chopped fibers.

As we see, the eight modes of long-term failure have just four "usual" critical plies. Of those, the UD and the chopped plies are the most important. The resin-rich liner is not a critical ply, as it cracks and fails in chemical service before the corrosion barrier. The corrosion barrier of chopped fibers – and not the resin-rich liner – is the critical ply in chemical service.

8.4   THE LONG-TERM LOAD-DEPENDENT FAILURES

The four long-term load-dependent modes of failure discussed earlier are applicable to plies, not to laminates. They correspond to critical crack densities, or accumulated damage, that develop in the plies. The durability of the critical ply is the time or the number of cycles required to grow the initially small crack densities to their critical failure values. The computation of the load-dependent ply durability requires knowledge of the regression equations linking the critical crack density to the number of loading cycles, as discussed later in this chapter. The crack density increases under cyclic loads until it reaches the critical value defining the ply durability and, by extension, the laminate durability. The failure criterion is the long-term crack density that fails the critical ply. The difficulty with this approach is the exact quantification, or measurement, of the crack densities or damage at the point of failure. The micro-crack densities and damage developing in the critical plies are difficult, if not impossible, to see and measure.

We avoid this problem by postulating a one-to-one inverse relationship between the critical damage and the ply residual, or long-term strength. The inverse relationship implies that large accumulated damages result in low residual ply strengths, as it should. Next, we consider the applied strains. Obviously the ply failure occurs when the applied strain equals its long-term, or residual, strength. Therefore, the criterion for long-term failure is

Applied strain = Residual strength

When the applied strain is high, the residual strength at failure is also high and the accumulated damage in the critical ply is low. The plies subjected to high strains fail sooner and display less damage (less crack densities) than those that fail later under low strains. Lower strains imply longer times to failure and more damage. This is easy to understand, considering that lower applied strains require more damage to fail a ply. The one-to-one inverse relation postulated here allows the description of regression equations in terms of applied strains or residual strengths, instead of critical damage. It is easier to work with strains instead of damage and crack densities. We will return to this fundamental topic in Chapter 12, with a slightly different approach based on the Paris law for crack growth.

Therefore, instead of ply "damage and crack density," we speak of "ply residual strength," or "ply long-term strength." The cyclic regression equations linking the number of cycles to failure and the ply long-term strength is

$$log\left(residual\ strength\right) = A - Gs \times logN \tag{8.1}$$

Equation (8.1) gives the number of cycles that grows the initially small ply damage—from the small applied strain—to the point that the *residual strength = applied strain*. That, per our adopted criterion, is the point of failure. The failure threshold is the short-term strength, that is, the applied strain that fails the ply in one cycle. Setting $N = 1$ in Equation (8.1), we have

$$log\left(threshold\right) = A \tag{8.2}$$

$$log\left(residual\ strength\right) = log\left(threshold\right) - G \times logN \tag{8.2A}$$

The regression Equation (8.2A) measures the number of cycles to grow the initially small crack density—from the applied strain—to a critical crack density at the number of cycles N. The critical crack density defines the failure point, that is, that point where the ply residual strength equals the applied strain. Substituting the residual strength for the applied strain in Equation (8.2A), we have

$$\log(strain) = A - G \times logN \tag{8.3}$$

Equation (8.3) is the regression equation used to predict the number of cycles to fail a ply subjected to cyclic strains. In this equation, the intercept $A = \log(\text{threshold})$ and the slope Gs are measured experimentally. The failure threshold is the crack density (expressed as strain) that fails the ply in short-term quasi-static tests, that is, for $N = 1$. As we see, the failure thresholds are in fact the short-term ply strengths.

Example 8.3

Explain the origins and differences between infiltration, weeping and loss of stiffness.

The UD plies crack easily and, for that reason, never have direct contact with corrosive chemicals. The infiltration threshold of UD plies has no practical relevance. The infiltration threshold is a property of plies of chopped fibers. The low glass content and random orientation of the chopped fibers do not crack easily and produce large resin-dominated infiltration thresholds.

The stiffness threshold is a property of UD plies. The laminate loss of stiffness comes from resin cracks running along the fibers in UD plies. The offending UD cracks have their origin as resin-fiber debonds from transverse or shear strains. The resin-fiber debonds initiated by these strains are subsequently accentuated by resin cracking. The long-term stiffness life of UD plies depends on the fiber-resin adhesion and the resin toughness. The chopped fibers are not critical in structural service, and their stiffness threshold is not relevant.

The weep threshold is an important property of both UD and chopped plies. The durability of industrial laminates in non-aggressive chemical service, such as sanitation, oil and others, comes from weeping not from infiltration failure. The resin toughness controls the weep threshold of sanitation pipes built with barriers of chopped fibers. In oil pipes, with no weep barrier of chopped fibers, the resin-fiber adhesion and the resin toughness control the weep threshold.

8.5   CRACK DENSITIES AND EQUIVALENT PLY STRAINS

The load-dependent laminate failures result from the accumulation of small cracks in the critical ply. The small cracks grow in number – not in size – and eventually reach a critical density at which they interact, coalesce and fail the ply. The computation of the ply durability requires knowledge of the critical crack density at the onset of coalescence.

Equation (8.3) gives the number of cycles N to grow the initially small crack densities to their critical value. The parameter A in Equation (8.3) is related to the ply threshold, defined as the short-term strength for each mode of failure. There is a threshold for every ply and every mode of failure. The thresholds are ply properties independent of the loading. In Chapter 1, we have listed the thresholds as ply properties, equal to any other, such as the moduli and expansion coefficients.

The experimental measurement of the thresholds consists essentially in straining the plies in short-term quasi-static loadings and measuring the failure strain. By definition, the ply fails on reaching the thresholds. The thresholds are nothing more than the ply short-term strengths. The four long-term load-dependent modes of failure have each their own failure thresholds that give rise to four long-term lives. By contrast, the long-term load-independent failures—chemical, anomalous, and abrasion—do not involve thresholds and critical crack densities.

Example 8.4

Explain the differences between design, allowable, critical, failure, and threshold strains.

The critical, failure, and threshold strains have the same meaning and designate the same thing. They are ply properties measured in short-term quasi-static lab test and indicate the ply short-term strength.

The allowable and design strains also have the same meaning and derive from the application of a short-term design factor to the critical strains:

$$design\ strain = \frac{threshold}{SF}$$

The product standard AWWA C950 for composite pipes used in water transmission suggests a short-term design factor SF = 1.8. When we say, for instance, that the weep threshold of a polyester barrier of chopped fibers is 0.80%, the allowable strain in the pipe hoop direction would be 0.80%/1.8 = 0.45%. The above short-term SF has no relation with the long-term SF computed from the unified equation (see Chapter 16).

8.6   THE FAILURE THRESHOLDS

The thresholds are short-term ply strengths marking the onset of the load-dependent failure modes. The plies or laminates statically strained below the thresholds never fail, because static strains do not grow cracks. Cyclic loads, however, promote crack growth and eventually fail the ply or laminate, even if their peak strains are less than the thresholds. The study of load-dependent durability consists in determining the time/cycles for the applied loading to fail the critical ply. The time/cycles to failure come from regression equations that plot as straight lines on log—log space. As any straight line, their description involves one failure point and a slope. The easiest failure points to measure are the short-term strengths that we have defined as the failure thresholds. The thresholds are, therefore, fundamental in the study of the long-term load-dependent ply durability. Each ply has four thresholds:

- The **infiltration threshold** is the critical crack density that allows the fast ingress of chemicals. The infiltration threshold separates the "slow diffusion regime" from the "high infiltration regime." Corrosion barriers operating above the infiltration threshold have low service lives from the fast ingress of corrosive chemicals.
- The **weep threshold** is the critical crack density that allows the leakage of fluids. The weep failure involves higher crack densities than those observed in infiltration.
- The **stiffness threshold** is the critical crack density leading to the onset of delamination and loss of laminate stiffness.
- The **rupture threshold** is the short-term rupture strength of the critical ply.

In discussing the failure thresholds, our attention focus primarily on the corrosion barrier of chopped fibers and the structural UD plies. It is instructive to say a few words about these plies. To begin with, they are the most important critical plies defining the long-term durability of composites. Apart from that, these plies have the following features:

- The UD plies consist of a large number of closely packed parallel continuous fibers that easily debond from the resin and initiate the cracks that cause loss of stiffness and weeping. Both the stiffness and the weep thresholds of UD plies are highly dependent on the fiber-resin bonding.
- By contrast, the corrosion/weep barrier is a collection of widely separated and randomly dispersed bundles of chopped fibers. The chopped fibers in the randomly dispersed bundles are parallel and closely packed, like those in the UD plies.

Both feature bundles with identical closely packed parallel fibers. Given this situation, we might conjecture that, perhaps, they might have the same infiltration and weep thresholds. This conjecture, however, is not true. The wide separation and random orientation of the short bundles of chopped fibers prevent the link-up and coalescence of the glass-resin debonds. The fiber-resin adhesion has no effect on the infiltration and weep processes. By contrast, the resin-fiber debonds will link-up and coalesce in UD plies, with strong effects on the infiltration and weep thresholds. With this background, we may safely conclude that the infiltration and weep thresholds of chopped fibers depend mostly on resin cracking, with a small influence from resin-fiber debonds. This situation reverses for the closely packed UD plies, which have infiltration, weep, and stiffness thresholds highly dependent on fiber-resin bonding. This simple reasoning explains the higher infiltration and weep thresholds of chopped fibers vis-à-vis those of UD fibers.

Like any short-term strength, the measurement of the failure thresholds involve the application of quasi-static loads to the critical ply and the annotation of the failure strain. The problem is not how to do the measurement, but the choice of the failure strain. Different people will disagree on the correct value to use. Some will say that such a strain has caused ply failure and others may say otherwise. What follows is my recommendation to promote a wise approach to the topic of threshold measurement.

The critical plies of interest are the UD and chopped fiber plies. The failure thresholds are the infiltration, weep, stiffness, and rupture. The isotropic chopped fibers have four thresholds. The orthotropic UD plies have eight thresholds, corresponding to the fiber and transverse directions. We can define, therefore, twelve thresholds, some of which have no practical interest.

In measuring the thresholds, the first thing to do is choose the short-term types and levels of damage representative of failure. The short-term strengths measured at such damages are the thresholds. We propose the following:

- The **infiltration threshold** is the strain at the onset of ply cracking. This is an obvious choice.
- The **weep threshold** is arbitrarily set at a certain level of crack densities. This arbitrariness may be a cause of dispute.
- The **stiffness threshold** derives from a combination of crack densities and the onset of delamination.
- The **rupture threshold**, of course, derives from the ply separation in two or more parts.

8.6.1 Thresholds of Fiber Loaded UD Plies

The UD plies do not develop cracks when loaded in the fiber direction. There are no infiltration, weep, and stiffness failures for fiber loaded UD plies. Rupture is the only mode of failure for UD plies loaded in the fiber direction. The UD plies loaded in the fiber direction lose some stiffness from fiber rupture, but this loss is not relevant. See the appendix on sudden death in Chapter 12. The rupture threshold of fiber loaded UD plies is the ply short-term tensile strength. The rupture threshold of fiber loaded UD plies is fiber dominated and has a value of $Tr = 3.0\%$.

TABLE 8.3

Rupture is the only mode of failure of fiber loaded UD plies. The rupture threshold in the fiber direction is fiber-dominated.

Failure thresholds of UD plies loaded in the fiber direction

Infiltration	—
Weep	—
Stiffness	—
Rupture	Tr = 3.0%

Table 8.3 lists the thresholds of UD plies loaded in the fiber direction.

8.6.2 Thresholds of Transverse Loaded UD Plies

The transverse loaded UD plies develop many micro-cracks that grow under quasi-static loads to merge and form large macro-cracks running parallel to the fibers. The formation of the first large crack marks the infiltration failure. The onset of large cracking is a valid marker of the infiltration threshold. The onset of cracking, however, is not a valid rupture criterion for transverse loaded embedded UD plies where the adjacent plies prevent rupture. The embedded UD plies under transverse quasi-static loads are capable of stretching and developing many large cracks without rupture. Eventually the density of macro-cracks becomes high enough to coalesce and allow the passage of fluids. This is the weep threshold, characterized by a critical density of macro-cracks. The density of large cracks increases with load increments until stabilization at a maximum value—known as CDS, or Characteristic Damage State – that triggers the onset of ply delamination. The CDS marks the ply failure by loss of transverse stiffness as well as of strength. The stiffness and rupture thresholds of transverse loaded UD plies coincide with the CDS (see Figure 8.2).

8.6.2.1 Infiltration

The infiltration threshold of transversely loaded UD plies is the ply short-term strength corresponding to the first macro-crack, measured by back extrapolation from acoustic emissions, loss of translucency, or loss of stiffness, as shown in the appendix of Chapter 13. The suggested values are Ti = 0.20% for polyesters and Ti = 0.30% for vinyl esters. The infiltration thresholds of transverse loaded UD plies have little relevance, since such plies do not have direct contact with chemicals.

8.6.2.2 Weep

The weep threshold of transverse loaded UD plies is the short-term ply strength corresponding to a pre-determined density of macro-cracks, loss

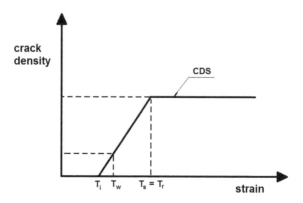

FIGURE 8.2 *Macro-crack density and the failure thresholds of transverse loaded UD plies. The infiltration threshold corresponds to the onset of cracking. The weep threshold is arbitrarily set at a certain crack density. The stiffness and rupture thresholds correspond to the same state of damage, known as CDS.*

of translucency, loss of stiffness, or acoustic count. See Figure 8.2 and the appendix of Chapter 13. The suggested weep threshold values of transverse loaded UD plies are Tw = 0.25% for polyesters and Tw = 0.40% for vinyl esters.

8.6.2.3   Stiffness

The stiffness threshold of transverse loaded UD plies is the short-term ply strength corresponding to the maximum and final density of macro-cracks known as CDS or Characteristic Crack Density. See Figure 8.2. The suggested stiffness thresholds are Ts = 0.40% for polyesters and Ts = 0.60% for vinyl esters.

8.6.2.4   Rupture

The rupture threshold of transverse loaded UD plies coincide with the stiffness threshold. See Figure 8.2. The suggested values are Tr = 0.40% for polyesters and Tr = 0.60% for vinyl esters.

Table 8.4 lists the thresholds of transverse loaded UD plies.

8.6.3   THRESHOLDS OF CHOPPED PLIES

Unlike UD plies, the chopped plies do not develop large cracks. Their durability analyses deals mostly with micro-damage or micro-cracks invisible to the naked eye. Rupture is the only clear-cut and undisputed failure of chopped fibers. failure points are controversial and require specialized equipment to measure.

TABLE 8.4

The stiffness and rupture thresholds of transverse-loaded UD plies describe the same state of damage and have the same value.

	Failure thresholds of transverse-loaded UD plies	
	Polyester (%)	Vinyl ester (%)
Infiltration	Ti = 0.20	Ti = 0.30
Weep	Tw = 0.25	Tw = 0.40
Stiffness	Ts = 0.40	Ts = 0.60
Rupture	Tr = 0.40	Tr = 0.60

TABLE 8.5

The stiffness failure is not relevant for chopped plies.

	Failure thresholds of chopped plies	
	Polyester (%)	Vinyl ester (%)
Infiltration	Ti = 0.30	Ti = 0.50
Weep	Tw = 0.80	Tw = 1.00
Stiffness	—	—
Rupture	Tr = 1.50	Tr = 2.50

Infiltration: The infiltration threshold of chopped plies is the short-term crack density measured by back extrapolating experimental data for acoustic emissions, loss of translucency, or loss of stiffness. See Figure 8.2 and Appendix 13.1. The suggested values are Ti = 0.30% for polyesters and Ti = 0.50% for vinyl esters.

Weep: The weep threshold of chopped plies is the short-term crack density measured by experimental acoustic emission, loss of translucency or loss of stiffness. See Figure 8.2 and Appendix 13.1. The suggested weep thresholds are Tw = 0.80% for polyesters and Tw = 1.0% for vinyl esters.

Stiffness: The stiffness threshold of chopped fibers has no practical relevance.

Rupture: The rupture thresholds of chopped plies are Tr = 1.50% for polyesters and Tr = 2.50% for vinyl esters.

Table 8.5 lists the thresholds of chopped plies.

8.7 THE WORK OF R. F. REGESTER

In November of 1968, R. F. Regester published a pioneering and perhaps the best paper ever in the history of composites for industrial chemical service, titled *Behavior of Fiber Reinforced Materials in Chemical Service* (see Ref. [16]). In his pioneering and extremely well-conducted research, Regester worked with typical laminates consisting of liners and corrosion barriers, immersed in solutions of sulfuric and hydrochloric acid. In a few cases, the immersed laminates did not have a liner and the corrosion barrier of chopped fibers saw a direct exposure to the acid attack. The resin used was a bisphenol A polyester, the best resin for chemical service in those days. The work consisted in measuring and reporting the depth penetrated by the chemica las well as its concentration profile in the corrosion barrier, as a function of time.

The complete work covers laminates immersed in water and caustic environments, not just acid. However, for the purposes of this book, an analysis of the reported acid results will suffice. For the complete work that includes water and caustic immersion, please see Ref.[16]. While conducting his groundbreaking work, Regester was not aware of the concepts of infiltration and weep thresholds. Such concepts were, of course, unsuspected in those long-ago days. However, a careful examination of his published results certainly points in their direction. What follows is my interpretation of the original findings reported by Regester in the distant year of 1968.

8.7.1 TIME OF EXPOSURE

As postulated in Chapter 11, the advancing front of corrosive chemicals in composite laminates is approximately flat. Furthermore, the penetrated plies remain in place to provide a shield that significantly slows down the penetration rate. The work of Regester gives confirmation to these theoretical proposals. Figures 8.3 and 8.4 show typical penetrant concentration in corrosion barriers immersed in 25% H_2SO_4 and 15% HCl solutions at 100°C, over a 6-month period. The reader will note the absence of resin-rich liners in these experiments.

8.7.2 GLASS CONTENT

The penetration rate of chemical species in pure resin castings is lower than in laminates. This is a consequence of the high permeability of the glass-resin interphase and the presence of voids in laminates. Furthermore, and for the same reasons, the concentration of the penetrated chemicals is higher in laminates than in resin castings. Both the chemical penetration rate and concentration increase with the glass content in laminates. Regester tested and confirmed these postulates, by exposing laminates with and without resin-rich liners to 25% sulfuric acid solution at 100°C. The test results showed the resin-rich liner slowing down down the penetration rate. However, the concentrations of SO_4 ions in the corrosion barrier and in the laminate structure do not depend on the liner presence (Figure 8.5).

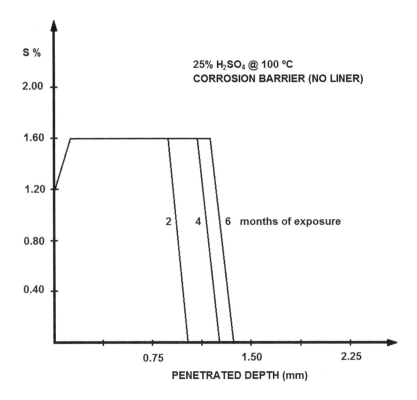

FIGURE 8.3 *The flat advancing front of H₂SO₄ in the corrosion barrier is clearly visible. Note the absence of the protective liner and the direct exposure of the corrosion barrier. The shielding effect significantly slows down the rate of chemical advance, indicating a long service life. The reader will note the lower acid concentration at the inner resin-rich skin (1.2%) compared with the 1.6% concentration in the corrosion barrier. As explained in Chapter 11, this is due to acid accumulation in voids and at the glass-resin interphase present in the corrosion barrier. The figure shows the large penetration rate observed in the first 2 months of exposure dropping off sharply with time. This is a consequence of the shielding effect.*

The work of Regester indicates the resin-rich liner not providing a similar barrier for the penetration of Cl ion from a 15% HCl solution at 100°C. As shown in Figure 8.6, the concentration and penetrated depth of Cl ions are independent of the presence of resin-rich liner. This finding seems to indicate that resin-rich liners cannot slowdown the penetration of chloride ions. This, of course, is not true. The liner will retard the ingress of all ions. The explanation for the null result reported by Regester is quite possibly the premature cracking of the bisphenol A liner in the presence of a 15% hydrochloric acid solution at 100°C. The cracked liner would certainly allow the free passage of the chloride ions.

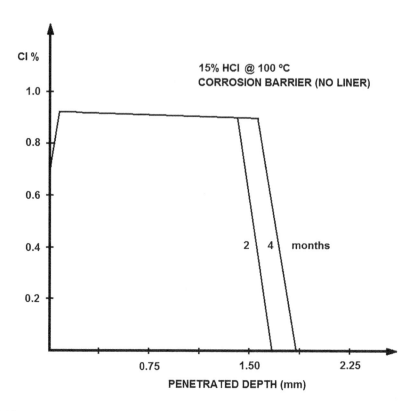

FIGURE 8.4 *The penetration of HCl follows the same pattern of H_2SO_4, except a little faster. The rapid penetration of HCl indicates the need of thick corrosion barriers.*

The foregoing discussion indicates the following hierarchy of plies with respect to chemical concentration and penetration rate:

- Liner: lowest chemical concentration and lowest penetration rate. It is unfortunate that most liners crackdown in service and are not reliable in the long term. Our analyses of laminate durability in chemical service—see Chapters 11 and 14—ignore the liner presence.
- Corrosion barrier: medium chemical concentration and medium penetration rate. The chopped fibers of the corrosion barrier are the critical plies in chemical service.
- UD structure: highest chemical concentration and highest penetration rate. The residual life of laminates in which the corrosive chemicals have penetrated the corrosion barrier is very short. The structural UD plies are not as good as the corrosion barrier in direct contact with chemicals.

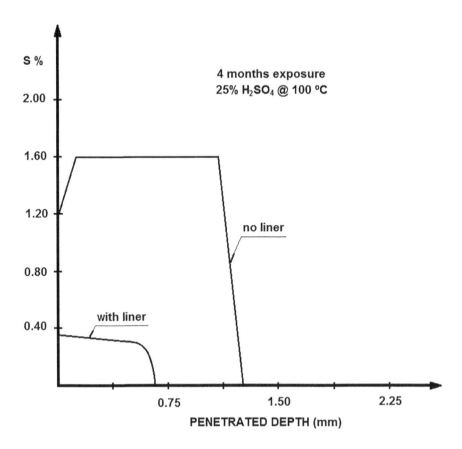

FIGURE 8.5   *The intact liner has a significant effect in reducing the short-term chemical penetration. This initial protection, however, becomes negligibly small in long-term service. Furthermore, the chemical concentrations in the corrosion barrier and in the laminate structural plies are independent of the liner presence.*

8.7.3   ACID CONCENTRATION

The concentration profile of the penetrated chemicals in immersed laminates varies with the ply construction and the concentration of the aggressive chemical solution. To confirm this, Regester tested corrosion barriers immersed in 5%, 25%, and 50% sulfuric acid solutions at 100°C for 4 months. The results in Figure 8.7 show the concentration of SO_4 ions within the laminate increasing with the concentration of SO_4 ions in the solution. The depth of penetration was greater at higher solution concentration, although the increase was more between 5% and 25% than between 25% and 50%. This occurs regardless of the liner presence.

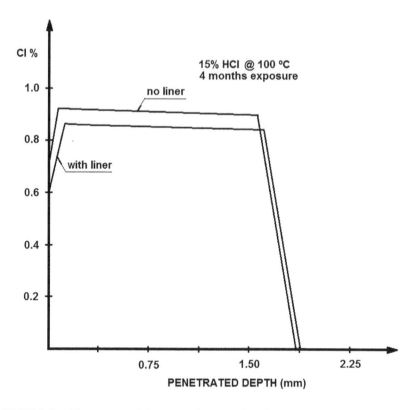

FIGURE 8.6 *The resin-rich liner cracks immediately on contact with the 15% HCl solution at 100°C. This cracking neutralizes the liner short-term protective effect. The reader should compare these graphs with those in Figure 8.5.*

8.7.4 TEMPERATURE

The effect of the operating temperature was determined by measuring the penetration rates at 100°C and 115°C for corrosion barriers exposed to 25% and 50% sulfuric acid solutions. Figure 8.8 shows that temperature increments from 100°C to 125°C substantially increased the penetrated depth. Furthermore, the concentration of SO_4 ions in the laminate increased with increasing temperature. These results indicate that increments in the operating temperature increase both the rate of penetration and the concentration of the offending chemicals. This is bad news for laminate durability in chemical service. The resin suppliers provide chemical resistance guides for their products, indicating the maximum operating temperature they can sustain in a variety of industrial chemicals.

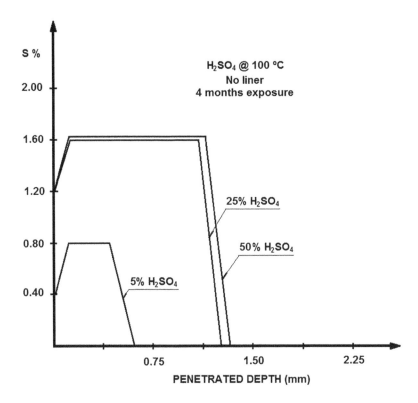

FIGURE 8.7 *Weak 5.0% solution of H_2SO_4 has lower penetrating power and gives smaller concentrations in the corrosion barrier than higher concentration acid solutions. This explains the long durability of composite pipelines in weak acid urban sewer service.*

8.7.5 Type of Cation

To determine if the anion penetration was independent of the cation, Regester tested NaCl and HCl solutions at the same temperature and concentration of Cl ions. The experiment lasted 2 months at 100C, and indicated the strong effect of the type of cation on the penetration rate of the anion. Regester observed that the larger effective ionic diameter of the easily hydrated Na ions substantially decreased their diffusion rate and slowed down the penetration of the chloride ions. To maintain electrical neutrality within the laminate, the slowest moving ion will determine the rate of diffusion for both ions. This explains the aggressive nature of HCl solutions when compared with those of benign NaCl. A similar reasoning applies to sulfuric acid and their salts.

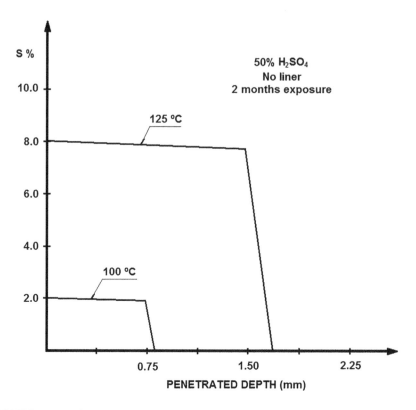

FIGURE 8.8 *The operating temperature has a significant effect on the chemical rate of penetration and concentration in the corrosion barrier. The resin suppliers provide chemical resistance guides for their products, indicating the maximum operating temperature they sustain in a variety of industrial chemicals.*

8.7.6 THE WORK OF REGESTER IN ACTUAL SERVICE

A careful analysis of the work of Regester leads to some interesting interpretations and conclusions. Tables 8.6 and 8.7 compare the sulfur and chloride concentrations in the test solutions and in the laminates.

Tables 8.6 and 8.7 provide the following information:

- The 25% sulfuric acid solution, with a sulfur content of 8.2%, saturates with a sulfur concentration of 0.3% in the resin-rich liner. This indicates a low solubility S = 0.3/8.2 = 3.7% of sulfuric acid in bisphenol A resins at 100°C.
- The sulfur concentration is five times higher—1.6/0.3 = 5—in the corrosion barrier than in the resin-rich liner. This is because of the high void content in the corrosion barrier, where the acid solution

TABLE 8.6

The table indicates low solubility of sulfuric acid in the liner and in the corrosion barrier. The higher concentration in the corrosion barrier is due to acid accumulation in voids and at the glass-resin interphase.

H_2SO_4 concentration in the acid solution (%)	Sulfur concentration in the acid solution (%)	Sulfur concentration in the liner (%)	Sulfur concentration in the corrosion barrier (%)
5	1.6	—	0.7
25	8.2	0.3	1.6
50	16.3	—	1.6

TABLE 8.7

The table indicates low solubility of hydrochloric acid in the corrosion barrier.

HCl concentration in the acid solution (%)	Chlorine concentration in the acid solution (%)	Chlorine concentration in the liner (%)	Chlorine concentration in the corrosion barrier (%)
15	14.6	—	0.8

gathers at the full 25% concentration. These results indicate also the possibility of higher acid solubility in the glass-resin interphase. High acid concentration at the glass-resin interphase is bad for strain-corrosion. For details, see Chapter 14.

- The 5% sulfuric acid solution gives a very low—0.7% sulfur—concentration in the corrosion barrier, which explains the apparent absence of strain-corrosion in underground sanitation pipes carrying acidic urban sewage. For details, see Chapter 14.
- The solubility of HCl in bisphenol A resins at 100°C is also very low.

We next discuss the results reported by Regester in light of the hypotheses proposed in Chapter 11.

- The chemical penetration in the corrosion barrier is slow and advances as a nearly flat concentration profile. This is in agreement with the proposals in Chapter 11.
- The chemical presence in the resin is a function of resin-chemical solubility and the solution concentration. This is in full agreement with Chapter 11.
- The penetrated plies that remains on the laminate surface provides a shield that slows down the rate of chemical advance. This is in agreement with Chapter 11.

- The liner retards the chemical penetration in the corrosion barrier. However, once penetrated, it gives no long-term protection other than the shielding effect. Although not explicitly mentioned, this conclusion is an obvious part of the work of Regester.

Let us say a few words regarding the liner and its usefulness. The foregoing discussion indicates that it is wise to ignore the long-term liner protection even in those cases when it remains intact throughout the laminate lifetime. Intact liners would at best delay the chemical penetration, while having no effect on its concentration in the corrosion barrier. It is safe and prudent to ignore the liner presence in all cases. Furthermore, most chemicals will quickly attack and crack the liner, destroying its protective function. See Figure 8.6 for the case of 15% HCl at 100°C. The premature liner destruction is not a cause of concern, since all laminates in chemical service operate below the infiltration threshold and the chemical concentration in the corrosion barrier does not depend on the liner presence.

Let us illustrate this reasoning in the case of underground sanitation pipes carrying acidic urban sewage. From Table 8.6, we see that a 5% solution of sulfuric acid—simulating the urban sewage—produces a very low and harmless acid concentration in the corrosion/weep barrier. With this background, we consider two scenarios.

- **Scenario 1.** Suppose sanitation pipes with no liners. The weep barrier operates in the following conditions:

    - *Weep barrier strained below the infiltration threshold.* The absence of cracking in this scenario assures a low acid concentration in the weep/corrosion barrier. The weep barrier has direct contact with the 5.0% acid solution and saturates at a harmlessly low concentration of just 0.70%. The absence of cracking assures a low acid concentration and a slow penetration rate. The pipe life is very large.
    - *Weep barrier strained above the infiltration and below the weep threshold.* The difference between this and the previous scenario is the faster and deeper acid penetration. The 5.0% concentration sulfuric acid that fills the small infiltration cracks has no effect on the resin and does little damage to the widely dispersed and randomly oriented chopped fibers. The pipe life, although less than in the previous scenario, is still very large.
    - *Weep barrier strained above the weep threshold.* The large cracking in this scenario is too severe, allowing the full strength 5.0% sulfuric acid to quickly inundate the weep barrier and attack the

UD structural fibers. The oriented UD fibers, unlike the random chopped fibers of the weep barrier, deteriorate fast in the presence of the full strength acid. The strain-corrosion rupture is fast in this case.

- **Scenario 2.** Suppose now sanitation pipes with intact liners. The weep barrier operates as in the previous scenario.

  - The intact liner delays the penetration of the acid solution, with everything else remaining as in the previous scenario. The strain-corrosion failure does not occur as long as the pipe operates below the weep threshold. The liner presence delays the strain-corrosion process but provides no assured long-term protection. The long-term strain-corrosion performance of pipes in sewer service is, for all practical purposes, independent of the liner presence. This conclusion is true of every long-term situation involving the liner.

The preceding discussion explains the apparent absence of strain-corrosion in underground sewage pipes operating below the weep threshold. For details, see Chapter 14.

Let us finalize this section clarifying a situation that had me puzzled for a long time. In simple words, I would put it this way:

> The concept of infiltration threshold is a fake proposition, since experience indicates a substantial resin embrittlement in the presence of chemical attack. The pristine infiltration threshold would be lost soon after the chemical comes in contact with the corrosion barrier.

All of the above is true and the resin embrittlement is real. However, only the resin penetrated by the chemical will show this embrittlement. The nonpenetrated resin, in the rest of the corrosion barrier, keeps its pristine properties and remains faithfully obedient to the original infiltration threshold concept. The concept of infiltration threshold holds for both aged and pristine corrosion barriers.

8.8   SHORT-TERM AND LONG-TERM SAFETY FACTORS

There are several short-term criteria to compare the ply strains with the failure thresholds. The most convenient is the maximum-strain criterion, which consists in a direct comparison of the ply strain components with the failure thresholds.

$$\frac{threshold}{ply\ strain} = SF$$

The maximum-strain criterion produce four short-term safety factors (SF), one for each mode of failure.

Other short-term criteria widely mentioned in the literature are the maximum-stress and the equations by Tsai-Wu and Tsai-Hill. The classical Tsai-Wu and Tsai-Hill quadratic criteria maintain their validity and meaning when analyzing for the short-term safety factors and the failure thresholds. After all, the failure thresholds are nothing more than short-term ply strengths. The four failure thresholds enter the quadratic equations in place of the short-term ply strength. The Tsai-Wu and Tsai-Hill equations produce short-term safety factors for infiltration, weep, stiffness, and rupture, in the same way as the maximum-strain criterion.

The reader should keep in mind that the above criteria—maximum-strain, maximum-stress, Tsai-Wu, and Tsai-Hill—apply to short-term failures only and, as such, cannot predict durability. In the realm of short-term failure analysis, the classical criteria are fully compatible with the failure thresholds. After all, we repeat, the failure thresholds are nothing more than short-term ply strengths. Things are different, however, in the realm of long-term failure caused by the accumulation of cyclic and static damage. The computation of the four long-term safety factors SF requires the unified equation discussed in Chapter 16.

This book deals only with the long-term safety factors. The short-term safety factors, when mentioned at all, derive from the maximum-strain criterion.

Example 8.5

Consider a ply of chopped fibers. Identify its load-dependent modes of failure, degradation mechanisms, failure criteria, and failure thresholds. This example deals only with load-dependent failures.

Modes of failure: In chemical service, infiltration. In sanitation service, weeping. Loss of stiffness and rupture are usually associated with UD plies, not with chopped plies.

Degradation mechanism: Damage accumulation from cyclic fatigue loads.

Failure thresholds: In chemical service, the infiltration threshold. In sanitation service, the weep threshold. The stiffness and rupture thresholds are usually associated with UD plies, not with chopped plies. The failure thresholds are the same as the well-known short-term strengths, critical strengths, and critical strains.

Fatigue thresholds: In addition to the failure thresholds, all plies have their own fatigue threshold. The fatigue threshold indicates a crack density (expressed as an equivalent strain) below which the growth of crack density is negligibly small. For any given ply, all thresholds (failure and fatigue) increase with increments in the resin toughness.

Short-term failure criteria: Maximum-strain, maximum-stress, Tsai-Wu, and Tsai-Hill.

Long-term failure criteria: The unified equation discussed in Chapter 16.

The object of this book is to present solutions to the durability of industrial laminates considering the eight modes of long-term failure and two critical plies—UD and chopped fibers. The arguments presented in the rest of the Part II are for the most part straightforward and easy to understand. The exception, perhaps, is the durability associated with the four load-dependent modes of failure, which involve the unified equation and requires some intellectual effort.

9 The Regression Equations of Failure

9.1 INTRODUCTION

The current method to estimate the load-dependent durability of laminates uses experimental cyclic regression equations linking the peak strain to the number of cycles to failure.

$$\log(\epsilon) = A_c - G_c \log(N) \qquad (9.1)$$

Equation (9.1) describes a failure line that includes all possible combinations of applied strain and number of cycles that fail the laminate at a given load ratio R and loading frequency. The regression parameters A_c and G_c are measured experimentally for each laminate and loading. The intercept A_c captures the laminate short-term strength. The slope G_c captures the rate of damage accumulation as a function of the load ratio R, the applied peak strain and the loading frequency. The importance of the loading frequency and the R ratio will become apparent in Chapters 16, 17 and 18.

The number of cycles N that fails the laminate is determined simply by entering the applied peak strain in Equation (9.1). This, basically, is how we currently compute the long-term rupture, stiffness, weeping, and infiltration durability of laminates when their failure Equation (9.1) is available. This very simple and straightforward solution is applicable only to those rare combinations of laminates and loadings for which the regression Equation (9.1) is available. Unfortunately, such equations of failure are difficult to obtain and are rarely available.

The determination of the A_c and G_c parameters requires expensive long-term testing for all possible combination of loadings and laminates. This, of course, is not feasible. We need a better tool, capable of solving the load-dependent long-term laminate durability when Equation (9.1) is not available. The unified equation, discussed in Chapter 16, is just the tool we need.

The few available failure Equations (9.1) apply mostly to laminates, not to plies. The reader may be willing to know what the difference is between a ply and a laminate regression equation. Let us clarify this topic.

- When applied to laminates, Equation (9.1) predicts the rupture durability for specific combinations of laminate construction and loadings. There are many such equations in the literature, used to predict the long-term rupture of structures such as, for instance, wind blades. The literature also lists a few laminate equations to predict the weep durability of oil pipes. The time variable, the load ratio R, and the loading frequency, not explicit in the laminate Equation (9.1), are accounted for in the specific values taken by the experimental slope G_c. In this book, we do not spend any time or effort discussing the laminate regression equations.
- When applied to plies, Equation (9.1) predicts the durability for all four load-dependent modes of failure, not just rupture. See Figure 9.1. As in the laminate equations, the experimental slopes G_c capture both the time variable and the load ratio R. For any given ply, different loadings produce different slopes G_c, which is a significant practical limitation. This chapter gives a detailed discussion of the ply regression equations.

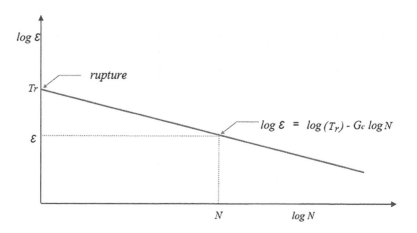

FIGURE 9.1A *Ply rupture life—The ply rupture durability N for any given peak strain ε comes from the rupture regression line. The rupture durability is final and marks the ultimate life of the ply. Any change on the load ratio R and test frequency (load path) would affect the slope G_c and require a new and specific regression equation. This is a problem. The complete solution of the load-dependent rupture durability would require an infinite number of regression equations. For any given ply, the regression parameter A_c varies with the resin toughness, but not with the loading. For the same ply and resin, all regression equations pass through the point $A_c = log(T_r)$.*

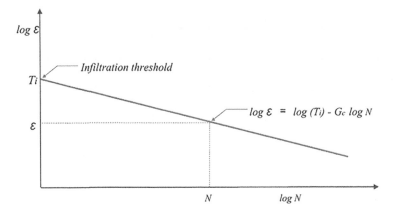

FIGURE 9.1B *Ply infiltration life*—*The infiltration durability of plies come from infiltration regression lines. After N cycles at a constant peak strain, the initially small crack density reaches a critical value that allows the rapid ingress of aggressive chemicals in the corrosion barrier. From this moment, the remaining service life is short. The attainment of the infiltration threshold raises a red flag and warns the engineer to get ready for either relining or replacing the equipment. In a few cases, as in small diameter pipes, no relining is possible and the equipment runs until it weeps or ruptures. The slope parameter G_c is a ply feature that varies with the load ratio R. The infiltration threshold Ti, however, is a ply property independent of the loading.*

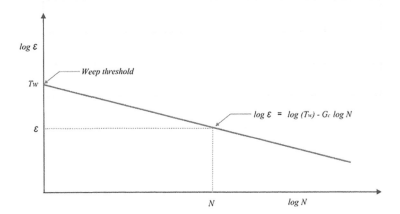

FIGURE 9.1C *Ply weep life*—*The weep durability of plies come from the weep failure lines. After N cycles under the peak strain ε, the initially small crack density reaches a critical value and weeping is inevitable. The weeping failure is not catastrophic and involves nothing more than leakage and loss of fluid. If such an occurrence is not objectionable, the pipeline may run until rupture. The slope G_c is a ply feature that varies with the load ratio R. The load-independent failure threshold T_w is a ply property that depends on the resin toughness.*

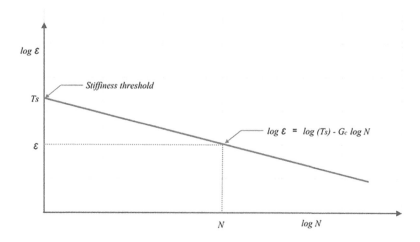

FIGURE 9.1D *Ply stiffness life—The stiffness durability of plies come from stiffness regression lines. The slope G_c is a ply feature that varies with the load ratio R. Unlike the infiltration and weep failures, the plies involved in stiffness failure have two regression lines, one for tensile and another for shear loadings.*

- A special family of ply regression equations are those derived from pure static ($R = 1.0$) and pure cyclic ($R = 0.0$) loadings. The slopes G_c (pure cyclic) and G_s (pure static) of the pure failure equations are ply properties fundamental in the development and application of the durability model proposed in this book. The pure slopes G_c and G_s are direct inputs to the unified equation discussed in Chapter 16. Furthermore, the pure regression equations predict the long-term ply strengths, which are also required in the unified equation. Later in this chapter, we discuss in detail the pure cyclic and the pure static ply regression lines.

Later in this chapter, we present a list of all ply regression equations available in the published literature. The available ply equations, although few in number and perhaps lacking in accuracy, form the empirical basis for the model proposed in this book to compute the load-dependent laminate durability. The proposed computation model makes use of idealized ply equations. To facilitate the presentation, we omit the idealization details. The reader will find the missing details in the Appendix 9.1.

9.2   THE PLY REGRESSION EQUATIONS

Let us leave aside the laminates and imagine Equation (9.1) applied to plies. The regression parameters A_c and G_c are measured experimentally for each ply and loading. The number of regression equations is infinite, as each ply would have a different equation for a different loading. This, of course, is an impractical situation.

The practical approach to such cases consists in interpolating between known regression lines. The interpolation is possible for the time-independent modes of failure such as infiltration, weeping, and stiffness. For the time-dependent rupture failures, involving fiber strain corrosion and the time variable, the accuracy of this interpolation is uncertain. The unified equation discussed in Chapter 16 provides a better solution than interpolation to all durability problems.

Given this background, we return to the ply failure equations. Our task is to explain the meaning of the regression parameters A_c and G_c. We start with A_c. By definition, the failure thresholds are the short-term ply strengths that fail the ply in the first loading cycle. Thus, by setting $N = 1$ in Equation (9.1):

$$\log(\text{threshold}) = A_c - G_c \log(1)$$

$$A_c = \log(threshold) \tag{9.1A}$$

As we see, the intercepts A_c of the ply regression lines derive directly from the failure thresholds, that is, from the short-term ply strengths.

$$A_c = \log(T_i) \quad (\text{Infiltration failure})$$

$$A_c = \log(T_w) \quad (\text{Weep failure})$$

$$A_c = \log(T_s) \quad (\text{Stiffness failure})$$

$$A_c = \log(T_r) \quad (\text{Rupture failure})$$

The above considerations define the regression parameters A_c as ply properties independent of the loading. For any given ply, the parameter A_c varies with the mode of failure, the resin toughness, and the operating conditions defined by temperature and moisture. However, A_c is a ply strength independent of the loading. This important conclusion will be apparent later. Let us consider next the slope parameter G_c.

The regression slopes G_c are not ply properties. They are ply features that vary with the loading and the mode of failure. Furthermore, the slopes might depend also on the resin toughness and the operating conditions of temperature and moisture. Any given ply has an infinite number of slopes. A full solution to the ply durability problem requires the measurement of a gigantic number of slopes and regression equations. Hercules himself would shy away from such a daunting task. See Figure 9.1.

9.3 SIMPLIFIED REGRESSION PLY LINES

The difficulty with the ply failure equations described in Figure 9.1 is their narrow applicability to specific conditions. Any change in resin, loading, operating conditions and mode of failure requires the experimental measurement of a specific and expensive failure line. This means an enormous number of experimental regression equations for any given ply. We need a way to reduce the number of required experimental failure equations with a minimum impact on their predictive accuracy. Let us see how to do that:

- The operating conditions pose a potentially serious complication. The development of specific equations for different temperatures and moisture condition is simply not feasible. Fortunately, this enormous effort is not necessary. As explained in Chapter 10, the regression equations developed for dry plies at room temperature are safe to use in any operating condition.
- The need of an experimental equation for each of the four modes of failure is also a problem. Again, fortunately, the new concept of fatigue threshold simplifies this complication to the determination of the failure equation for just one mode. Having one experimental equation, for any mode of failure, suffices to determine the other three. For chopped plies, the infiltration, weep, and stiffness lines derive from the experimental rupture equation. For transverse loaded UD plies, the weep, stiffness, and rupture equations derive from the experimental infiltration line. See Figure 9.2.
- The unified equation, discussed in Chapter 16, solves the problem posed by a wide variety of different loadings.

The above simplifications are central in the development of our load-dependent durability models. They save a lot of experimental work with acceptable loss of accuracy. However, we still need one specific set of equations for every resin. This difficulty we address later.

9.3.1 DERIVING THE SIMPLIFIED REGRESSION EQUATIONS

Let us now discuss the connection between the four load-dependent regression equations. We have an intuitive feeling for the existence of such a

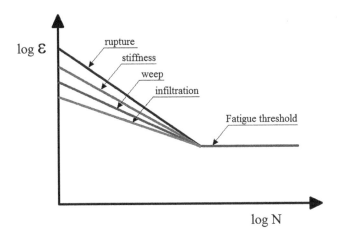

FIGURE 9.2   *The intercepts A_c of the simplified infiltration, weep, stiffness and rupture lines derive from the short-term failure thresholds. The slopes G_c of each line derive from the slope of any one of the four lines. See the text. The four regression lines originate at their short-term failure thresholds – the short-term strengths – and meet at the long-term fatigue threshold – the long-term fatigue strength. For any given ply, the fatigue threshold is the same for all modes of failure and all resins. See the text. Given that the thresholds are known ply properties, the gigantic problem of determining a specific equation for each load-dependent mode of failure reduces to measuring the slope of just one regression line. As explained in the text, for any given ply, changing the resin toughness shifts all short-term thresholds – infiltration, weep, stiffness and rupture – while keeping the long-term fatigue threshold unchanged. The reader should bear in mind that the lines depicted in the figure refer to the same ply and the same loading. Such a ply, as we know, has fixed thresholds. Different loadings would produce different slopes, with all lines originating and ending at the fixed thresholds, as explained in the text.*

connection. We can imagine the ply damages starting at some pristine state and accumulating to reach the critical state at failure. The damage path originates at the same pristine state and evolves in the same way passing through all damage states corresponding to the four modes of failure. First, the infiltration state of damage, followed by the weeping and stiffness states until the ultimate damage state corresponding to rupture.

Now imagine the ply subjected to an extremely small load, requiring an extremely long time to failure. The accumulated damage from such a long-duration load is very large and the ply residual strengths reduce to extremely low values. In this small-load and long-duration situation, the differences between the weep, infiltration, stiffness, and rupture long-term strengths become negligibly small. In the very long run, the long-term ply strengths corresponding to the four failure modes coincide and the four regression lines converge. The residual strength at the point of convergence is the fatigue threshold of the ply. The short-term thresholds define one end of the regression lines. The long-term fatigue threshold defines the other end (Figure 9.2).

The state of damage of the ply at the long-term fatigue threshold is so high as to obliterate the difference in resin toughness. For any given ply, the point of convergence in Figure 9.2 is resin-independent, that is, is the same for all resins. Let us introduce an analogy to clarify this statement. Imagine a young Olympic athlete and an average person. There is an enormous difference in performances when both are young. However, when both are 90 years old, the average citizen has a good chance to beat any Olympic gold medalist. In their youth, this would be unthinkable. Not so in their old age. The aging process is a great leveler. The fatigue threshold, therefore, is a long-term ply property independent of the mode of failure and the resin toughness.

Figure 9.2 shows the four regression lines originating at the short-term thresholds and converging at the long-term fatigue thresholds. The regression equations of the four lines in Figure 9.2 are as follows:

$$log\epsilon = log\left(T_r\right) - G_c^r logN \quad \left(Rupture\ line\right)$$

$$log\epsilon = log\left(T_w\right) - G_c^w logN \quad \left(Weep\ line\right)$$

$$log\epsilon = log\left(T_i\right) - G_c^i logN \quad \left(Infiltration\ line\right)$$

$$log\epsilon = log\left(T_s\right) - G_c^s logN \quad \left(Stiffness\ line\right)$$

The experimental measurement of one slope suffices to compute all others. Let us say we measure G_c^r, the rupture slope. The weep, infiltration, and stiffness slopes follow from simple geometric reasoning. The following expressions derive from inspection of Figure 9.2:

$$G_c^w = \frac{log\left(T_w/T_0\right)}{log\left(T_r/T_0\right)} \times G_c^r$$

$$G_c^i = \frac{log\left(T_i/T_0\right)}{log\left(T_r/T_0\right)} \times G_c^r$$

$$G_c^s = \frac{log\left(T_s/T_0\right)}{log\left(T_r/T_0\right)} \times G_c^r$$

In the above equations, T_0 is the long-term fatigue threshold. The other symbols are self-explanatory.

The rupture regression lines are the only ones practical to measure experimentally. This is because the end-points of rupture failures are uncontroversial and easy to observe. The end-points of the other modes of failure are either controversial or not so easy to detect. Of the four regression lines in Figure 9.2, only one—the rupture—is actually measured. The remaining three lines derive from it. The exception to this statement is the regression lines of transverse loaded UD plies which derive from the infiltration, not the rupture, regression equation.

The least controversial and easiest to observe failure end-point of transverse loaded UD plies is the first crack, which marks the onset of infiltration. The failure end-points of the other modes—weep, stiffness, and rupture—of transverse loaded UD plies are controversial and difficult to detect. As a result, all regression slopes of transverse loaded UD plies derive from the infiltration line. In analogy with the preceding discussion, these slopes are

$$G_c^w = \frac{log\left(T_w/T_0\right)}{log\left(T_i/T_0\right)} \times G_c^i$$

$$G_c^r = \frac{log\left(T_r/T_0\right)}{log\left(T_i/T_0\right)} \times G_c^i$$

$$G_c^s = \frac{log\left(T_s/T_0\right)}{log\left(T_i/T_0\right)} \times G_c^i$$

The short-term failure thresholds in Figure 9.2 are ply properties that may shift up and down, depending on the resin toughness, but stay the same for any given resin and regardless of the loading. The long-term fatigue threshold has the same value for all loadings and all resins. For any given ply, different loadings give lines of different slopes, all starting and ending at the fixed thresholds.

Example 9.1

The British Plastics Federation reported the rupture slope of pure cyclic loaded (R = 0) polyester chopped plies as $G_c^r = 0.075$. This rupture cyclic slope is valid for polyester chopped plies. The rupture slopes of chopped plies at other R ratios are not available.

Let us compute the pure weep and infiltration regression slopes of polyester chopped plies. The discussion is limited to pure loadings, that is, to R = 0. The short-term failure thresholds of polyester chopped plies are:

$$T_w = 0.80\%$$

$$T_i = 0.20\%$$

$$T_r = 1.50\%$$

$$T_0 = 0.05\%$$

The weep and infiltration slopes for R = 0 are

$$G_c^w = \frac{log(T_w / T_0)}{log(T_r / T_0)} \times G_c^r$$

$$G_c^w = \frac{log(0.80 / 0.05)}{log(1.50 / 0.05)} \times 0.075 = 0.061$$

$$G_c^i = \frac{log(T_i / T_0)}{log(T_r / T_0)} \times G_c^r$$

$$G_c^i = \frac{log(0.20 / 0.05)}{log(1.50 / 0.05)} \times 0.075 = 0.031$$

Example 9.2

Let us derive the four cyclic regression equations of polyester chopped plies at the load ratio R = 0.8. The ply parameters are those in Example 9.3. To compute the desired equations, we need the cyclic rupture slope at R = 0.8. As a rule, such a slope is not available. However, for illustration purpose, let us assume $G_c^r = 0.050$ for R = 0.8. With this information, the weep and infiltration slopes for R = 0.8 are

$$G_c^w = \frac{log(T_w / T_0)}{log(T_r / T_0)} \times G_c^r$$

$$G_c^w = \frac{log(0.80 / 0.05)}{log(1.50 / 0.05)} \times 0.050 = 0.041$$

$$G_c^i = \frac{\log\left(T_i / T_0\right)}{\log\left(T_r / T_0\right)} \times G_c^r$$

$$G_c^i = \frac{\log\left(0.20 / 0.05\right)}{\log\left(1.50 / 0.05\right)} \times 0.050 = 0.020$$

Having the slopes, the failure equations at R = 0.8 are

$$\log\left(\epsilon\right) = \log\left(1.5\right) - 0.050\log\left(N\right) \quad \left(Rupture\right)$$

$$\log\left(\epsilon\right) = \log\left(0.8\right) - 0.041\log\left(N\right) \quad \left(Weep\right)$$

$$\log\left(\epsilon\right) = \log\left(0.2\right) - 0.020\log\left(N\right) \quad \left(Infiltration\right)$$

Figure 9.3 shows the desired rupture and weep lines. Different load ratios R produce lines of different slopes, all passing through the fixed thresholds.

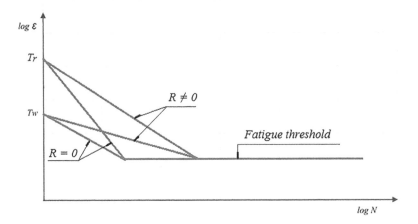

FIGURE 9.3 *For any given ply, different load ratios R give different regression slopes. The thresholds, however, are ply properties that remain unchanged for the same ply, regardless of the loading. The regression lines start at their specific short-term thresholds and finish at the long-term fatigue threshold. For any given ply, the experimental measurement of one regression slope suffices to determine all others. The reader will note the line slopes decreasing with increments in the R ratio. The slope is maximum at R = 0.0 (pure cyclic) and decreases to $G_c = 0$ at R = 1.0.*

The simplification introduced by the concept of fatigue threshold eliminates the need to develop expensive regression equations for each of the four modes of failure. All we need is one equation, for one mode of failure. The others derive from it. This is an enormous simplification. Furthermore, as explained in Chapter 10, the operating conditions (temperature and moisture) have a small and negligible effect on the regression equations. This is another major simplification, allowing the extension of test results obtained on dry plies at room temperature to all operating conditions. The only question remaining is the resin effect on the regression lines, which we address in the next sub-section.

Example 9.3

Suppose a sanitation pipe operating at a load ratio R = 0.80. The following information is available for the critical ply of polyester chopped fibers.

$T_r = 1.5\%$ $\left(Rupture\ threshold\right)$

$T_i = 0.20\%$ $\left(Infiltration\ threshold\right)$

$T_w = 0.80\%$ $\left(Weep\ threshold\right)$

$G_c^r = 0.050$ $\left(Rupture\ regression\ slope\ for\ R = 0.80\right)$

$T_0 = 0.05\%$ $\left(Fatigue\ threshold\right)$

$\varepsilon = 0.20\%$ $\left(Operating\ peak\ strain\right)$

We wish to compute the number of cycles to (a) rupture, (b) weep, and (c) infiltrate a pipe subjected to the operating peak strain 0.20%. The reader will notice the measured rupture slope $G_c^r = 0.050$ for this specific loading R = 0.8. For any loading, the regression slope is always less than that for the pure cyclic loading $G_c^r = 0.075$.

This problem is easy to solve with the known rupture regression equation for R = 0.80. The normal situation is the absence of such equation, which dictates the use of the unified equation.

From Example 9.6, the regression lines of chopped fibers for R = 0.8 are

$log\epsilon = log\left(1.5\%\right) - 0.05logN$

$log\epsilon = 0.176 - 0.05logN$ $\left(Rupture\right)$

$$log\epsilon = log(0.80\%) - 0.041logN$$

$$log\epsilon = -0.097 - 0.041logN \quad (Weep)$$

$$log\epsilon = log(0.20\%) - 0.020logN$$

$$log\epsilon = -0699 - 0.020logN \quad (Infiltration)$$

The reader will recall that the informed rupture equation comes from measurements on plies of chopped fibers submitted to the load ratio R = 0.80. This information is seldom available in the real world. In this case, the given rupture regression equation is directly applicable to the problem at hand. The weep and infiltration equations, derived from the given rupture equation, are also directly applicable.

The number of cycles to infiltration is

$$log0.20 = -0.699 - 0.020logN$$

Solving this equation, we have

$$N(infiltration) = 1 \; cycle$$

The pipe infiltrates on the first cycle, which is understandable since the peak operating strain (0.20%) is equal to the infiltration threshold. This infiltration is perfectly acceptable for pipes in water service and is not a failure.

The number of cycles to weep is

$$log0.20 = -0.097 - 0.041logN$$

$$N(weep) = 4.81 \times 10^{14} \; cycles$$

The number of cycles to rupture is

$$log0.20 = 0.176 - 0.05logN$$

$$N(rupture) = 3.15 \times 10^{17} \; cycles$$

This numerical example illustrates the enormous power of the regression equations, when available. The problem is their unavailability. The unified equation applies in the absence of the regression equations. See Chapters 16 and 18.

9.3.2   THE RESIN EFFECT

We are finally ready to discuss the effect of resin toughness on the ply slopes G_c. The connection between the slopes of different resins come from a simple inspection of Figure 9.2. As already discussed, the long-term fatigue threshold (T_0) is the same for all resins, while the short-term thresholds vary with the resin. From Figure 9.2 we have

$$\left(G_c^r\right)_{VER} = \frac{\log\left(T_r / T_0\right)_{VER}}{\log\left(T_r / T_0\right)_{POL}} \times \left(G_c^r\right)_{POL}$$

Where the subscripts VER and POL refer to plies of vinyl ester and polyester resins, respectively. The above equation computes the slope of any resin from the easy to measure ply thresholds and the measured slope of another resin. This is a significant simplification, indicating that knowledge of just one ply slope, developed for any resin, suffices to compute the ply slopes of any other resin.

The experimental evidence, discussed in the Chapter 10, indicates a small increase in the ply rupture slope G_c^r with increments in the resin toughness. Let us confirm this statement. From the above equation we have

$$\left(G_c^r\right)_{VER} = \frac{\log\left(2.50 / 0.05\right)_{VER}}{\log\left(1.50 / 0.05\right)_{POL}} \times \left(G_c^r\right)_{POL} = 1.15 \times \left(G_c^r\right)_{POL}$$

As we see, the large difference in toughness between vinyl ester and polyester resins implies in a small 15% variation in the rupture slope. The reader will understand that the slopes mentioned in this section hold for any loading or R ratio. In particular, they are valid for the pure cyclic slopes at R = 0.0. We say this to facilitate the discussions presented in Section 9.5.

Example 9.4

Let us solve the problem of Example 9.3, assuming the sanitation pipes with a weep barrier of vinyl ester. The infiltration threshold of vinyl ester chopped plies is Ti = 0.50%, the weep threshold is Tw = 1.00%, the elongation at rupture is Tr = 2.50% and the fatigue threshold is T_0 = 0.05%. The operating strain is 0.30%, higher than the 0.20% assumed in Example 9.3.

The rupture slope of the vinyl ester chopped ply for R = 0.80 is

$$\left(G_c^r\right)_{VER} = 1.15 \times \left(G_c^r\right)_{POL}$$

$$\left(G_c^r\right)_{VER} = 1.15 \times 0.050 = 0.058$$

The weep and infiltration slopes for the vinyl ester chopped ply are

$$G_c^w = \frac{\log\left(T_w / T_0\right)}{\log\left(T_r / T_0\right)} \times G_c^r = \frac{\log\left(1.00 / 0.05\right)}{\log\left(2.50 / 0.05\right)} \times 0.058 = 0.044$$

$$G_c^i = \frac{\log\left(T_i / T_0\right)}{\log\left(T_r / T_0\right)} \times G_c^r = \frac{\log\left(0.50 / 0.05\right)}{\log\left(2.50 / 0.05\right)} \times 0.058 = 0.034$$

The regression lines for the vinyl ester chopped ply are

$$log\epsilon = \log\left(2.5\%\right) - 0.058logN$$

$$log\epsilon = 0.398 - 0.058logN \quad \left(Rupture\right)$$

$$log\epsilon = \log\left(1.00\%\right) - 0.044logN$$

$$log\epsilon = -0.044logN \quad \left(Weep\right)$$

$$log\epsilon = \log\left(0.50\%\right) - 0.034logN$$

$$log\epsilon = -0.301 - 0.034logN \quad \left(Infiltration\right)$$

The number of cycles to infiltration is

$$log0.30 = -0.301 - 0.034logN$$

$$N\left(infiltration\right) = 3.3 \times 10^6 \, cycles$$

The number of cycles to weep is

$$log0.30 = -0.044logN$$

$$N\left(weep\right) = 7.6 \times 10^{11} \, cycles$$

The number of cycles to rupture is

$$log0.30 = 0.398 - 0.058logN$$

$$N(rupture) = 7.5 \times 10^{15} cycles$$

This example illustrates the enormous superiority of vinyl ester resins vis-à-vis polyesters.

9.4 THE PURE REGRESSION EQUATIONS

So far, we have been discussing ply regression equations for any loading, that is, any R ratio. The pure regression equations, i.e., those equations in which $R = 1.0$ (pure static) or $R = 0.0$ (pure cyclic) are the topic of this section.

9.4.1 PURE STATIC RUPTURE OF FIBER LOADED UD PLIES ($R = 1.0$)

Mark Greenwood developed the applicable regression equations for this case. See Chapter 12.

$$log(\epsilon) = 0.400 - 0.077log(hours) \quad \text{Boron-free glass.}$$

$$log(\epsilon) = 0.347 - 0.130log(hours) \quad \text{E glass.}$$

The long-term static strengths of UD plies tensile loaded in the fiber direction are

$$logS_S = 0.400 - 0.077log(X) \quad \text{Pure static strength. Boron-free glass.}$$

$$logS_S = 0.347 - 0.130log(X) \quad \text{Pure static strength. E glass.}$$

In the above, X is the long-term expected durability in hours.

9.4.2 PURE CYCLIC RUPTURE OF FIBER LOADED UD PLIES ($R = 0.0$)

Guangxu Wei developed the applicable regression line for this case. See Chapter 12.

$$log(\epsilon) = 0.480 - 0.089log(N)$$

The long-term cyclic strength in the fiber direction is

$$\log S_C = 0.480 - 0.089 \log(Y) \quad \text{Pure cyclic strength}$$

Where Y is the number of cycles expected in X hours.

9.4.3 PURE CYCLIC INFILTRATION OF TRANSVERSE LOADED UD PLIES ($R = 0.0$)

Guangxu Wei measured the infiltration pure line of transverse loaded poly-ester UD plies. For details, see Chapter 18.

$$\log(\epsilon) = \log 0.20 - 0.034 \log(N) \quad \text{Infiltration, pure cyclic, polyester}$$

The long-term infiltration strength is

$$\log S_C = \log 0.20 - 0.034 \log(Y) \quad \text{Pure cyclic, polyester}$$

Where Y is the number of cycles expected in X hours.

9.4.4 PURE CYCLIC SHEAR RUPTURE OF UD PLIES

The UD shear regression equations are useful in the study of long-term stiffness failure. These equations are not available.

9.4.5 PURE CYCLIC RUPTURE OF CHOPPED FIBERS ($R = 0.0$)

The British Plastics Federation developed the pure cyclic rupture equation for polyester chopped fibers.

$$\log(\epsilon) = \log(1.50) - 0.075 \log(N) \quad \text{Rupture, pure cyclic, polyester}$$

The long-term cyclic rupture strength is

$$\log S_C = \log(1.50) - 0.075 \log(Y) \quad \text{Long-term pure cyclic strength}$$

In the above equation, Y is the number of cycles expected in X hours.

9.4.6 PURE STATIC RUPTURE OF CHOPPED FIBERS ($R = 1.0$)

The British Plastics Federation developed the pure static rupture regression equation for polyester chopped fibers:

$$\log(\epsilon) = \log(1.50) - 0.045 \log(\text{hours}) \quad \text{Pure static rupture, polyester}$$

The long-term static strength is

$$\log S_S = \log(1.50) - 0.045 \log(X) \quad \text{Pure static strength}$$

In the above equation, X is the target lifetime in hours. The static slope is resin-independent.

There remains one final question to address, regarding the thickness of the embedded UD ply. In this regard, we quote from Ref. [20], taking the liberty of slightly editing the original text to comply with the terminology in this book.

> The transverse cracking process in the general cross-ply laminates is influenced by several parameters. The first effect to consider is the so-called constraint effect. As demonstrated by Parzini et al. the thickness of embedded UD plies will influence the thresholds. For thin plies larger thresholds are observed. At a certain minimum thickness of about 0.5 mm, a constant value of the threshold is reached, which is the threshold value for the UD ply. UD plies embedded in laminates and thinner than 0.5 mm have larger thresholds than isolated plies.

Taking into consideration that most UD plies embedded in commercial laminates are thinner than 0.5 mm, the ply thickness might be a topic of considerable discussion. A conservative approach to this discussion consists in ignoring the ply thickness effect and measuring the transverse UD thresholds in embedded plies thicker than 0.5 mm. That way we capture a conservative short-term transverse strength.

Section 9.5 lists the pure regression equations of commercial interest.

9.5   TABULATED PURE REGRESSION EQUATIONS

The generic forms of the regression equations are:

$$\log(\epsilon) = A_C - G_C \log(N) \quad (\text{Pure cyclic})$$

$$\log(\epsilon) = A_S - G_S \log(\text{hours}) \quad (\text{Pure static})$$

The intercept parameter for the cyclic loadings is

$$A_C = \log(threshold)$$

Where "threshold" indicates the short-term failure threshold – the short-term ply strength – applicable to the specific mode of failure.

In Sub-sections 9.3.1 and 9.3.2, we developed a very powerful protocol to compute the cyclic slope G_c of any mode of failure and any resin from just one experimentally measured slope. When the rupture slope is available, the other slopes for the same resin are:

$$G_c^w = \frac{log(T_w/T_0)}{log(T_r/T_0)} \times G_c^r$$

$$G_c^i = \frac{log(T_i/T_0)}{log(T_r/T_0)} \times G_c^r$$

$$G_c^s = \frac{log(T_s/T_0)}{log(T_r/T_0)} \times G_c^r$$

When the infiltration slope is available, the other slopes for the same resin are:

$$G_c^w = \frac{log(T_w/T_0)}{log(T_i/T_0)} \times G_c^i$$

$$G_c^r = \frac{log(T_r/T_0)}{log(T_i/T_0)} \times G_c^i$$

$$G_c^s = \frac{log(T_s/T_0)}{log(T_i/T_0)} \times G_c^i$$

The resin effect on the regression equations is twofold. First, the resin toughness directly controls the short-term thresholds that define the regression parameter A_c. Second, the resin toughness affects the ply slope G_c as shown in the equation below.

$$\left(G_c^r\right)_{VER} = \frac{log(T_r/T_0)_{VER}}{log(T_r/T_0)_{POL}} \times \left(G_c^r\right)_{POL}$$

Where the subscripts VER and POL stand for tough vinyl esters and brittle polyesters, respectively. So far, the discussions in this Section are valid for any loading, or R ratio.

As we see, the knowledge of the failure thresholds and one cyclic slope suffices to compute the ply regression equations for all modes of failure and all resins subjected to loadings of the same R ratio. Our goal in this Section is the tabulation of the pure regression equations. Therefore, going forward from this point, we focus on the pure cyclic slopes developed for R = 0.0.

The following Tables 9.1a through 9.1d list all pure regression equations of plies of commercial interest. The pure equations are useful to compute the ply durability – and by extension the laminate durability – under pure cyclic or pure static loads. Such pure loading cases are rare in real life. Storage tanks provide examples of pure static loading. I cannot think of a single application involving pure cyclic loading. Most engineering applications involve the simultaneous presence of cyclic and static loads. The pure equations per se are useless.

In real life, when the cyclic and static loads act simultaneously, we take one of two choices:

- We can measure experimentally the regression equation for that specific loading and laminate. This is unfeasible, given the infinite combinations of loadings and laminates.
- We can use the unified equation described in Chapter 16. This is the best choice.

TABLE 9.1

Pure rupture failure equations. The fiber dominated static long-term rupture equations have the same slope for all resins.

Rupture

UD plies, fiber direction, static boron-free glass	$\log(\epsilon) = 0.400 - 0.077 \times \log(hour)$
UD plies, fiber direction, static E glass	$\log(\epsilon) = 0.347 - 0.130 \times \log(hour)$
UD plies, fiber direction, cyclic any glass	$\log(\epsilon) = \log(3.00) - 0.089 \times \log(N)$
UD plies, transverse direction, cyclic polyester	$\log(\epsilon) = \log(0.40) - 0.051 \times \log(N)$
UD plies, transverse direction, cyclic vinyl ester	$\log(\epsilon) = \log(0.60) - 0.061 \times \log(N)$
Chopped plies, static polyester	$\log(\epsilon) = \log(1.50) - 0.045 \times \log(hour)$
Chopped plies, static vinyl ester	$\log(\epsilon) = \log(2.50) - 0.045 \times \log(hour)$
Chopped plies, cyclic polyester	$\log(\epsilon) = \log(1.50) - 0.075 \times \log(N)$
Chopped plies, cyclic vinyl ester	$\log(\epsilon) = \log(2.50) - 0.086 \times \log(N)$

The unified equation is applicable to plies embedded in any laminate subjected to any loading combination. It differs from the laminate regression equations in two points. First, it applies to a critical ply, not to the entire laminate. Second, it works by interpolation between ply equations developed for a few selected loadings, that is, for loadings defined by a few chosen R ratios. The ply equations developed for non-pure loadings are as those depicted in Figure 9.1. We need quite a few of those to make accurate interpolation. For details, see Chapters 17 and 18. The high cost and difficulty to determine a large number of regression equations, at several R ratios, is probably the only weak point of the unified equation. I personally do not see the composites industry investing the time and money to develop the required experimental regression equations. For a deeper discussion of this topic, see Chapter 18.

Be it as it may, the pure static and pure cyclic equations tabulated in Tables 9.1–9.4 provide the essential inputs needed in the unified equation and are central to the study of the load-dependent laminate durability. The fact that previous investigators have developed all pure failure equations is a blessing. We should be thankful and appreciative of the insightful work by Mark Greenwood, Guangxu Wei, and the British Plastics Federation.

Example 9.5

Explain the difference between the ply general regression lines depicted in Figure 9.1 and the pure regression lines discussed in the previous section.

TABLE 9.2
The pure stiffness failure equations of practical interest are those of transverse loaded UD plies. The chopped plie and the UD plies loaded in the fiber direction are not critical in stiffness failure.

Stiffness

UD plies, fiber direction, static boron-free glass	—
UD plies, fiber direction, static E glass	—
UD plies, fiber direction, cyclic any glass	—
UD plies, transverse direction, cyclic polyester	$\log(\epsilon) = \log(0.40) - 0.051 \times \log(N)$
UD plies, transverse direction, cyclic vinyl ester	$\log(\epsilon) = \log(0.60) - 0.061 \times \log(N)$
Chopped plies, static polyester	—
Chopped plies, static vinyl ester	—
Chopped plies, cyclic polyester	—
Chopped plies, cyclic vinyl ester	—

TABLE 9.3

The pure weep failure equations of interest are those of chopped and transverse loaded UD plies. The UD plies loaded in the fiber direction do not develop resin cracks leading to weep failure. Statically loaded chopped plies do not grow cracks and do not cause weep failure.

Weep

UD plies, fiber direction, static boron-free glass	—
UD plies, fiber direction, static E glass	—
UD plies, fiber direction, cyclic any glass	—
UD plies, transverse direction polyester	$\log(\epsilon) = \log(0.25) - 0.039 \times \log(N)$
UD plies, transverse direction vinyl ester	$\log(\epsilon) = \log(0.40) - 0.051 \times \log(N)$
Chopped plies, static polyester	—
Chopped plies, static vinyl ester	—
Chopped plies, cyclic polyester	$\log(\epsilon) = \log(0.80) - 0.061 \times \log(N)$
Chopped plies, cyclic vinyl ester	$\log(\epsilon) = \log(1.00) - 0.066 \times \log(N)$

TABLE 9.4

The pure infiltration failure equations of interest are those of chopped plies. The UD plies never have contact with corrosive chemicals. Statically loaded chopped plies do not grow cracks and do not cause infiltration failure.

Infiltration

UD plies, fiber direction, static boron-free glass	—
UD plies, fiber direction, static E glass	—
UD plies, fiber direction, cyclic any glass	—
UD plies, transverse direction polyester	$\log(\epsilon) = \log(0.20) - 0.034 \log(N)$
UD plies, transverse direction vinyl ester	$\log(\epsilon) = \log(0.30) - 0.044 \times \log(N)$
Chopped plies, static polyester	—
Chopped plies, static vinyl ester	—
Chopped plies, cyclic polyester	$\log(\epsilon) = \log(0.30) - 0.040 \times \log(N)$
Chopped plies, cyclic vinyl ester	$\log(\epsilon) = \log(0.50) - 0.051 \times \log(N)$

The lines in Figure 9.1 are generic representations of regression equations for any ply and any loading. Those lines, as explained, compute the number of cycles to ply failure under specific R ratios and load paths. The A_c parameters in Figure 9.1 are the same as those in the pure regression equations. In fact, the A_c parameters are short-term ply strengths independent of the loading. The cyclic slopes G_c in Figure 9.1 vary with the loading – the R ratio – and are not the same as those in the pure regression lines. In fact, the highest possible slopes are those of the pure cyclic lines. See Figure 9.3. The ply regression lines in Figure 9.1 provide the data to compute the interaction parameter G_{sc}. See Chapters 17 and 18.

The pure regression equations hold for R = 0.0 (pure cyclic) and R = 1.0 (pure static). The slopes G_c and G_s of the pure regression equations are direct inputs to the unified equation.

Example 9.6

The short-time (30 seconds) quality control hydrostatic pressure test currently mandated on all sanitation and oil pipes is performed at pressures above the pipe rating and slightly less than the weep threshold. For example, a pipe designed to operate at 100 PSI may have a weep threshold of 200 PSI and be pressure tested at 175 PSI, just slightly short of the weep threshold. This is a safe procedure as long as the pipes operate under static pressure. However, in the presence of cyclic loads, there is a legitimate concern that the damage incurred in such a high-pressure test might reduce the weep durability.

In fact, from the concepts introduced in this chapter, any pressure excursion above the nominal would enlarge the crack density and reduce the cyclic durability. The engineer should carefully ponder any pressure spike above the rated pressure. See the numerical Examples 9.7 and 9.8.

As explained in Chapter 20, the impermeable pipes with aluminum foil do not require these quality control pressure tests. The impermeable pipes are a welcome low cost solution to eliminate the expensive and time-consuming quality control pressure testing and at the same time assure a weep-free pipe. For details, see Chapter 20.

Example 9.7

Consider an oil pipeline of ± 55 UD vinyl ester plies with weep threshold $T_w = 0.40\%$. The peak operating transverse strain is $\varepsilon = 0.20\%$. The pipe is factory tested for 30 seconds at a transverse strain of 0.30% as part of the quality control program. Let us quantify the effect of such a pressure test on the expected weep durability. We consider two scenarios:

1. Pure static condition
2. Pure cyclic conditions

Pure static service: The transverse test strain 0.30% is less than the weep threshold 0.40%. Since static loads do not grow cracks, the test damage does not affect the long-term weep failure. The pipeline operates as if it had never been pressure tested. In the specific case of pure static pressure, the factory pressure test is no cause of long-term concern.

Pure cyclic service: The transverse test strain of 0.30% increases the crack density and reduces the durability of the pipeline in the presence of cyclic loadings. This is because the in-service damage growth starts from a higher level corresponding to the 0.30% test strain. The complete evaluation of the durability in this case, consisting of a combination of static and cyclic loadings, requires the use of the unified equation introduced in Chapter 16. However, for illustration purposes, we offer in this Example a solution for the simple case of pure cyclic load. From Table 9.3, the pure cyclic weep equation of transverse loaded vinyl ester UD plies submitted to the peak operating strain of 0.20% is

$$\log(0.20) = \log(0.40) - 0.051 \times \log(N_1)$$

From this equation, we obtain $N_1 = 800\ 000$ cycles. This is the number of cycles to weep a healthy, pristine and not pressure-tested pipe. The crack density in this case starts to grow from the small value corresponding to the applied peak operating strain 0.20%.

We next compute the number of cycles to grow the damage – crack density – to the level produced by the test pressure 0.30%. To do this we assume the pipe weeps at 0.30%. The slope of the weep regression equation in this case is

$$G_c^w = \frac{\log(T_w / T_0)}{\log(T_r / T_0)} \times G_c^r = \frac{\log(0.30 / 0.05)}{\log(0.60 / 0.05)} \times 0.061 = 0.044$$

The failure equation for this hypothetical case is

$$log(0.20) = log(0.30) - 0.044 log N_2$$

Solving the above equation, we obtain $N_2 = 10\ 000$ cycles. This is the number of cycles for a healthy pipe submitted to the nominal strain 0.20% to attain the damage state corresponding to the test strain of 0.30%.

The number of cycles required to weep the damaged pipe is the difference between N_1 and N_2. Therefore,

$$N = N_1 - N_2 = 800\ 000 - 10\ 000 \approx 800\ 000\ cycles$$

We have come to the unexpected conclusion that the factory hydrostatic test conducted at pressures 50% higher than the rated pressure—below the weep threshold—has a negligible effect on the pipe cyclic durability. The number of cycles to weep remains practically unchanged after the pressure test.

Example 9.8

Determine the spike pressure (strain) that would reduce the pipe weep durability to half its original value. Assume the pipe parameters and operating conditions of Example 9.7.

The weep threshold is $T_w = 0.40\%$ and the fatigue threshold is $T_0 = 0.05\%$. The peak operating strain is $\varepsilon = 0.20\%$ and the pipe operates under pure cyclic loading.

The number of cycles to weep the pristine pipe is, as before

$$log(0.20) = log(0.40) - 0.051 logN_1$$

$N_1 = 800\ 000$ cycles.

We wish to determine the short duration spike strain that would weep the pipe in

$N_2 = N_1/2 = 400\ 000$ cycles

Let ω designate the unknown spike strain. The weep slope of the hypothetical regression equation is

$$G_c = \frac{log(\omega / 0.05)}{log(0.60 / 0.05)} \times 0.061$$

The unknown spike strain ω comes from

$$log(0.20) = log(\omega) - \frac{log(\omega / 0.05)}{log(0.60 / 0.05)} \times 0.061 \times log(400\ 000)$$

Solving the above equation, we have

$\omega = 0.38\%$

This unexpected and amazing result shows that the short-duration spike strain that reduces the pipe life to 50% of its original value is 0.38%, just 5.0% short of the weep threshold 0.40%. Our analysis seems to indicate that swing pressures up to 90% of the weep threshold are not overly detrimental to the pipe durability.

Example 9.9

This example was included as an afterthought, in the hope that corny repetition may help clarify long established and hard to abandon concepts.

The test protocol described in ASTM D 2992 A defines the method to establish the cyclic weep regression line of composite pipes. The arrangements of the cyclic protocol are essentially identical to those of the sister method ASTM D 2992 B to establish the static weep regression line. Both methods subject pipe segments to different pressure levels (static in procedure B, cyclic in procedure A) to measure the time or number of cycles to weep. In both cases, the pipe specimens are over-strained to shorten the test time.

The regression equation generated by the ASTM D2992 A protocol is

$$log\epsilon_{hoop}^{pipe} = A_{weep}^{pipe} - G_{weep}^{pipe} \times log \tag{9.2}$$

The ply regression equation proposed in this chapter is

$$log\epsilon_2 = A_c - G_c logN \tag{9.3}$$

Let us highlight the differences between the above equations:

First difference

- The ASTM Equation (9.2) measures the number of cycles to weep the pipe. Actually, it measures the travel time of water in traversing the cracked pipe wall. Equation (9.2) holds only for pipes cracked above the weep threshold.
- Equation (9.3) measures the number of cycles to grow the initially small crack density to the critical density corresponding to the weep damage.

Second difference

- Equation (9.2) considers the pipe hoop strain.
- Equation (9.3) considers the UD ply transverse strain.

Third difference

- Equation (9.2) defines failure at the onset of weeping.
- Equation (9.3) defines failure at the weep threshold.

Fourth difference

- To illustrate the fourth difference, we resort to an aircraft analogy.
- The ASTM Equation (9.2) predicts the time for the aircraft to hit the ground when out of fuel.
- Equation (9.3) predicts the time for the aircraft to run out of fuel.

These are the basic differences between the classical and the new approach to weep durability. The airplane analogy is especially compelling. The important thing to know when flying any aircraft is the time to run out of fuel, not the time to hit the ground when out of fuel.

APPENDIX 9.1

THE IDEALIZED RUPTURE EQUATION

The experimental rupture regression line is not strictly linear as presented in this chapter. The upper left end of the experimental line, corresponding to high strains and few cycles, flattens out to reflect the short-term strength of the ply. Furthermore, the lower right end of the experimental line also flattens out in response to the fatigue limit.

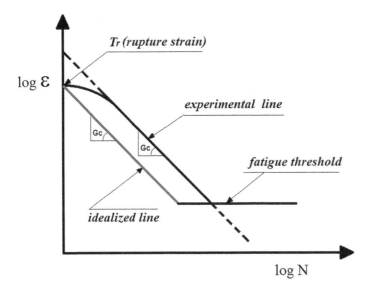

FIGURE 9.4 *The experimental rupture regression line does not capture the short-term ply strength and the long-term fatigue threshold. The idealized rupture line captures both the short-term rupture strength and the long-term fatigue threshold. The two lines have identical slopes.*

Figure 9.4 captures the above statements. The experimental rupture regression line predicts higher short-term and lower long-term failure strains than the actual material strength. Furthermore, the experimental line does not capture the short-term ply strength and the long-term fatigue threshold.

The experimental rupture regression line does not capture the thresholds and is not good for design. The alternative idealized line proposed in Figure 9.4 captures the thresholds. The idealized equation takes the form

$$log(\epsilon) = A_C - G_C log(N)$$

Where A_c and G_c are the regression parameters, N is the number of cycles to rupture, and ϵ is the applied peak strain. The idealized rupture lines originate at the short-term rupture strength (the failure threshold) and runs parallel to the experimental line to meet the fatigue threshold. The idealized regression parameter A_c derives directly from the rupture short-term strength:

$$A_c = log(short - term\ strength)$$

The two lines are parallel and have the same slope. The rupture slope G_c is the only experimental parameter measured in regression tests. The slopes for the other modes of failure—infiltration, stiffness and weep— derive from the rupture slope G_c, as explained in this chapter.

APPENDIX 9.2

THE FATIGUE THRESHOLD

The fatigue threshold derive from the concept of infinite used in engineering practice. In the engineering sense, the term infinity defines the remoteness at which extremely distant causes cease to have appreciable local effects on the problem or affair at hand. As an example, the remote planet Mars has a negligible gravitational effect on earth. The Mars distance from Earth is, therefore, infinite, even if we know this is not strictly correct. The same concept holds for plies subjected to low cyclic peak strains that produce extremely long cyclic lives.

Let us assume two identical plies made of different resins. The rupture regression equations for these plies vary with the resin toughness, as explained in the text:

$$log\epsilon = A_1 - G_1 logN$$

$$log\epsilon = A_2 - G_2 logN$$

The above lines have different slopes and meet at some point. By definition, the long-term fatigue threshold is the long-term ply strength at the point where the two lines meet. This, of course occurs at a very large number of cycles. The small fatigue threshold *"To"* and the corresponding very large number of cycles *"No"* derive from solving the following system in two equations and two unknowns.

$$logT_0 = A_1 - G_1 logN_0$$

$$logT_0 = A_2 - G_2 logN_0$$

The above system consists of two rupture regression lines for the same type of ply made with different resins. From the discussion presented in the text, we have learned that only one such a line is available at this time, derived for brittle polyester resins. The lack of a second equation, derived for a different resin, prevents the correct computation of the long-term fatigue threshold. All we need is one additional experimental regression line measured on plies of vinyl ester or any other flexible resin. However, we do have empirical evidence from laminate tests (laminate, not ply), suggesting the "infinite" number of cycles $N_0 = 10^{20}$ at the fatigue threshold for pure cyclic loading. See the work of Prof. Mandel on laminates, briefly described in the Examples 10.5 and 10.6. Taking $N_0 = 10^{20}$ as the "infinite" number of cycles at the fatigue threshold for pure cyclic loading, we have the following suggestions for the pure cyclic ultimate long-term ply strengths corresponding to the fatigue thresholds.

Transverse Loaded UD Plies. The fatigue threshold of transverse loaded polyester UD plies, with a short-term rupture strength $T_r = 0.40\%$, a slope $G_c = 0.051$ and $N_0 = 10^{20}$ is

$$logT_o = log0.60 - 0.051log10^{20}$$

From the above equation, we have $T_0 = 0.04\%$.

Chopped Plies. The fatigue threshold of polyester chopped plies, with a short-term strength $T_r = 1.50\%$, a slope $G_c = 0.075$ and $N_0 = 10^{20}$ is

$$logT_o = log1.50 - 0.075log10^{20}$$

From the above we have $T_0 = 0.05\%$

Fiber Loaded UD Plies. The pure rupture cyclic regression equation of UD plies loaded in the fiber direction, with a short-term strength $T_r = 3.00\%$ and a slope $G_c = 0.089$ is

$$logT_o = log3.00 - 0.089log10^{20}$$

From the above equation, we have $T_0 = 0.05\%$

Table 9.5 summarizes the above findings

TABLE 9.5
The table indicates a convergence of all fatigue thresholds. Based on these values, we propose a fatigue threshold $T_0 = 0.05\%$ for all plies, all resins and all loadings, pure or otherwise. The No value will change with the load ratio R, while the To = 0.05% remains unchanged.

Pure fatigue thresholds (T_0) for No = 10^{20} cycles

	Transverse loaded UD plies (%)	Chopped plies (%)
Polyester	0.04	0.05
Vinyl ester	0.04	0.05
Fiber loaded UD plies, any resin	0.05	

The fatigue threshold is not the same as the fatigue limits described in the appendix of this book. The values of the theoretically possible – see the Appendix – fatigue limits should vary with the ply construction and the resin matrix. The fatigue threshold, however, is the same for all resin matrices and plies. The fatigue threshold is a mathematical idealization describing the long-term convergence point of all failure lines. We might think of the fatigue threshold as a theoretical lower bound of the fatigue limit.

Note: On meeting the fatigue limit, or the fatigue threshold, the regression lines turn horizontal. This is required to avoid the paradox of tougher plies failing sooner than brittle ones.

10 Effects of Temperature, Moisture and Resin Toughness

10.1 INTRODUCTION

This chapter deals with the effects of resin toughness, fiber sizing, environmental moisture, and operating temperature (OT) on the durability of composites. The discussions will highlight the strong and expected positive influence of the resin toughness, as well as the small roles of the environmental moisture and operating temperatures. The approach is intuitive and qualitative, as we try to avoid as much as possible the complex details that shed little or no light on the subject matter of this book. The role of resin-glass interphase/interface is barely touched, just enough to convey the basic ideas.

Increments in the resin toughness have a strong positive effect on the ply failure thresholds. As an example, the infiltration failure of the corrosion barrier of chopped plies increases from 0.20% for brittle polyesters to 0.50% for tough vinyl esters. The weep, stiffness, and rupture thresholds also show similar improvements. The improved failure thresholds have an important commercial impact in favor of tough versus brittle resins.

By contrast, increments in the environmental moisture and operating temperature have a modest, although positive, impact on the failure thresholds. The important point in this regard is that, however small, the increased failure thresholds are assurances of better—not of worse—ply performance at higher operating temperatures and moisture conditions. This improved performance justifies the use of the failure thresholds measured at room temperature in dry plies to all moisture conditions and OTs. Such a simple finding eliminates the need of collecting expensive experimental data and provides an enormous simplification to the study of the durability of composites. The reader should not erroneously assume from this statement that the environmental moisture and temperature have no effect on the durability of laminates. This impression is not true. In spite of their small positive impact on the failure thresholds, both the moisture and operating temperature have a strong, sometimes decisive, role in the residual strains. The residual strains, as we know, play an important part in the study of load-dependent durability. A careful engineer will not ignore the effects of the thermal and hydric residual strains.

We start this chapter discussing the effects of temperature, moisture, and resin toughness on the failure thresholds of UD plies. The effects on the plies of chopped fibers will follow.

10.2   EFFECTS OF TEMPERATURE AND MOISTURE ON THE FAILURE THRESHOLDS OF UD PLIES

The failure thresholds are short-term ply strengths measured at room temperature in a dry atmosphere. The question we seek to answer is … Are these thresholds valid at any temperature and moisture? We know from Chapter 1 that some ply properties, such as the expansion coefficients and moduli, are not sensitive to changes in temperature and moisture. Is this insensitivity also valid for the short-term strengths?

Catherine Woods and Walter Bradley answered this question in a remarkable paper titled "Effect of Seawater on The Interfacial Strength of E Glass/graphite Epoxy." See Ref. [18]. These two authors computed the micro-strains on unidirectional (UD) plies assuming the rectangular fiber array shown in Figure 10.1. The reasoning and conclusions that follow derive from the published data and findings of the above authors, assuming UD plies with a rectangular fiber array.

The resin shrinkage from crosslinking and cooling down pulls the fibers together, placing the points B (Figure 10.1) in radial compression. At the same time, the surrounding fibers constrain the resin shrinkage toward the point A, placing the mass within the dotted circle under tensile strains (Figure 10.1).

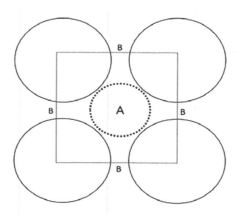

FIGURE 10.1   *The square array of fibers indicates resin residual tension within the dotted circle A and fiber compression at the points B. These residual strains are relieved with water intake and increments in the operating temperature.*

At room temperature and in a dry atmosphere, the resin-rich region within the dotted circle A is in tension, while the points B are in compression. Suppose now we immerse the UD ply, initially at room temperature and dry, in hot water. In this condition, the immersed ply saturates with water and takes a higher equilibrium temperature. The increased temperature and high water intake have the same effect on the ply, since both expand the resin. Whatever we say about water intake is equally valid for temperature rise.

The residual radial compressive strains at the points B act to prevent glass-resin debonds. As we have just seen, the resin shrinkage from cooling down and crosslinking place these points in radial compression and therefore increase the resin-fiber bond strength. That is a surprising conclusion. Our instinct would lead us to expect the cure/cooling shrinkages to decrease, not to increase, the resin-fiber bond strength. Proceeding with the argument, we see that increments in the operating temperature, or in the water intake, will expand the resin and reduce the residual fiber compression at the points B. This will lower the fiber-resin bond strength and is detrimental to the infiltration threshold. We have our first conclusion:

Increments in water intake and in operating temperature (OT) lower the infiltration threshold of transverse loaded UD plies.

Let us now check the resin strains within the dotted circle A. As we have seen, the curing and thermal shrinkage put this region in tension, which is bad for cracking. However, increments in the operating temperature or in the water intake will expand the resin and alleviate this tension, thereby retarding the crack initiation. This is good for the weep and stiffness thresholds. We have our second conclusion:

Increments in water intake and in operating temperature (OT) improve the weep and stiffness thresholds of transverse loaded UD plies.

In conclusion, increments in the operating temperature OT or in the water intake Δm have the positive effect of retarding the resin cracking and the negative effect of facilitating resin-fiber debonding. This situation is bad for infiltration, and good for stiffness and weeping. Table 10.1 summarizes these conclusions.

10.3   AN IMPORTANT PRACTICAL CONSEQUENCE

Figure 10.2 shows the static weep regression lines of oil pipes measured at two temperatures. Comparing the two lines, we conclude that higher test temperatures extrapolate to lower HDB (Hydrostatic Design Basis) values. Later in this chapter, we will explain the reason for the increased line slope at higher temperatures. For the moment, we see that higher temperatures

TABLE 10.1

Changes in the operating temperature and in the water intake expand/contract the resin in the same way and have similar effects. Anything said about changes in operating temperature is also applicable to water pickup. The findings in this table indicate that the failure thresholds measured on dry UD plies at room temperature are valid at any temperature and moisture. The error introduced from this simplification is conservative and safe for stiffness and weep, while slightly risky for infiltration. Infiltration, however, is not a relevant mode of failure for UD plies.

	Increased Δm	**Increased OT**
Infiltration threshold (debonding)	Decrease	Decrease
Weep threshold (resin cracks)	Increase	Increase
Stiffness threshold (resin cracks)	Increase	Increase

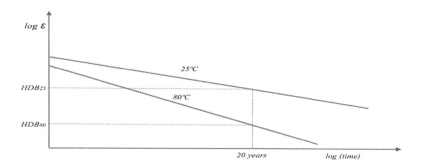

FIGURE 10.2   *The slopes of the classical static weep lines increase with the operating temperature. This unimpeachable experimental conclusion extrapolates to lower HDB's and falsely indicates poorer performance of pipes operating at higher temperatures. The weep threshold concept, on the other hand, indicates improved pipe performance at higher temperatures.*

produce lower HDB values in composite pipes. This unquestionable experimental evidence, showing unmistakable and uncontroversial lower HDBs at higher temperatures, have an enormous economic impact on the use of composite pipelines. The measured lower HDBs force the designers to increase the pipe thicknesses in high temperature services. The cost penalty associated with the thicker pipe is enormous, not only in terms of increased materials usage, but also in terms of the formidable quality control and qualification tests required at the high operating temperature.

The theory developed in this book is in disagreement with the above conclusion. First, as explained in Chapter 13, we reject the concept of extrapolated HDB. Second, as shown in Table 10.1, the weep threshold of UD plies increases, not decreases, at higher OTs. Higher weep thresholds mean increased pipe performance and safer operation at higher temperatures, in direct opposition to the classic HDB conclusion.

We have reached an interesting conceptual conflict. On the one hand, the currently accepted HDB concept points to lower pipe performance at higher operating temperatures. On the other hand, the weep threshold concept proposed in this book indicates otherwise. Conflicts like this require experimental evidence.

10.4   EXPERIMENTAL EVIDENCE ON THE WEEP THRESHOLD

In 1994, a group of French scientists published some interesting experimental results on the short-term weeping of composite pipes. In essence, the French group pressure tested several pipe specimens made with different resins at several temperatures. All tested specimens had identical diameter, thickness, and laminate construction. The resin and the temperature were the only variables. The objective of the experiment was to quantify the effects of resin toughness and operating temperature on the pipe's short-term weeping. The tested pipe specimens had no weep barrier (oil pipes), and the strains were correctly measured transverse to the fibers. It is interesting to speculate on the results of these experiments, had the French group been aware of the concept of weep threshold back in 1994. Their test design would have been simpler to carry out, and their results a lot more useful and meaningful. Unfortunately the experiments focused entirely on the short-term pipe weeping, not on the weep threshold.

The following discussion comes from the test results reported by the French group. We discuss the quasi-static loading first and the cyclic loading later. Figure 10.3 illustrates the reported test results of pipe specimens subjected to quasi-static pressurization at room temperature. A close inspection of Figure 10.3 indicates the following:

- The infiltration threshold T_i is lower for the brittle pipes, as expected.
- The weep threshold T_w is also lower for the brittle pipe, as expected.
- The short-term weep pressure, however, is higher for the brittle pipes, which is not expected.

The same experiment, conducted at higher temperatures with pipe specimens made with the same resin, gave similar results. Increments in resin toughness or in the operating temperature delay the pipe cracking and produce higher weep thresholds. This improvement is expected for both tougher resins and at higher temperatures. However, the lower short-term weep pressure of the tougher and hotter pipes remains a puzzle.

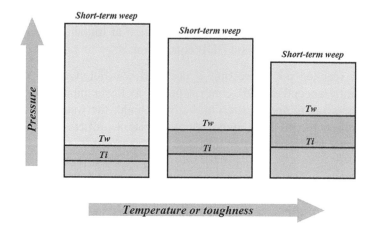

FIGURE 10.3 *The infiltration and weep thresholds are higher for the tougher pipes, as expected. The short-term weep pressure, however, is lower, which is quite puzzling.*

The point in need of explanation is … Why the pipe specimens tested at higher temperature— or made of tougher resins—weep at lower short-term pressure, while having higher infiltration and weep thresholds? This apparent paradox is easily resolved by looking into the weep mechanism at the ply level. We proceed to describe this mechanism for embedded UD plies subjected to quasi-static tensile loads. The weep mechanism for chopped fibers is discussed later.

10.5   A CLOSER LOOK AT THE WEEP FAILURE OF OIL PIPES

The following discussion applies to UD plies in oil pipes subjected to quasi-static pressure increments. The test specimens have no weep barrier and no cyclic pressure is applied.

Note 10.1: We believe a digression is required at this point to clarify an important issue related to crack growth in composite materials. Cracks grow in two ways. First, they grow from monotonic increments in the applied static load. This is the "quasi-static" loading. Second, they grow from cyclic loads. This is the "fatigue" loading. The reader should bear in mind that non-cyclic loads grow crack densities—or damage—only if monotonically increased. The present discussion is applicable to quasi-static, non-cyclic, loadings. The discussion of fatigue damage will follow shortly.

The UD plies embedded in oil pipes subjected to quasi-static pressure increments develop a multitude of small penny-shaped micro-cracks in radial planes parallel to the fibers. These micro-cracks grow in number and in size in response to pressure increments. The penny-shaped micro-cracks grow in size, in this case, because there are no fibers crossing their path either in the fiber or in the thickness directions. Eventually, the growing micro-cracks reach the plies immediately above and below, which arrest their growth in the thickness direction. The crack growth continues, however, in the fiber direction. As the pressure increases, the initially small penny-shaped micro-cracks cease their growth in the thickness direction and grow parallel to the fibers, turning into long macro-cracks. If the pressure increment ceases, the cracks stop their growth and become stationary. The cracks will resume their growth if the pressure is again increased. The formation of the first macro-crack running parallel to the UD fibers in quasi-static pressure tests marks the infiltration threshold of the ply.

Note 10.2. There are three cases of self-similar crack growth in composites. One is ply delamination, not discussed in this book. Other is the macro-crack growth along the UD fibers, described above. The growth is possible in these cases because the cracks find no fibers crossing their path. The third and last case of self-similar crack growth is laminate strain-corrosion, discussed in Chapter 14.

The length, opening, and density of macro-cracks in embedded UD plies (i.e., plies in laminates) increase with the transverse tensile load. Eventually, the growing cracks become so plentiful and so long as to cross with similar cracks in adjacent plies. Such crossing points form passages that allow the water to move from one ply to the other. That is the way crack coalescence works in UD laminates. The weep threshold is the transverse tensile strain at the onset of such coalescence. The pipes never weep if strained below this threshold. Above the threshold, the pipes certainly weep.

The measurement of the weep threshold is controversial, since the onset of crack coalescence is hard to guess. In practice, we may define the weep threshold as the strain that develops an arbitrarily chosen number of macro-cracks per unit length. The controversy lies in choosing this critical crack density. Some people might say two cracks per inch, while others would say three cracks per inch. Be it as it may, the weep threshold is a fundamental parameter in the study of pipe durability.

Note 10.3: The time to weep adopted as a failure criterion in the ASTM D 2992 B test protocol measures the water travel time in traversing pipe walls cracked beyond the weep threshold. The time to weep depends on the pipe's wall thickness, the magnitude of the transverse strain, and the crack pattern. The time to weep is not relevant in the study of the durability of composites.

We next discuss how the resin toughness, the environmental moisture and the operating temperature affect the weep time of pipes strained beyond the weep threshold. The discussion deals with the travel time of water in pipes cracked above the weep threshold, and not with the weep threshold itself. We return to the experiment performed by the French group in 1994. The tested pipe specimens had identical wall thicknesses and construction. The applied pressure was quasi-static. The number of crossover points, those where the macro-cracks in adjacent plies intersect, increases with the macro-crack density and length. Longer cracks are more likely to intersect than shorter ones. The same applies to higher crack densities, that is, the presence of many macro-cracks also lead to many crossover points. The conditions most favorable to the shortest travel times are high densities of long macro-cracks. The reader will notice we have sometimes used the term "crack" to denote the long "macro-crack" that develops along the fibers in UD plies.

Experimental evidence—backed by careful reasoning—indicates that brittle plies develop many short cracks, as opposed to the few long cracks observed in tough plies. The crucial question is which condition produces pipes with shorter weep times, few long cracks, or many short cracks? The experiments conducted by the already mentioned French group indicate that UD plies with few long cracks produce pipes of shorter weep times. A reasonable explanation for this observation is as follows:

The many short macro-cracks formed in the brittle pipe interact to form tortuous pathways ending mostly in blind alleys that stop the water flow. The advancing water accumulates in these alleys and takes a long time to find its way through the complex maze in the brittle pipe wall. The many short cracks produce long weep times in brittle pipes. By contrast, the travel time is shorter in the straight pathways formed by the few long macro-cracks present in the tough pipe. This situation is analogous to driving a car through densely populated areas, where the travel time is shorter if we take one long expressway instead of driving through many short neighborhood streets.

We complete the present discussion with a visual comparison of the crack patterns that develop in oil pipes with UD plies of different resin toughness—or at different temperatures. Figure 10.4 shows the outside surface of tough and brittle pipes, with the ply micro-cracks running in the fiber direction. We see the tough UD plies on the right with fewer and longer cracks than the brittle UD plies on the left. From our preceding discussion, the tough pipes weep sooner than the brittle ones. This, in short, is the explanation for the bizarre experimental result reported by the previously mentioned French researchers in 1994.

In conclusion, all else being equal, the tough pipes weep sooner and have regression lines that plot below those of similar brittle pipes. The same conclusion holds for pipes at higher temperatures. This explains the steeper line slopes of tougher – or hotter – pipes shown in Figures 10.2 and 10.5. However, as explained in Chapter 13, the regression weep lines are meaningless for

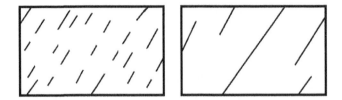

FIGURE 10.4   *Macro-cracks in UD plies. The few long cracks present in tough pipes (right) intercept many similar long cracks in adjacent plies. The opposite occurs with the many short cracks present in brittle pipes (left). The few long cracks in tough UD plies provide direct pathways for the advancing water and produce pipes with short weep times.*

long-term predictions. The true indicator of long-term weep performance is the weep threshold, which is higher for tougher resins or hotter pipes. Figure 10.5 shows the details. The real and valid weep threshold concept indicates improved pipe performance at higher temperatures, in direct conflict with the false indications of the weep regression lines. Such a compelling argument leads to an inescapable conclusion.

- The weep performance of oil pipes increases with the operating temperature, in direct contrast with the predictions of the ASTM D2992 B test protocol.

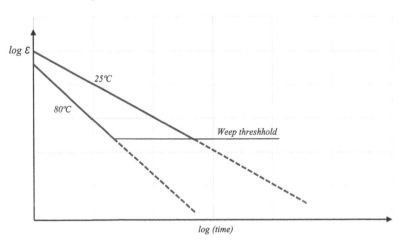

FIGURE 10.5   *The static weep lines of pipes at lower temperatures plot above those at higher temperatures. This undisputed experimental fact falsely indicates poorer pipe performance at higher operating temperatures. However, the true indicators of long-term weep performance, the weep thresholds, improve at higher temperatures. This amazing, unexpected and compelling argument suggests that pipes designed for room temperature service can safely operate at higher temperatures.*

This amazing conclusion has enormous economic significance. The use of room temperature pipes in high-temperature service would bring the following benefits:

- The expensive and difficult qualification and quality control tests performed at higher temperatures are no longer required.
- The pipes designed for 25°C can operate at higher temperatures.
- The resin heat distortion temperature (HDT) is the only limitation on the upper working temperature.

Example 10.1

Compare the fatigue crack propagation in plies and in resin castings.

Like all homogeneous materials, neat resin castings do not arrest crack growth and develop a single large crack under cyclic loads. By contrast, the fibers present in composite plies arrest the crack growth and force the development of many small cracks.

Let us compare the crack growth in resin castings of different toughness. The brittle casting cracks sooner and grows a single crack faster than the tough casting. The conclusion is as follows:

- The brittle resin casting has a lower cracking threshold and produce a larger crack than the tough casting.

Now we compare the crack growth in plies of different toughness. The brittle plies crack sooner and produce a large density of small cracks. The tough plies crack later and develop fewer cracks. The conclusion is as follows:

- The brittle ply has a lower cracking threshold and produce shorter cracks than the tough ply.

The cracking starts sooner and the crack densities grow faster in brittle than in tough plies. In the specific case of transverse loaded UD plies, in which the cracks grow along the fibers, brittle plies produce larger crack densities of shorter lengths than tough plies. This example highlights the difference between cracks evolving in castings and plies.

Example 10.2

This numerical example discusses a few weep test results reported in the literature for ±55 epoxy oil pipes.

Table 10.2 shows experimental static weep parameters As and Gs for ±55 epoxy oil pipes at three temperatures. The static slopes increase from Gs = 0.030 at 25°C to Gs = 0.102 at 82°C. A similar slope variation occurs in pipes of different resin toughness. This is in agreement with our previous discussions.

Let us compute the extrapolated HDB at 25 years and 50 years at the reported temperatures. The following equation applies:

$$log \epsilon = A_S - G_S \times log\left(hours\right)$$

For 25 years at 25°C

$$log\left(HDB\right) = -0.096 - 0.030 \times log\left(25 \times 365 \times 24\right)$$

$$HDB = 0.55\%$$

For 25 years at 65°C

$$log\left(HDB\right) = -0.032 - 0.073 \times log\left(25 \times 365 \times 24\right)$$

$$HDB = 0.38\%$$

For 25 years at 82°C

$$log\left(HDB\right) = 0.087 - 0.102 \times log\left(25 \times 365 \times 24\right)$$

$$HDB = 0.35\%$$

Similar computations lead to the HDB values at 50 years. The results are in Table 10.2.

TABLE 10.2

Temperature dependence of the static weep parameters A_s and G_s for ±55 epoxy oil pipes. The weep thresholds of epoxy UD plies are not available. The weep thresholds listed refer to ±55 polyester UD plies in the hoop direction under a 2:1 loading. The weep thresholds of epoxy UD plies should be higher. The slope increase with increments in the operating temperature is in agreement with our previous discussions. We do not question the relative positions and slopes of the regression lines. What we question is the validity of the extrapolated HDB as a weep predictor. The extrapolated HDB concept ignores the weep threshold and has no use in the analysis of pipe durability.

Temperature (°C)	A_s	G_s	HDB (25 years)	HDB (50 years)	Weep threshold (Hoop)
25	−0.096	0.030	0.55	0.54	0.80
65	−0.032	0.073	0.38	0.36	>0.80
82	0.087	0.102	0.35	0.32	>0.80

The regression lines extrapolate to lower HDB values at higher temperatures. The classical approach mandates a pipe derating for higher temperature services. However, in line with our previous discussions, the weep thresholds actually improve at higher temperatures and indicate otherwise. We do not have the weep thresholds of epoxy UD plies. Their values, however, should be higher than those reported in Table 10.2 for polyester UD plies. The weep threshold concept indicates improved long-term weep performance of composite pipes at higher operating temperatures.

10.6   FAILURE THRESHOLDS OF CHOPPED PLIES

In the preceding sections, we discussed the resin and environmental effects on the failure thresholds of UD plies. This section checks these effects on the failure thresholds of chopped plies. The quality of the fiber-resin bonding has great importance in the performance UD plies, as it controls the onset and growth of the large macro-cracks that weep the pipes. The fiber-resin bonding is not as relevant in plies of chopped fibers. The wide separation and random orientation of the fiber bundles in chopped plies minimize the effect of debond cracks in favor of resin cracks. As a result, the infiltration and weep thresholds of chopped plies are resin-dominated ply properties, with a small role played by the fiber-resin bonding. The four failure thresholds of chopped plies increase with increments in the resin toughness.

From the previous discussion, we conclude – as for UD plies – that the resin plasticization from increments in the operating temperature and moisture conditions improve all thresholds. As we did in the case of UD plies, we propose the use of the failure thresholds measured on dry chopped plies at room temperature at any operating temperature and moisture condition. The error introduced by this proposal is conservative and safe for chopped fiber and UD plies.

The discussions in this chapter have provided solid and compelling arguments supporting the use of the failure thresholds measured on dry plies at room temperature at any operating temperature and moisture condition. This is a tremendous simplification to the study of the long-term durability of composites.

10.7   WEEP TIMES OF SANITATION PIPES

The weep times of sanitation pipes with weep barriers of different resins are similar to those of oil pipes with UD plies. See Section 10.5. Figure 10.6 shows the cross section of a chopped ply in a pressurized pipe. Note the many short cracks in the brittle ply (left) in comparison with the few long cracks in the tough ply. Note also the irregular crack pattern in the brittle ply, forming many branches ending up mostly in blind alleys. This leads us to conclude that, as in oils pipes, tough sanitation pipes have lower weep pressures than brittle ones. The weep threshold, however, improves with increments in the resin toughness and at higher temperatures. The same arguments and

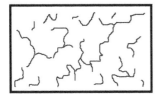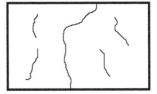

FIGURE 10.6   *Macro-cracks in chopped plies. Cross section view of two plies of chopped fibers, showing the crack patterns that develop in brittle and tough resins. The brittle resin on the left forms many short cracks that end up mostly in blind alleys. The tough resin on the right forms few long cracks that provide a direct pathway for the water.*

conclusions reached earlier for oil pipes and UD plies are also valid for sanitation pipes and chopped plies.

10.8   IMPROVING THE FAILURE THRESHOLDS

This section discusses a few ideas to improve the resin-glass bonding and the failure thresholds. The arguments develop around the concepts of interface and interphase (Figure 10.7). The interface is the glass-resin contact surface. The interphase is a thin transition layer around the fibers, created when the matrix dissolves and blends with the fiber sizing. The interphase forms a cushion around the fibers.

The interface and interphase control the onset and growth of fiber-resin cracks that govern the weep and stiffness failures of UD plies. It is amazing that such a tiny fraction of the UD plies controls the stiffness life of composite structures such as aircraft parts and wind blades. The glass-resin

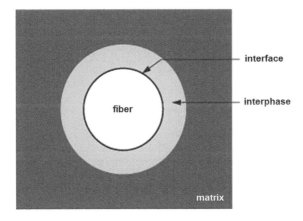

FIGURE 10.7   *The interface controls the glass-resin bonding. The interphase and resin matrix control the rate of crack growth.*

bonding strength and the resin toughness are matters of enormous commercial importance.

The following discussion is limited to UD plies statically loaded in the transverse direction. The discussion of cyclic loadings will follow later.

The total peak transverse strain drives the fiber-resin debonds and the subsequent crack growth in the UD ply. The total peak strain includes all strains, except those at the fiber level, which are part of the ply properties captured in the characterization process. See Appendix 10.3. Figure 10.8 shows the fiber-resin interphase modeled as springs, to illustrate its role in transferring loads from the resin matrix to the fiber interface.

The following arguments focus the resin-rich region A within the dotted circle, as well as the points B near the fibers. See Figure 10.1. The decisive importance of the strains at these locations is apparent. The strains within the dotted circle A control the resin crack propagation. The strains at the fiber surface B initiate and grow the fiber-resin debonds. We seek answers to the following question:

What are the best glass sizings and resin matrices for improved infiltration, stiffness, and weep thresholds in UD plies?

The answers to this apparently complex question are surprisingly simple:

- The film former present in the fiber sizing should endure high elongation and have low stiffness to avoid premature ruptures and transfer the least possible stress to the interface. To achieve this, it should be readily soluble in the resin matrix to provide a low stiffness, tough interphase. Figure 10.8 models the flexible interphase as low

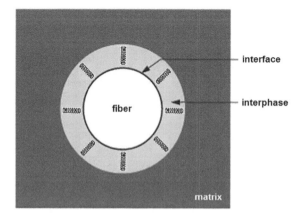

FIGURE 10.8 *The interphase modeled as springs illustrates the load transfer from the resin matrix to the fiber surface.*

stiffness springs. There is good experimental evidence support-ing this idea. Excellent fiber-resin adhesion and large infiltration thresholds have been reported for transverse loaded UD plies with extremely flexible film formers. See Refs. [19, 20, 22].

- The interface should have the best possible adhesion to the glass fibers, perhaps by increasing the silane loading or through the use of hydroxyl-rich film formers.
- The resin matrix should have the highest toughness possible to accommodate large strains before cracking and also to develop few cracks. The resin modulus should not be too low, since such a move would be detrimental to the HDT. The vinyl ester resins combine high toughness and high HDT.
- The resin matrix should also have good adhesion to the glass fibers. Excellent glass adhesion is one of the reasons for the distinct superi-ority of the hydroxyl-rich vinyl ester resins vis-à-vis polyesters.

Reducing the ply thickness and the glass loading also improve the infil-tration, stiffness, and weep thresholds of UD plies. The discussion of such refinements is beyond the scope of this book.

10.9   CYCLIC LOADINGS

Here, we finally come to the cyclic loadings. The discussion that follows is applicable to all load-dependent failure modes and, although focused on UD plies, is valid for chopped plies as well.

Figure 10.9 shows a typical static weep line defining the failure envelope of composite pipes. Note the sloped regression part of the line turning hor-izontal on reaching the weep threshold. All pressure/strains above the weep

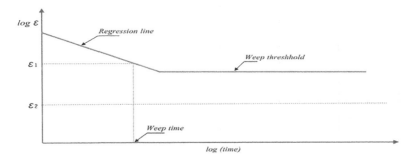

FIGURE 10.9   *Static weep envelope of typical pipes illustrating the failure times for two strains. The high strain ε_1 weeps the pipe at the time predicted by the regres-sion line. The low strain ε_2 is less than the weep threshold and never weeps the pipe.*

threshold fail the pipe in the times predicted by the sloped part of the regression line. No pressure/strain below the threshold ever weeps the pipe. As examples, consider the two static strains ϵ_1 and ϵ_2 shown in Figure 10.9. The strain ϵ_1 is higher than the weep threshold and intercepts the sloped regression line at the indicated weep time. The strain ϵ_2 is lower than the weep threshold and never intercepts the failure envelope. The avoidance of weep failure in the presence of static loadings is easy. Simply stated, if the applied static strain is less than the weep threshold, the pipe never weeps.

This situation changes in the presence of cyclic loads. The peak cyclic strain produces initial crack densities that are, of course, less than the threshold, otherwise they would cause instant failure. However, unlike the static case, the cyclic load grows the small crack densities to their critical values. As explained in Chapter 9, the regression equation to predict the number of cycles to fail the critical ply is

$$\log(\epsilon) = A_c - G_c \log(N) \tag{10.1}$$

Where ε is the peak strain and N is the number of cycles to failure.

In the previous sections, we discussed the effects of resin toughness and environmental condition on the ply failure thresholds that define the static parameters As. In this section, we will do the same thing for the cyclic regression parameters Ac and Gc. We begin with Ac.

As explained earlier, by taking N = 1 in Equation (10.1), we obtain $A_c = \log(threshold)$. The static As and cyclic Ac parameters are the same entity, both defined by the ply short-term strengths, also known as failure thresholds. The failure thresholds, as already explained, change with the resin toughness and the environmental conditions. We recall that increments in the resin toughness substantially increase the ply short-term strengths. As an example, the weep threshold of transverse loaded UD plies varies from 0.20% for brittle polyesters to 0.50% for tough vinyl esters. As for the effect of the environmental conditions on Ac, we have already shown that changes in temperature and moisture have a negligible effect on the failure thresholds. Therefore, the Ac parameters in Equation (10.1) do not change with the environmental conditions. For a list of the failure thresholds of commercial importance, see Tables 8.3–8.5.

We have the following first conclusions:

- The operating conditions (temperature and moisture) have no significant effect on Ac.
- The resin toughness has a strong and easily quantifiable effect on Ac.

We now turn our attention to the cyclic rupture slope Gc. The rupture regression line contains the short-term rupture threshold Tr and the postulated long-term fatigue threshold T_0. The slope Gc of the rupture regression line is easy to obtain from inspection of Figure 9.2:

$$G_c = \frac{log\left(T_r/T_0\right)}{logN}$$

Recalling the discussions in the Appendices 9.1 and 9.2 of Chapter 9, the pure fatigue threshold is arbitrarily set as $T_0 = 0.05\%$ when $N = 10^{20}$ for all resins and all plies. Entering these values in the above equation, we have

$$G_c = 0.065 + 0.050 \times log\left(T_r\right) \tag{10.2}$$

Equation (10.2) gives a good estimation of the rupture slope Gc of any ply of any resin. The resin toughness enter the equation by way of the rupture threshold Tr. Equation (10.2) is not exact, since we have arbitrarily set the fatigue threshold at $T_0 = 0.05\%$ for all plies and all resins.

Given that the environmental conditions have a negligible effect on the thresholds, we offer a second conclusion:

- The cyclic slopes Gc are not affected by the environmental conditions.

Back to Equation (10.2), the cyclic slope Gc is clearly affected by the ply rupture threshold Tr, which varies with the resin toughness. We have a third conclusion:

- The resin effect on the cyclic slope Gc is easily quantifiable through the rupture threshold Tr.

The above conclusions lead to the following simplifications:

1. The environmental conditions—temperature and moisture—have no effect on the cyclic regression parameters Ac and Gc. The ply regression equations measured at room temperature in a dry atmosphere are valid in all environmental conditions.
2. The resin toughness affects the rupture slope Gc as quantified in Equation (10.2).

The above simplifications are fundamental in the study of the load-dependent cyclic durability of composites. They simplify the analysis and significantly reduce the cost involved in measuring a plethora of expensive regression equations.

Example 10.3

Compute the rupture slopes of transverse loaded UD plies and chopped plies using Equation (10.2). Compare the results with those in Table 9.1.

The rupture regression slopes computed from Equation (10.2) $G_c = 0.065 + 0.050 \times log(T_f)$ are as follows:

$G_c = 0.065 + 0.050 \times log(0.60) = 0.054$ *Transverse loaded UD, vinyl ester*

$G_c = 0.065 + 0.050 \times log(0.40) = 0.045$ *Transverse loaded UD, polyester*

$G_c = 0.065 + 0.050 \times log(1.50) = 0.074$ *Chopped, polyester*

$G_c = 0.065 + 0.050 \times log(2.50) = 0.085$ *Chopped, vinyl ester*

$G_c = 0.065 + 0.050 \times log(3,00) = 0.089$ *UD ply loaded in the fiber direction*

The above numbers are slightly different from the values in Table 9.1. The reason for the discrepancy is our arbitrary definition of $T_0 = 0.05\%$ for all resins and all plies. In the absence of experimental evidence, however, Equation (10.2) is very useful, as it allows the computation of any regression line – for any ply and any resin – from the easy to measure failure thresholds.

Figure 10.10 shows the pure cyclic rupture lines of transverse loaded UD plies made with different resins. The vinyl ester line starts at Ac = log(0.60). The polyester line starts slightly below, at Ac = log(0.40). The lines meet at the fatigue threshold corresponding to $N_0 = 10^{20}$. The lines in Figure 10.10 are pure, because we have taken $N_0 = 10^{20}$ which is the number of cycles corresponding to the fatigue threshold of pure cyclic loadings.

Example 10.4

Compute the opening rate of macro-cracks in embedded UD plies under cyclic loading.

The crack magnitude of the openings in embedded UD plies derive from the ply thickness and the global laminate strain. The global strain and the ply thickness are constant. Therefore, the crack openings do not change under cyclic loadings. The macro-cracks in UD plies under cyclic loads grow in length, not in opening.

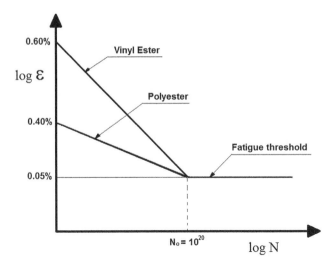

FIGURE 10.10   *Pure regression lines of transverse loaded UD plies made with different resins. Note the higher Ac = log (0.60) of the tough vinyl ester and the lower Ac = log (0.40) of the brittle polyester. The resin toughness affects both the Ac and Gc regression parameters. The two lines converge at the fatigue threshold $N_0 = 10^{20}$ and $T_0 = 0.05\%$.*

Example 10.5

The reader may be curious to see at least one laminate regression equation. There are many such equations in the literature, developed for a variety of laminates intended for several structural applications, like wind blades and others. Figure 10.11 shows two laminate regression equations, experimentally developed by the legendary John Mandel. See Ref [31].

The experiments leading to Figure 10.11 involved the pure fatigue testing of epoxy and polyester laminates of identical construction [±45/0/±45/0/±45]. The R ratio is R = 0.1, which we take as pure cyclic. We have a few comments on Figure 10.11.

As we have repeatedly warned, the testing of laminates produces specific results applicable to the tested laminate and loading. The experimental slopes of the regression lines in Figure 10.11 are specific to the tested laminate and represent neither the transverse nor the longitudinal regression slopes of UD plies.

The laminate regression equations in Figure 10.11 expressed on log-log space are as follows:

$$\log(\epsilon) = 0.784 - 0.147 logN \quad (Epoxy)$$

$$\log(\epsilon) = 0.456 - 0.128 logN \quad (Polyester)$$

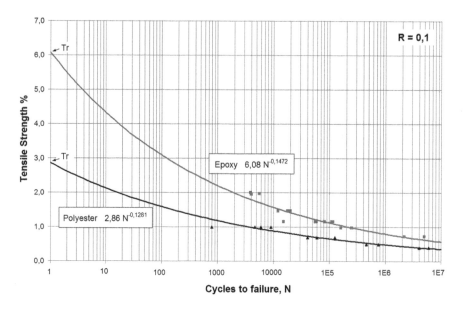

FIGURE 10.11 *Pure fatigue regression equations for epoxy and polyester laminates of equal construction. On log–log space, the two equations plot as straight lines. Drawing by Gabriel Gonzalez.*

By definition, the fatigue threshold of the laminate is the point where the above equations meet. Entering To and No in the above system, we have

$$\log\left(T_0\right) = 0.784 - 0.147 logN_0$$

$$\log\left(T_0\right) = 0.456 - 0.128 logN_0$$

Solving the above, we have:

No = 10^{20}

To = 0.007%

These coordinates define the fatigue threshold of the tested laminate. This is the only fatigue threshold I could find in the literature for any laminate or ply. The reader will note the discrepancy between the pure fatigue threshold measured on the specific laminate of this example and the proposed ply values in Table 9.2. The laminate measured fatigue threshold, To = 0.007%, is less than the proposed To = 0.050%. This is a surprise. In the long-term, the cyclic regression lines of all plies and laminates, regardless of the resin matrix, should converge to approximately the same fatigue threshold.

For lack of better data, and for consistency, we assume the pure fatigue threshold listed in Table 9.2 as applicable to all laminates and all plies, regardless of the resin matrix. Having the fatigue threshold and the short-term strength, we can estimate the pure regression rupture slope of any laminate as

$$G_c = \frac{\log\left(T_r/T_0\right)}{\log N_0}$$

Where Tr is the laminate short-term rupture strength. The values T_0 and N_0 are, of course, the coordinates of the laminate pure fatigue threshold. Taking $T_0 = 0.05\%$ and $N_0 = 10^{20}$, we have

$$G_c = 0.065 + 0.050 \times \log\left(T_r\right) \tag{10.3}$$

Equation 10.3 gives the pure cyclic rupture slope Gc of the laminate tested by Prof. Mandell for any resin. Equation 10.3 is the same as Equation (10.2), which gives the regression slope of plies. The reason for this equality lies is the assumption that all plies and laminates converge to the same fatigue threshold. If we were to reject this convergence and adopt the fatigue threshold measured for the laminate, the slope would be

$$G_c = 0.108 + 0.050 \times \log\left(T_r\right) \tag{10.3A}$$

Example 10.6

Use Equation 10.3A to compute the pure regression rupture slopes of the laminates in Example 10.5.

For the epoxy laminates with Tr = 6.08%:

$$G_c = 0.108 + 0.050 \times \log\left(T_r\right)$$

$$G_c = 0.108 + 0.050 \times 0.784 = 0.147$$

For the polyester laminates with Tr = 2.86%:

$$G_c = 0.108 + 0.050 \times \log\left(T_r\right)$$

$$G_c = 0.108 + 0.050 \times 0.456 = 0.130$$

The power of Equation (10.3A) is apparent when we wish to simulate the effect of different resins on the fatigue durability of laminates. The only parameter needed to do a reasonably accurate computation of the regression line is the easy to measure short-term laminate rupture strength Tr. Let us illustrate this by assuming a resin for which the measured laminate short-term strength is Tr = 3.55%.

From Equation 10.3A the pure regression slope is

$$G_c = 0.108 + 0.050 \times log\left(T_r\right)$$

$$G_c = 0.108 + 0.050 \times 0.550 = 0.136$$

The laminate cyclic regression equation for this specific resin is

$$log\epsilon = A_C - G_C \times log\left(N\right)$$

$$log\epsilon = log\left(3.55\right) - 0.136 \times log\left(N\right)$$

$$log\epsilon = 0.550 - 0.136 \times log\left(N\right)$$

The reader will appreciate the significance of this numerical example and the far-reaching capability of Equations 10.3 and 10.3A. We have derived the regression equation of a laminate simply by measuring its short-term tensile strength.

Example 10.7

Table 10.2 reports the measured static weep regression parameters As and Gs of epoxy±55 oil pipes subjected to 2:1 loadings. From these parameters, we write the weep regression equations (to predict the weep times) at the tested temperatures as follows:

$$log\epsilon = -0.096 - 0.030 \times log\left(hours\right) \quad \left(25C\right)$$

$$log\epsilon = -0.032 - 0.073 \times log\left(hours\right) \quad \left(65C\right)$$

$$log\epsilon = +0.087 - 0.102 \times log\left(hours\right) \quad \left(82C\right)$$

The above equations indicate the intercepts (thresholds) varying from As = − 0.096 at 25°C to As = + 0.087 at 82°C. The weep tests on oil pipes indicate a tremendous effect of the operating temperature on the regression parameter As, in direct contradiction with the conclusions in this chapter. Let us explain this contradiction.

There is no contradiction. The above weep regression equations refer to the water travel time through cracked pipes. This travel time has nothing to do with the onset of weeping and the weep threshold. The weep equations predict the travel times of water in cracked pipes and have nothing to do with the ply regression lines. There is no contradiction.

APPENDIX 10.1

CRACKS IN BRITTLE PLIES

The macro-crack density is different in embedded as opposed to isolated UD plies. For one thing, the isolated plies rupture at the formation of the first transverse macro-crack. The embedded plies, on the other hand, do not break after the first crack and develop multiple cracking. In embedded UD plies, the macro-cracks take an elongated elliptical shape and grow parallel to the fibers.

The crack growth in the fiber direction does not go on forever. There are at least two mechanisms arresting their growth:

- First, there is a decrease in the stress intensity factor K_I as the crack length increases. For details, see any book on fracture mechanics. Eventually, the UD macro-cracks become too long in comparison with the ply thickness and this brings the value of K_I close to zero. The low stress intensity factor K_I stops the crack growth.
- Second, there is the mutual shielding of the crack tips approaching each other from opposite directions. As the crack tips approach each other, their high stress fields first combine to accelerate the crack growth. Later, once the tips pass each other, the stress field drops and the crack growth stops. The shielding effect is clear in Figure 10.12.

Figure 10.12 illustrates the arrest process observed on a trunk of eucalyptus tree left for years to dry in the sun. Observe the arrest of the approaching cracks close enough to have their tips shielded. The shielding effect from the high crack density present in brittle plies prevents the development of long cracks. The opposite situation occurs in tough plies. Of course, the same reasoning applies to different temperatures. This argument explains why brittle and cold plies develop many short cracks, when compared to tough and hot ones.

FIGURE 10.12   *The crack growth stops as they approach each other from opposite directions. The photo shows macro-cracks on a tree trunk left to dry in the sun.*

APPENDIX 10.2

THE RESIN-GLASS INTERPHASE

The resin matrix dissolves the glass sizing forming an interphase layer around the fibers. The interphase is a blend of the fiber sizing with the resin matrix. All phenomena involving the immediate glass-resin vicinity, such as fiber blooming and debonding, involve the interphase, not the resin matrix itself. As an example, the chemical attack on the glass fibers depends on the solubility of the aggressor product in the interphase, not in the resin. Furthermore, the interphase controls the fiber-resin debonds that trigger the infiltration, weep, and stiffness failures.

The interphase has never been properly studied and characterized. Perhaps the failure concepts introduced in Chapter 8, together with the laminate strain-corrosion discussed in Chapter 14, will change this situation.

APPENDIX 10.3

MICRO-STRAINS AT THE FIBER LEVEL

This appendix explains the reason for excluding the micro-strains at the fiber level from the analysis of laminate durability. The residual micro-strains at the fiber level derive from:

- Resin cure shrinkage
- Differential resin-fiber thermal shrinkage
- Resin expansion from water intake, counteracting the thermal and cure shrinkages.

A careful analysis of the above causes indicates equal values for the residual micro-strains in all plies made with the same resin. Different resins produce different micro-strains, but the same resin produces the same micro-strains. The process of ply characterization automatically captures these residual micro-strains, which form a uniform and fixed baseline that is always the same for the same resin. The residual micro-strains shift the failure thresholds by the same amount and for that reason are not a concern. This is a welcome simplification, since these micro-strains are too complex and not easy to compute.

11 Durability of the Corrosion Barrier

11.1 INTRODUCTION

The engineering community is well aware of the long service lives of composite equipment in a variety of harsh environments. It would take no more than a brief tour of any modern pulp bleaching or chlorine–alkali facility to realize the importance of composites in services where metal corrosion is a major issue. However, after many years of good service, there is still no reliable method to predict the durability of the corrosion barriers in such cases. The many studies and test methods developed over the years to evaluate laminates immersed in aggressive chemicals, while certainly useful to establish suitability, are not adequate to quantify the durability. This is understandable in view of the overwhelming complexity of the problem. The many combinations of resin types, working temperatures and other process variables, coupled with a myriad of aggressive chemicals make the prediction of laminate durability in aggressive environments a most daunting endeavor.

The concept of infiltration threshold, detailed in Chapter 8, is very useful in the analysis of laminates in chemical services. By definition, the chemical attack on the corrosion barrier is load-independent, as long as the operating strain is less than the infiltration threshold. This load-independence eliminates the strain as a variable, and allows the determination of the corrosion parameters by simple immersion tests. This is a tremendous simplification. In this chapter, we introduce a simple method to predict the durability of corrosion barriers in chemical services.

11.2 TWO TYPES OF AGGRESSIVE CHEMICALS

We define as aggressive all chemicals that penetrate the laminates and lessen their performance. The damaging potential of the aggressive chemicals derives from their (a) ability to penetrate the laminates, (b) solubilize in the resin, and (c) interact with the fibers and the resin matrix. Some chemical species have low penetrating power and do damage mostly on the equipment surfaces. Others pervade the entire laminate and interact with all plies. This chapter classifies the chemicals according to their ability to react with the resin.

- **Reactive:** This group includes most chemicals that are in industrial use. The resin matrix captures the reactive species and considerably slows down their rate of advance. The reactive chemicals take a long time to traverse the corrosion barrier and, hence we call them non-penetrating. The damage from the non-penetrating chemicals is limited to the plies near the exposed surface of the laminate. All reactive chemicals, such as acids and oxidizers fall in this category.
- **Non-reactive:** This group includes the chemicals that do not interact chemically with the resin matrix. The absence of chemical interaction allows their easy penetration in the entire laminate. The molecular size, shape, and polarity of non-reactive chemicals control their penetration rate in polymers. There is no chemical reaction involved in this penetration, which is essentially friction dependent. The most important non-reactive chemicals are water and solvents that, as a rule, have high penetration rate and saturate the entire laminate in a short time. Being a short-term process, the damage from the penetrating chemicals—with the exception of water—is outside the scope of this book. The reader will remember that the penetrating chemicals are the only ones capable of reaching all fibers in all plies.

The terms reactive, aggressive, and non-penetrating are synonymous when referring to chemicals in contact with composites. Likewise, the terms non-reactive, non-aggressive, and penetrating chemicals are interchangeable in every situation.

Figure 11.1 shows the concentration profiles of three non-penetrating chemicals in a typical laminate designed for corrosion service. The chemical concentrations—marked on the vertical axis—are larger in the corrosion barrier than in the liner. This is because of chemical accumulation in voids and at the resin–glass interphase present in the corrosion barrier, and absent in the neat resin liner. The chemical concentration in the resin itself is the same both in the liner and in the corrosion barrier. The reader will understand that the excess chemical accumulates at the glass–resin interphase and in voids, not in the resin itself. The following five points are clear from Figure 11.1:

1. The interaction with the resin slows down the chemical penetrating rate and flattens the advancing front. The flatness of the advancing front is a feature of all reactive chemicals in contact with composite laminates. Other factors being equal, the higher the reactivity of the penetrating species, the flatter is the advancing front. For example, the highly reactive chlorine dioxide displays a very neat and flat front when advancing in the corrosion barrier.

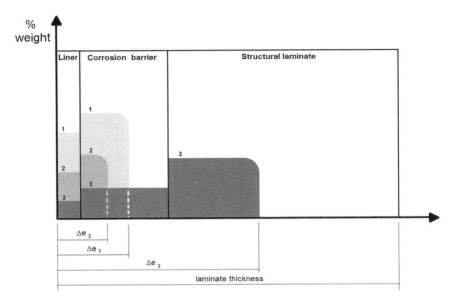

FIGURE 11.1   *The flat advancing fronts of reactive chemicals allow the accurate measurement of the penetrated depth* Δe. *The liner retards the chemical advance with no effect on the concentration itself. The durability analysis of the corrosion barrier ignores the liner presence.*

2. The liner has no effect on the chemical concentration in the corrosion barrier. The only benefit from the liner presence is the slowing down, or retardation, of the penetration rate. In this book, the estimation of the durability ignores the liner presence. For further details, see the work of Regester in Chapter 8.

3. The chemical damage to the corrosion barrier depends on (a) the operating temperature, (b) the chemical concentration, (c) the penetrated depth, and (d) the resin–chemical interaction.

4. The rate of penetration of a reactive chemical does not relate to its solubility, or concentration, in the resin. The concentration profile of product 3 indicates a low solubility and a high rate of penetration.

5. The concentration profile of product 3 indicates a partial penetration of the structural plies. This situation is dangerous because the design of the structural plies contemplates stiffness, not the slowing down of chemical penetration or resistance to chemical attack. The only chemicals allowed in the structural plies are the non-reactive solvents and, of course, water.

The damage from aggressive products depends on their chemical concentration in the penetrated laminate, as well as on the penetrated depth. The concentration, of course, varies with the solubility of the chemical in the resin. The penetrated depth varies with the exposure time, the operating temperature, and other variables. Figure 11.1 shows the concentration profiles and the penetrated depths of three products in a standard laminate at a given time t. Given enough time, all aggressive chemicals advance past the corrosion barrier and reach the structural plies. When this happens, the maintenance crew shuts down the equipment and takes action to replace the penetrated corrosion barrier. The time for the advancing front to reach the structural plies defines the durability of the corrosion barrier and the service life of the equipment. The maintenance engineer carefully monitors the advancing front, replacing the penetrated corrosion barrier with a fresh one to make sure that no aggressive chemical ever reaches the structural plies. In applications where this is not possible, as in small diameter pipes, the chemicals would continue their advance and do serious damage to the structural plies. This is the only instance of non-penetrating chemicals causing premature rupture failure of laminates. As a rule, the fresh corrosion barrier introduced in the relining operation keeps the aggressive chemicals away from the structural plies to assure they have no effect on the long-term rupture failure of the equipment.

Example 11.1

The 5.0% solution of sulfuric acid present in urban sewage is not damaging to polyesters. The liner and the corrosion barrier of sanitation pipes are fully preserved. How could the strain-corrosion process occur in sanitation pipes?

The full preservation of the liner and the weep barrier depresses the acid concentration in the laminate and protects the structural plies. This statement is correct as long as the weep barrier is not cracked. However, should the pipe be strained above the weep threshold, the 5.0% acid solution would gain direct access to the structural UD plies and cause fast strain-corrosion rupture. The sanitation pipes operating below the weep threshold do not fail from strain-corrosion. For details, see Chapter 14.

The rate of chemical penetration depends on its concentration, diffusivity, and reactivity in the corrosion barrier. Some resins produce equipment with long service lives not because of their inherent chemical resistance, but from their low solubility and permeability to the aggressive chemicals. This explains the superior performance of highly cross-linked resins and the outstanding effect of post-cure on the durability of corrosion barriers.

11.3 NON-REACTIVE CHEMICALS

The non-reactive species, such as water and solvents, penetrate the laminates by squeezing their way into the openings between resin molecules. This friction-controlled diffusion is relatively fast and saturates the entire laminate in a short time.

Example 11.2

Compute the solubility of ethanol in resin castings in the following scenarios:

1. Resin casting immersed in 100% pure ethanol.

   The ethanol saturation in resin castings immersed in 100% solutions at room temperature is 14.0%. The solubility of ethanol in resin at room temperature from a 100% pure alcohol solution is:

   $$solubility = \frac{ethanol\ concentration\ in\ the\ resin}{ethanol\ concentration\ in\ the\ solution} = \frac{14\%}{100\%} = 0.14$$

2. Resin casting immersed in a 50% aqueous ethanol solution.

   The ethanol saturation in resin castings immersed in 50% solutions at room temperature is 6.0%. The ethanol solubility in resin from a 50% aqueous solution at room temperature is:

   $$solubility = \frac{ethanol\ concentration\ in\ the\ resin}{ethanol\ concentration\ in\ the\ solution} = \frac{6\%}{50\%} = 0.12$$

   This example illustrates the solubility of chemical species in resin matrices. The work of Regester, discussed in Chapter 8, illustrates the low solubility of sulfuric acid in bisphenol A polyester resins.

Even if non-reactive, the penetrated solvents degrade the structural capability of laminates, especially at high operating temperature. The degradation from some solvents is so high as to preclude the use of composites in structural services, even at room temperature. This is the case of ethanol. The mechanical degradation from the 14.0% absorption of ethanol weakens the laminates and limits their use to non-structural, room temperature services, as in underground storage tanks. Composite materials do not work well in pressurized or high-temperature ethanol services. The recently developed impermeable pipes (Chapter 20) do not allow the ingress of solvents and neutralize their effect. The impermeable pipes open new opportunities for composites in solvent services at high pressure and high temperature.

Water is a very special solvent. First, it has a small molecule. Second, its low solubility (1.0%) in the resin has no relevant effect on the laminate properties. The small water molecules penetrate all plies and have a presence everywhere at all times both in the resin matrix and at the glass–resin interface. The water presence is unavoidable. Even laminates operating in dry atmospheres are loaded with significant amounts of residual moisture carried in the resin and the glass from their manufacturing processes. Water is, therefore, a penetrating species that can reach all plies and is always present in all composite laminates.

The effect of water on the long-term laminate durability is detailed in Chapters 10, 12, and 20. Chapter 10 argues that water has no effect on the weep, infiltration, and stiffness durability. Chapter 12 discusses the water effect on the fiber strain-corrosion and rupture durability. Chapter 20 describes the anomalous failure of sand-cored composite pipes in water service.

The laminate damage from solvents other than water is a short-term phenomenon. This book deals only with long-term failures, which excludes the action of solvents other than water. Example 11.3 summarizes all we have to say about solvents and their effect on durability:

Example 11.3

Describe the durability of composites in solvent services.

The non-reactive chemicals, such as water and solvents, can easily diffuse into the resin matrix. Their penetration is fast and they soon reach all plies. The penetrated solvents plasticize the resin and lower their heat distortion temperature (HDT). As a simple example, we mention the modest water pickup of 1.0%, which is enough to reduce the resin HDT by a significant 10°C. The penetrated solvents also reduce the mechanical properties of the laminate. This reduction may range from moderate and acceptable, as in the case of water, to excessive and unacceptable.

The applicability of composites depends on the solvent solubility in resin castings at room temperature. The decision to allow or deny direct contact of solvents with composites requires nothing more than a simple resin casting short-term immersion test at room temperature. This is not a long-term issue and therefore is outside the scope of this book. The recent introduction of the impermeable pipe technology (Chapter 20) has completely solved the composite–solvent issue, which is no longer a concern.

Let us focus on the water effect. Water can be harmful to laminates in two ways. First, like all solvents, it lowers the resin HDT and has a slight effect on the mechanical properties, as explained. Second, it involves a continuous and sustained strain-corrosion attack on the glass fibers, which leads to long-term laminate rupture. Water is the only known chemical capable of (a) fully penetrating the laminate and (b) strain-corroding the glass fibers in all plies.

The rest of this chapter deals with the durability of corrosion barriers in reactive chemicals, the discussions that follow focus exclusively on reactive chemicals and do not involve solvents.

11.4   DURABILITY OF THE CORROSION BARRIER

The chemical durability of any laminate derives from the rate of penetration of the aggressive chemical in the critical corrosion barrier. The chemical interaction slows down the diffusion rate of reactive species in polymers. As a result, Fick's equations describing the penetration of non-reactive water or solvents are not valid for reactive chemicals. In fact, there is no model to describe the rate of advance of reactive chemicals in corrosion barriers. The best we can do in such cases is to recognize the flatness of the advancing front and describe the penetrated depth Δe empirically as a function of corrosion parameters. The following equation describes the penetrated depth Δe:

$$\log(\Delta e) = A + B\log(c) + C\log(t) + \frac{D}{T}$$

The above equation computes the penetrated depth Δe of reactive chemicals in corrosion barriers. Section 11.5 explains the parameters and variables in the above equation. The important fact to keep in mind at this point is the requirement of a low rate of penetration of reactive chemicals in corrosion barriers. This is an obvious requirement. Had it been otherwise, the reactive chemicals would quickly penetrate and destroy the laminate, and the composite equipment would not be a viable solution. The long service lives of corrosion barriers in highly aggressive chemical services attest to the slow nature of this penetration process.

11.5 MEASURING THE PENETRATED DEPTH Δe

In non-abrasive services, the penetrated depth Δe remains on the laminate surface as a shield that slows down the chemical advance. This shielding effect is very important in protecting the corrosion barrier. The penetrated depth of the corrosion barrier loses its mechanical properties and is not relevant in structural design. The standard procedure is to ignore and exclude the entire corrosion barrier in all computations involving aggressive chemicals. This is so even if we know that the non-penetrated part of the corrosion barrier, as well as the structural laminate, retain their pristine conditions. (Figure 11.2).

A good way to measure the penetrated depth Δe is by optical means on cut edges of immersed laminates. To highlight the penetrated depth, the cut edges are polished and rubbed with a contrasting dye. This may require highly contrasting dyes for proper detection (Figure 11.3).

Example 11.4

Discuss and compare the corrosion advance in metals with the deterioration of the corrosion barrier in composites.

The general corrosion of metals offers good examples of non-penetrating processes. The impermeable metallic lattice prevents the chemical penetration and

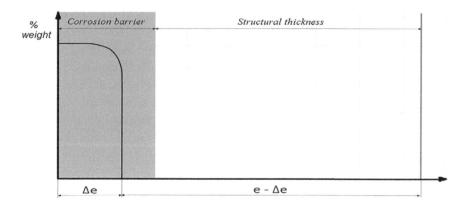

FIGURE 11.2 *The aggressive chemical penetrates slowly in the corrosion barrier. The penetrated depth Δe loses its structural capability. The non-penetrated part of the laminate, represented as e – Δe, is not affected by the chemical presence.*

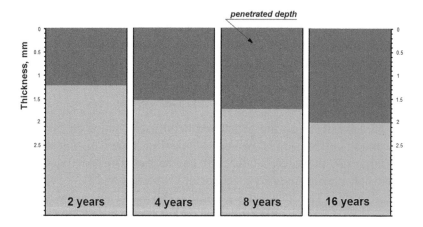

FIGURE 11.3 *Cross-section of a laminate exposed to chemical services over 16 years. Observe the non-linear advance of the aggressive chemical. The flat advancing front is clearly visible. (Drawing by Gabriel Gonzalez.)*

the lost metal thickness measures the degradation. The corroded and removed metal ceases to play any structural role and the remaining metal keeps its properties intact. Metallic structures lose load capacity in corrosive environments from loss of thickness.

A similar process occurs in composites, except the corroded/penetrated depth remains on the laminate surface. The penetrated depth loses mechanical properties and, as in metals, has no load-bearing contribution. The penetrated depth remains in place, providing an important shield that slows down the chemical advance. The non-penetrated part of the corrosion barrier keeps its properties intact.

11.6   MEASURING THE CORROSION PARAMETERS

The corrosion parameters A, B, C and D in Equation (11.1) compute the depth of the corrosion barrier penetrated by the chemical. Their measurement involves the immersion of representative laminates in the test fluid and annotating the penetrated depth Δe as a function of the exposure time t, the chemical concentration c, and the temperature T. The least squares method applied to the collected data produce a statistical estimation of the corrosion parameters A, B, C, and D. This is a typical multiple linear regression problem in which the mean value of the dependent variable Δe is estimated as a function of three independent variables, concentration, temperature, and time. This quantitative method to predict the durability of the corrosion barrier is a major advance from the current qualitative ASTM C 581 test method. The quantitative multiple regression test method described above differs from the qualitative ASTM C581 in the following:

- The test coupons used in the new method have no liner. The liner is an "add-on" feature that may give a small boost to the service life, but guarantees no long-term protection. It is safer to ignore the liner as a predictor of service life.
- The new method is quantitative and computes the penetrated depth Δe.
- The Barcol hardness and the visual appearance of the immersed coupons are irrelevant in the new method.

Table 11.1 summarizes these differences.

11.7   SIMPLIFIED EQUATIONS

The complete equation to predict the penetrated depth Δe in the corrosion barrier is:

$$\log(\Delta e) = A + B\log(c) + C\log(t) + \frac{D}{T} + E\varepsilon \tag{11.1}$$

Where
  Δe is the penetrated depth
  A, B, C, D, and E are experimental corrosion parameters
  c is the concentration of the chemical solution
  t is the time variable
  T is the absolute temperature
  ε is the strain, or elongation, in the corrosion barrier.

TABLE 11.1

Although similar in execution, the two test methods are fundamentally different in meaning. The immersion test proposed in ASTM C 581 method is a good estimator of resin suitability—not a predictor of durability.

	ASTM C 581 test method	Proposed test method
Scope	Evaluates suitability	Predicts durability
Criteria	Hardness	Penetrated depth Δe
	Coupon appearance	
	Medium appearance	
	Retention of flexural strength	
Significance	Suggests suitability	Predicts durability
Coupons	Liner on both surfaces	Liners are not used
	Two 450 g/m² mats	Coupons made of chopped glass
Procedure	Immersion of coupons	Immersion of coupons
	Specimens tested for flexural strength at regular intervals	Penetrated depth "Δe" measured at regular intervals
Calculation	Tabulates/constructs graph of hardness as a function of time	The penetrated depth Δe depends on the exposure time, the chemical concentration
	Tabulates/constructs graph of flexural strength as a function of time	and the operating temperature using a multiple regression technique
Interpretation	Resin is suitable if flexural strength and hardness level off	The regression equation predicts the penetrated depth and the service life
	Other indicators are the visual appearances of the laminate and media	Visual appearances and hardness are irrelevant

Since all corrosion barriers operate below the infiltration threshold, the strain parameter $E = 0$. With this simplification, the complete equation reduces to

$$\log(\Delta e) = A + B\log(c) + C\log(t) + \frac{D}{T} \tag{11.2}$$

The laboratory tests to measure the parameters A, B, C, and D in Equation (11.2) are too complex. A possible simplification would be testing for the most aggressive concentration of every chemical. This conservative approach would save a lot of testing effort. Such a simplification would eliminate the concentration variable c from Equation (11.2) and lead to

$$\log(\Delta e) = A + C\log(t) + \frac{D}{T} \tag{11.3}$$

A further simplification is still possible, by testing every chemical at its maximum commercial operating temperature. This would eliminate the temperature variable T from Equation (11.3) and lead to

$$\log(\Delta e) = A + C \log(t) \tag{11.4}$$

The simplified Equation (11.4) finds application in services where the chemical concentration and the operating temperature never change, as in the storage of ClO_2 and in the computation of the residual life of equipment aged after long service exposure. For details on the residual life of aged equipment, see Appendix 11.3.

The ultimate simplification to the general Equation (11.1) occurs in abrasive services that removes the penetrated material from the laminate surface and eliminates their shielding effect. In such cases, Equation (11.1) takes a simple linear relationship:

$$\Delta e = A \times t$$

11.8   ESTIMATING THE SERVICE LIFE

The actual experimental values of the corrosion parameters A, B, C, and D, are not available. However, to illustrate their application, we will assume the following values:

$$A = -0.219 \quad B = +0.206 \quad C = +0.321 \quad D = -97.760 \quad E = 0.000$$

The above values are valid for Δe in millimeters and t in months. The strain parameter was taken as $E = 0$, since the operating strain is less than the infiltration threshold. Let us use Equation (11.2) to compute (a) the service life of a standard 2.0 mm corrosion barrier and (b) the required thickness of a corrosion barrier to last a specified time.

Example 11.5

Estimate the service life of a standard 2.0 mm corrosion barrier. The equipment is shut down for maintenance when the penetrated depth reaches the value $\Delta e = 2.0$ mm. The service life is obtained by entering $\Delta e = 2.0$ mm and the above corrosion parameters in Equation (11.2). The service life is estimated for a chemical of concentration c = 0.5. The operating temperatures are 80°C and 27°C.

For the operating temperature 27°C

$$\log(2.0) = -0.219 + 0.206 \times \log(0.5) + 0.321 \times \log(t) - \frac{97.76}{273 + 27}$$

Solving for the time t, we obtain $t \approx 673$ months (56 years) at the operating temperature 27°C:

For the operating temperature 80°C

$$\log(2.0) = -0.219 + 0.206 \times \log(0.5) + 0.321 \times \log(t) - \frac{97.6}{273 + 80}$$

Solving for the time t, we obtain $t \approx 472$ months (39 years) at the operating temperature 80°C.

Example 11.6

In this example, we design a special corrosion barrier.

Our task is to determine the thickness of a corrosion barrier to last 25 years operating at 90°C in a chemical with concentration c = 0.3. Per Equation (11.2), the penetrated depth after 25 years at 90°C is

$$\log(\Delta e) = -0.219 + 0.206 \log(0.3) + 0.321 \log(25 \times 12) - \frac{97.76}{273 + 90}$$

$$\Delta e = 1.58 \, \text{mm}$$

The strain parameter was taken as E = 0, since the tensile strain in the corrosion barrier is less than the infiltration threshold.

Example 11.7

In 2001, Mark Greenwood published a remarkable paper reporting extensive creep-rupture tests on pultruded rods immersed in several aggressive chemicals. The objective of the study was to compare the performance of two glass compositions on the long-term rupture lives of UD plies. The tested rods were identical in every respect except for the different glass compositions. This worked example shows our interpretation of the test results.

The rod specimens were subjected to constant tension while immersed in the corrosive fluid. It must be realized that creep tests performed in corrosive media combine the effects of two distinct modes of attack. For meaningful results, the investigator should judiciously separate the effects of the penetrating water from those of the non-penetrating chemical. Mark Greenwood, in his 2001 paper, did not separate these effects. Instead, his conclusions were based on the classical assumption that the reactive chemicals fully penetrated and degraded the entire rod. The classical assumption of full penetration is not correct. As we have seen, the reactive chemicals have their effect limited to a small surface depth Δe.

We make use the published data to determine the penetrated depth Δe. The radius of the rods remained unchanged throughout the test. This leads to two equations

$$\sigma_r^0 = \frac{F_r^0}{\pi R^2} \quad \sigma_r^t = \frac{F_r^t}{\pi R^2} \tag{11.5}$$

Where σ_r^0 and σ_r^t are the tensile strengths of the rods prior to and after immersion, respectively. Note the rod radius R remains unchanged, and the tensile strength σ_r^t drops in the same proportion as that of the measured rupture force F_r^t. This is the classical interpretation.

Our interpretation is quite different. We say that the drop in tensile strength occurs in response to (a) a decrease in the structural radius of the rod equal to the penetrated depth and (b) the water strain-corrosion of the core fibers. The water damage to the core fibers is independent of the chemical damage to the outer fibers. The separation of these simultaneous effects involves an elaborate analysis not shown here. The separation of the two effects produces Equation (11.6):

$$\Delta e = R \left\{ 1 - \left[(retention) \times \left(\frac{t_r}{t_0} \right)^{K_s} \right]^{\frac{1}{2}} \right\} \tag{11.6}$$

Where

Δe is the penetrated depth
R = 3,175 mm is the original radius of the rod
Retention is the reported tensile strength retention for the rods at the rupture
   time t_r
t_r is the reported time to rupture
t_0 = 0.1 hour is the time to rupture in short term tensile tests
K_s = 0.0854 is the slope of the static regression line for the rods made of E
   glass in water
K_s = 0.0654 is the slope of the static regression line for the rods made of
   boron-free glass in water.

Equation (11.6) computes the penetrated thickness Δe from the measured strength "retention." We illustrate its use for rods tested in acid. According to Mark Greenwood, the strength retentions at 50 years for rods creep tested in acid are 0.90% and 12.10% for E glass and boron-free glass, respectively. The time to rupture in Equation (11.6) is 50 years, or $t_r = 438,000$ hours.

For the E glass rods in acid, the penetrated depth from Equation (11.6) is

$$\Delta e = 3.175 \left\{ 1 - \left[(0.009) \times \left(\frac{438000}{0.1} \right)^{0.0854} \right]^{\frac{1}{2}} \right\} = 2.60mm$$

For the boron-free glass rods in acid, the penetrated depth from Equation (11.6) is

$$\Delta e = 3.175 \left\{ 1 - \left[(0.121) \times \left(\frac{438000}{0.1} \right)^{0.0654} \right]^{\frac{1}{2}} \right\} = 1.35mm$$

As shown, the penetrated depth Δe is small considering the long exposure time of 50 years. The penetration on the E glass rods is 2.60 mm after 50 years in acid, compared to 1.35 mm for the boron-free rods. For equal performance after 50 years, the rods of E glass should have diameters $2 \times (2.60 - 1.35) = 2.50$ mm larger than the rods of boron-free glass. This small difference in the penetrated depths suggests that the glass composition is not a driving factor in the durability of corrosion barriers. Furthermore, the tested rods were made exclusively of UD fibers with no corrosion barrier to slow down the chemical penetration.

The classical interpretation of this test leads to the scary conclusion that the rods of E glass, regardless of diameter, lose all their strength (retention of 0.90%) after 50 years of immersion in acid. This conclusion is so unlikely that Mark Greenwood interjected an apologetic comment in his 2001 paper:

"Considering the relatively good performance of isophthalic polyester laminates reinforced with traditional E glass, the results from this study appear to be excessively severe. However, the manufacturing method typically used for storage tanks that would hold a strong acid uses a liner or barrier layer that would significantly delay the adverse effects of the acid on the structural layers of the chemical storage tank."

The above comment is implicit recognition of chemical attack limited to the surface plies. The inner plies remain intact. Table 11.2 summarizes our interpretation of the test results for other chemicals reported in Ref. [2] for other chemicals.

This chapter introduced a new method to quantify the durability of corrosion barriers. The basic assumptions underlying the new method lead to the following conclusions:

1. The reactive, non-penetrating chemicals, determine the durability of the corrosion barrier, independent of the applied loading.
2. The ASTM C 581 test method is good to assess resin suitability, not the durability of composites in corrosive environments.
3. The durability of boron-free and regular E glass in corrosive environments are essentially the same. The glass composition has a strong effect on fiber strain-corrosion and the rupture life, not on the corrosion durability.
4. The resin matrix governs the chemical durability of the corrosion barrier.

TABLE 11.2

Comparison of stress-rupture test results on pultruded rods of e glass and boron-free glass. The penetrated depth Δe is small in both cases, given the time of 50 years and the absence of corrosion barriers.

Medium at 23°C	Type of Glass	Strength Retention at 50 Years	Δe at 50 Years	Comments
5% salt solution	E glass	0.271	zero	The *hydrolytic* stability of of boron-free glass is superior to that of E glass
	Boron-free glass	0.368	zero	
1.0N HCl	E glass	0.009	2.60 mm	After a 50 years *immersion* in acid, the rods of E glass lose (2.60 – 1.35 = 1.45 mm) more thickness than those of boron-free glass
	Boron-free glass	0.121	1.35 mm	
Cement extract	E glass	0.148	0.83 mm	After 50 years *immersed* in cement extract, the rods of E glass lose (0.83 – 0.57 = 0.26 mm) more thickness than those of boron-free glass
	Boron-free glass	0.248	0.57 mm	

APPENDIX 11.1

DERIVING EQUATION (11.1)

This appendix derives Equation (11.1). The derivation will take into consideration the concentration of the chemical in the solution, the exposure time, the total strain, and the operating temperature. The rate of penetration increases with the concentration (c), and decreases with the reactivity (r) of the chemical. Furthermore, it decreases with the shielding effect from the penetrated material that remains on the laminate surface. These considerations lead to the differential equation:

$$\frac{d(\Delta e)}{dt} = k \frac{D \times S \times c^a}{r \times (\Delta e)^s} \tag{11.7}$$

Where
  Δe is the penetrated depth.
  t is the time of exposure
  k is a factor to accommodate the units of time, thickness, etc.
  D is the coefficient of diffusion of the chemical into the corrosion barrier
  S is the solubility of the chemical in the resin
  c is the concentration of the chemical solution
  r is the reactivity (aggressiveness) of the chemical

a is the activity of the chemical in aqueous solution
s is the shielding effect from the penetrated material.
The following relationships amplify Equation (11.7).

$$D = D_0 \times e^{\left(\frac{-E_D}{RT} + b\varepsilon\right)} \qquad S = S_0 \times e^{\left(\frac{-E_S}{RT}\right)} \qquad r = r_0 \times e^{\left(\frac{-E_R}{RT}\right)}$$

Where
E is the activation energy of each process
R is the gas constant
T is the absolute temperature
ε is the tensile strain
b is a parameter that depends on the glass-resin bonding

The above relationships recognize the effect of temperature on diffusivity, solubility, and reactivity. The effect of the tensile strain on diffusivity is also recognized. Entering the above expressions in Equation (11.7), we have

$$\frac{d(\Delta e)}{dt} = k \frac{D_0 \times S_0 \times c^a \times e^{b\varepsilon}}{r_0 \times (\Delta e)^s} \times e^{-\left(\frac{E_D - E_R + E_S}{RT}\right)}$$

or

$$\frac{d(\Delta e)}{dt} = k \frac{D_0 \times S_0 \times c^a \times e^{b\varepsilon}}{r_0 \times (\Delta e)^s} \times e^{-\left(\frac{E}{RT}\right)}.$$

The coefficients D_0, S_0 and r_0 take different values depending on the resin–chemical interaction. Although system-dependent, these coefficients are independent of the temperature and the exposure time. Integrating the above equation, we have

$$\Delta e = \left[k \times \frac{D_0 \times S_0}{r_0} \times (s+1) \right]^{1/s+1} \times c^{a/s+1} \times e^{b\varepsilon/s+1} \times t^{1/s+1} \times e^{-\left(\frac{E}{RT(s+1)}\right)}, \qquad (11.8)$$

On log–log space, Equation (11.8) becomes

$$\log(\Delta e) = \left(\frac{1}{s+1}\right)\log\left[k \times \frac{D_0 \times S_0}{r_0} \times (s+1)\right] + \left(\frac{a}{s+1}\right)\log(c) + \left(\frac{b\varepsilon}{s+1}\right)\log e$$

$$+ \left(\frac{1}{s+1}\right)\log(t) - \left(\frac{E}{RT(s+1)}\right)\log e.$$

By definition, the permeability P of a given solvent–laminate system is the product of the coefficient of diffusion D and the solubility S. Since most engineers have an intuitive feel for permeability, we enter the permeability $P = D \times S$ in Equation (11.8).

$$\log(\Delta e) = \left(\frac{1}{s+1}\right)\log\left[k \times \frac{P_0}{r_0} \times (s+1)\right] + \left(\frac{a}{s+1}\right)\log(c) + \left(\frac{b \times \log e}{s+1}\right) \times \varepsilon$$

$$+ \left(\frac{1}{s+1}\right)\log(t) - \left(\frac{E \times \log e}{R(s+1)}\right) \times \frac{1}{T}$$

Or

$$\log(\Delta e) = A + B\log(c) + C\log(t) + \frac{D}{T} + E\varepsilon \qquad (11.1)$$

Where

$$A = \left(\frac{1}{s+1}\right)\log\left[k \times \frac{P_0}{r_0} \times (s+1)\right] \quad B = \frac{a}{s+1}$$

$$C = \frac{1}{s+1} \quad D = -\frac{E \times \log(e)}{R(s+1)} \quad E = \frac{b \times \log e}{s+1}$$

Equation (11.1) computes the penetrated depth from the corrosion parameters A, B, C, D and E. The strain parameter always takes the value $E = 0$, since the corrosion barrier operates below the infiltration threshold in all cases.

Note 11.1: We believe the reader will not confuse the corrosion parameters D and E in the above equations with the diffusivity D and the Arrhenius activation energy E.

Example 11.8

This example considers a few limiting scenarios in the application of Equation (11.1).

$r_0 = 0$: This is the case of inert solvents and water, which penetrate the entire laminate. The parameter A and the penetrated depth Δe in this case are infinite.

$s = 0$. This situation occurs in abrasive services that removes the penetrated depth Δe. The shielding effect is lost, exposing the laminate to direct attack at all times. In these cases, Δe increases linearly with time.

$s = \infty$: These are the cases where the corrosion barriers are fully insulated from the corrosive chemical, as in the impermeable pipes with aluminum foil. The chemical penetration will not occur and Δe is zero.

$P_0 = 0$: In this scenario, the corrosion barrier is impermeable to the chemical. Examples of real life situations approaching this are the low permeability corrosion barriers of glass flakes or thermoplastic liners.

$b = 0$: The parameter b reflects the effect of the tensile strain on the diffusivity. This effect is small in corrosion barriers operating below the infiltration threshold. Therefore, we propose that $b = E = 0$ when the corrosion barrier operates below the infiltration threshold.

$a \neq 1$: The response of chemicals in water solution is not linear with respect to concentration. Some systems show sharp increases while others show a flat response in activity as the concentration changes. The exponent a accounts for this, with $a > 1$ representing slow response and $a < 1$ representing fast response. The linear case occurs, of course, when $a = 1$.

APPENDIX 11.2

DAMAGES AND DEFECTS IN LINERS AND CORROSION BARRIERS

Defective Liners: We begin this appendix discussing the effect of defective liners on the equipment durability. The protection provided by healthy liners is at best temporary. It is true that intact resin-rich liners slow down the penetration of chemicals in laminates and prolong the equipment life. It is also true that liners easily crack or strain-corrode and do not last very long in chemically aggressive environments. Furthermore, as shown in Chapter 8, the liner presence does not depress the chemical concentration in penetrated corrosion barriers.

In conclusion, although good at extending the service life of corrosion barriers, the liners are not an overriding necessity. In this book, we take a conservative approach and ignore the liner presence in discussing the durability of composites in chemical service. Given this background, we may ask what actions to take in the presence of defective, cracked or damaged liners in new equipment. Should we repair the damages and manufacturing defects in the liners?

The liner defects/damages arise in two ways:

- From cuts, gouges, contamination, voids, pits, air bubbles, osmotic blisters, porosity, or abrasion. Such defects/damages are restricted to the liner itself and do not penetrate the fiber-reinforced corrosion barrier. These liner defects/damages have a negligibly small effect on the service life of the equipment and do not need repair.

- Liner cracks from impact loads, or overstraining the corrosion barrier. These cracks in the liner —in fact all cracks—need repair, as they may seed the initiation of the strain-corrosion process.

We can safely ignore all liner defects or damages from entrapped air, insects and other foreign objects and debris such as ceiling droppings or loosened bristles from laminating brushes. Likewise, we can also ignore damages from osmotic blisters and manufacturing flaws like pits, voids, porosity, and others. Such liner defects and flaws are not overly detrimental to the durability of the corrosion barrier and have negligible effects on the life of the equipment. In fact, we can safely ignore all manufacturing defects limited to the liner, and not penetrating into the corrosion barrier. The liner defects pose no threat.

However, the presence of hairline cracks in the liner, from overstrained corrosion barriers or impact loads may be dangerous. Malmo (Ref. [9]), Norwood and Millman (Ref. [5]) investigated the cracking of gelcoats bonded to substrates of chopped fibers. Although specific to gelcoats, the work developed by these investigators is fully applicable to liners bonded to corrosion barriers. The following captures the essence of the work of these two authors:

> All hairline cracks observed in liners, regardless of cause—excessive laminate deformation, cyclic or impacting loads—have their origin in the underlying corrosion barrier of chopped fibers. The small and imperceptible damages in the corrosion barrier transfer to the unreinforced liner that develops the hairline cracks. This hairline cracking indicates a possible damage to the corrosion barrier and a danger situation.

We complete the above statement explaining that hairline cracks in the liner are early warnings of possible dangerous strain-corrosion situations. For details, see Chapter 14. The use of dye penetrants helps the identification of hairline cracks in the liner that require repair. In conclusion, the only objectionable defects in liners are the hairline cracks. All other defects are tolerable.

Defective Corrosion Barriers: We now move to discuss the effect on durability from defective corrosion barriers. The most common manufacturing defects found in the corrosion barriers are dry spots, entrapped foreign material, voids and delaminations. The dry spots and foreign material are easily spotted and corrected for in the manufacturing process and, for these reasons, are not a matter of concern. The voids present in corrosion barriers derive from entrapped air forming disc-shaped bubbles between plies, parallel to the laminate surface. They are similar to small delaminations. In fact, we can think of delaminations as very large and flat air bubbles. Both defects – voids and delaminations – are disc-shaped, run between plies, have small thicknesses and are parallel to the laminate surface.

The reader will note the absence of crazing and cracks in the corrosion barrier. These defects are typical of homogeneous materials, like liners, and never occur in fiber-reinforced chopped plies. We proceed to discuss the effect of these defects on the equipment durability in chemical service.

From the discussions in this Chapter, we recall that the chemical durability is the time for the aggressive chemicals to traverse the corrosion barrier. The advancing front of the chemical, we also recall, is approximately flat and reaches any given depth simultaneously. Suppose now the existence of a flat void, or delamination, in the path of the advancing chemical. The penetrating chemical will reach this defect and fill it up before moving on. From the penetration point of view, the void offers no resistance to the advance of the chemical. The aggressive molecules that enter the void move ahead of the rest of the advancing front by a quantity equal to the void thickness. This is because the rate of advance in the void is extremely high when compared to the slow diffusion process in the rest of the laminate. We have a conclusion:

The voids/delaminations reduces the effective thickness of the corrosion barrier by a quantity equal to their own thicknesses. Thick voids are more detrimental than thin delamination defects.

In addition to the above, the voids also develop ugly and frightful blisters in the corrosion barrier. Such blisters, however, are at a depth penetrated in the entire corrosion barrier, and therefore already discarded. The important thing in evaluating the residual life of any equipment is the penetrated depth in the corrosion barrier, not its visual aspects.

Considering that voids are very thin, they have a small effect on the durability of the corrosion barrier. As an example, consider a void that is 0.20 mm thick. This is a thick air bubble entrapped in the corrosion barrier. The compaction effort in the manufacturing process should reduce the void thicknesses to less than 0.20 mm. Anyway, the presence of such an unusually thick void represents a reduction in the effective thickness of the corrosion barrier of just $0.20/2.00 = 10\%$. This is a negligible effect.

In conclusion, the presence of voids (air bubbles) and delaminations in the corrosion barriers are not overly detrimental to the equipment durability in chemical service. The exceptions to this statement are the strain-corrosion and the anomalous failures discussed in Chapters 14 and 20, respectively. The long-term strain-corrosion and anomalous failures develop from voids and delaminations. As for the chemical failure, discussed in this chapter and in this appendix, the voids and delaminations seem to have an irrelevant effect.

The reader should compare the discussions in this appendix with the maximum allowable defects in liners and corrosion barriers in ASTM D 2567. This ASTM standard specifies the maximum allowed size (3.0 mm) and density (4/in²) of "spherical", not flat, voids in the corrosion barrier. The important feature in voids are their thickness, not their sizes. Besides, it is hard to imagine 3.0 mm spherical voids in 2.0 mm corrosion barriers.

APPENDIX 11.3

RESIDUAL LIFE OF AGED LAMINATES

This appendix estimates the residual life of aged equipment from measurements of their penetrated depths or residual stiffness. The analysis covers both the load-dependent and the load-independent modes of failure of laminates aged in service while operating in stable environments.

Load-dependent Modes of Failure. The load-dependent modes of failure ignore the penetrated depth (assumed zero) and focus on the time or number of cycles to reach the failure thresholds. In the absence of chemical attack, the loss of stiffness is the only ageing variable to measure. However, as discussed in Appendix 12.1, the sudden-death phenomenon assures a practically constant stiffness throughout the laminate life for all load-dependent failures. The constancy of the laminate stiffness in such services rules out any possibility of linking the residual life of the equipment to its residual stiffness. This is true for all equipment operating in nonaggressive chemicals, or those with properly maintained corrosion barriers. For details, see Appendix 12.1. In summary, there is no way to determine the laminate residual life in load-dependent cases. All load-dependent laminates face a sudden death.

Load-independent Modes of Failure. The most important cases of load-independent failures involve the chemical penetration and attack on equipment storing or carrying aggressive products. In the absence of proper maintenance, the aggressive chemical traverses the corrosion barrier and advances into the structural plies, causing the laminate to lose stiffness. As discussed in this chapter, the complete equation to estimate the depth penetrated by the aggressive chemical is

$$\log(\Delta e) = A + B \log(c) + C \log(t) + \frac{D}{T} + E\varepsilon \tag{11.1}$$

In the specific case of aged equipment operating in steady conditions, time is the only ageing variable and the above equation reduces to

$$\log(\Delta e) = K + C \log(t)$$ (11.9)

Where

$$K = A + B \log(c) + \frac{D}{T} + E\epsilon = constant$$

In Equation (11.9), the parameter K captures the effects of variables such as chemical concentration, resin type and operating temperature. The parameter C captures the shielding effect from the penetrated material that remains on the laminate surface.

The exponential form of Equation (11.9) is

$$\Delta e = kt^C$$ (11.10)

Where $K = logk$

From Equation (11.10), the penetrated depth in aged equipment operating in stable environments is exponential in the time variable. The estimation of the penetrated depth as a function of time requires two parameters. The k parameter captures the resin type and the operating conditions, while the C parameter captures the shielding effect.

The removal of the penetrated depth Δe in abrasive services eliminates the shielding effect, making s = 0 and C = 1. Therefore, the penetrated depth in abrasive services varies linearly with time as

$$\Delta e = kt$$ (11.11)

See Figure 11.4.

In non-abrasive service, the determination of the parameters K (or k) and C requires the measurement of two values of the penetrated depth, taken at different times. The parameters K (or k) and C enter Equation (11.9) to compute the equipment durability as defined by the original laminate thickness "e" and the minimum allowable thickness $(e)_{cri}$. From Equation (11.9) we have

$$\log(e - e_{cri}) = K + C \times log(durability)$$ (11.12)

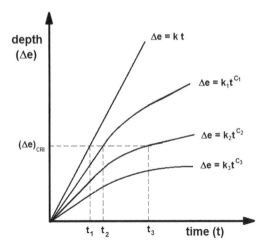

FIGURE 11.4 *The penetrated depth of laminates operating in stable conditions varies exponentially with time. The k parameter captures the effects of the resin type and operating conditions. The C parameter captures the shielding effect. The exponential relation turns linear in abrasive service, where the shielding s = 0 and C = 1. The graph shows the times (durability) to penetrate a critical depth (Δe)$_{cri}$.*

Where

e is the pristine, or original, laminate thickness

e_{cri} is the minimum allowable laminate thickness

The residual life of the equipment is

$$\left(\text{residual life}\right) = \left(\text{durability}\right) - \left(\text{time in service}\right)$$

The estimation of the residual life requires two parameters and the measurement of the penetrated depth at two inspection times. This is necessary to compute the parameters K and C. A single measurement of the penetrated depth, taken at one inspection time, holds only in abrasive service where the shielding effect is not active.

An alternate form of Equation (11.12) is

$$\log\left(S - S_{cri}\right) = K + C \times \log\left(\text{durability}\right) \tag{11.13}$$

Where S is the pristine laminate stiffness and S_{cri} is the minimum allowed stiffness, both measured in the hoop direction in cylindrical tanks. Again,

Equation (11.13) requires two measurements of the laminate residual modulus, taken at different inspection times.

The following reasons favor the visual measurement of the penetrated depth versus the stiffness:

- The residual stiffness/modulus method assumes the full chemical destruction of the penetrated depth, which is not true in most cases. The assumption of full destruction gives penetrated depths less the real values and overestimate the residual life. Besides, this method is destructive.
- The method involving the measurement of the penetrated depth is accurate and non-destructive

Example 11.9

Compute the residual life of a storage tank regularly relined to replace the penetrated corrosion barrier.

The regular replacement of the corrosion barrier prevents the chemical penetration in the structural plies. The measurement of the penetrated depths have no meaning in this case. Furthermore, in the absence of chemical attack, the sudden-death phenomenon assures a constant laminate stiffness throughout the tank life. The measurement of the residual stiffness is also useless in this case.

Example 11.10

Compute the residual life of a storage tank in service for 10 years with no relining. Two measurements, taken at different inspection times, indicated the following depths of chemical penetration.

Inspection time	Penetrated depth
60 months (5 years)	4.0 mm
120 months (10 years)	6.0 mm.

The minimum allowable laminate thickness is $(e)_{cri} = 8.0$ mm. The original laminate thickness is e = 15.0 mm.

The equation to compute the penetrated depth is

$$\log(\Delta e) = K + C\log(t)$$

Entering the measured data in the above equation, we have

$$\log(4.0) = K + C \times \log(60)$$

$$log(6.0) = K + C \times log(120)$$

Solving the above system, we obtain

$K = 0.397$ and $C = 0.183$.

The tank durability comes from Equation (11.12)

$$log(e - e_{cri}) = K + C \times log(durability)$$

$$log(15 - 8) = 0.397 + 0.183 \times log(durability)$$

The durability is 280 months (23 years)

The residual life is

$$residual\ life = (durability) - (time\ in\ service)$$

$$residual\ life = 23 - 10 = 13\ years.$$

In most cases, the revalidation process to evaluate the residual life applies to badly damaged or very old equipment. The revalidation of healthy or middle-aged equipment is not a common occurrence. The question asked in such cases is

"Can the tank survive in service until the next plant shutdown"

The pressing need of an urgent answer gives the inspector no opportunity to make two separate inspections. The conservative answer offered in such cases come from just one measurement of the penetrated depth. The inspector assumes the absence of shielding in this case, i.e., s = 0. This gives C = 1 and a fast linear chemical penetration, as in abrasive services.

$$\Delta e = kt$$

The computation of k comes from a single measurement of the penetrated depth.

Example 11.11

Estimate the residual life of the equipment described in Example 11.10 assuming a single measurement taken after 5 years in service.

The measured penetrated depth at 5 years is 4.0 mm, which gives k = 0.80 mm/year.

The durability is

$$durability = \frac{15-8}{0.80} = 8.75 \text{ years}$$

The residual life is

residual life = 8.75 – 5 = 3.75 years.

From Example 11.10, the correct durability is 23 years and the residual life is 23 – 5 = 18 years.

Example 11.12

Estimate the residual life of the equipment described in Example 11.10 for a single measurement taken after 10 years in service.

The measured penetrated depth at 10 years is 6.0 mm, which gives k = 0.60 mm/year.

The durability is

$$durability = \frac{15-8}{0.60} = 11.67 \text{ years}$$

The residual life is

residual life = 11.67 – 10 = 1.67 years.

From Example 11.10, the correct durability is 23 years and the residual life is 23 – 10 = 13 years.

12 Long-Term Fiber Rupture

12.1 INTRODUCTION

As discussed in Chapter 8, there are five long-term load-dependent modes of laminate failure:

- The *infiltration failure* from crack densities large enough to allow the ingress of fluids.
- The *weep failure* from crack densities large enough to allow the leakage of fluids.
- The *stiffness failure* from damages large enough to affect the modulus of elasticity.
- The *rupture failure* from fiber damage densities large enough to break the laminate.
- The *strain-corrosion failure* from the combined action of chemicals and mechanical loads.

This chapter deals exclusively with the long-term rupture from fiber damage under pure static or pure cyclic loads acting alone. The long-term infiltration, weeping and stiffness failures under pure loadings are the subject of the next chapter. The strain-corrosion failure is the topic of Chapter 14. solution of The long-term load-dependent failures involving the simultaneous action of cyclic and static loads require the unified equation discussed in Chapter 16.

We begin with the study of fiber rupture under pure static loads. The pure cyclic case follows in the sequence.

12.2 GENERALIZED STRAIN-CORROSION OF GLASS FIBERS

There are two types of static long-term ruptures of laminates, both caused by strain-corrosion. The first type – known as "static fatigue" or fiber strain-corrosion – comes from a generalized water degradation of all fibers in all plies. The second type – known as laminate strain-corrosion – comes from a localized chemical attack on small areas of the laminate. This chapter deals with the generalized fiber strain-corrosion. For a comprehensive discussion of the localized laminate strain-corrosion, see Chapter 14.

Water is the only known chemical capable of (a) reaching the entire laminate and (b) reacting with the glass fibers. Other chemicals are either (a) not reactive with the fibers or (b) not capable of reaching all plies. The ubiquitous water molecules penetrate the entire laminate and is present at the glass surface of all plies at all times. Water is the only known chemical species capable of causing generalized long-term fiber strain-corrosion. The high mobility of the water molecules assures that all plies reach hydric equilibrium with the moisture in the atmosphere, or in the soil, or in the chemical solutions in a short time. Furthermore, the manufacturing processes of the fibers and resins leave behind small residual moisture that makes the water presence in laminates unavoidable.

The process of fiber strain-corrosion involves the deterioration of tensile loaded glass fibers in the presence of water. This book will not discuss the mechanism of water attack. Such a discussion would add nothing to our goal, which is the estimation of laminate durability. For details on the mechanism of fiber strain-corrosion, see references 27, 28 and 29.

To understand the process of fiber strain-corrosion, let us consider a single fiber tensioned by the static force F, as in Figure 12.1. The static force F produces small cracks in pre-existing manufacturing flaws present at the fiber surface. The small cracks open under load and expose fresh glass surfaces to water attack. This results in a crack growth, which expose new material, etc., until the fiber ruptures under the static force F. Figure 12.1 illustrates the tensioned open cracks growing in time. The damage from the simultaneous action of a static strain and chemical attack (in this case, by water) characterizes the strain-corrosion process. The reader will note the crack grows in the absence of cyclic loads. We have just explained the process of fiber strain-corrosion.

The theory of fracture mechanics recognizes cyclic loads as the only driver of crack growth. Strain-corrosion, driven by static loads, is the exception. We repeat. Strain-corrosion is the only known process capable of growing

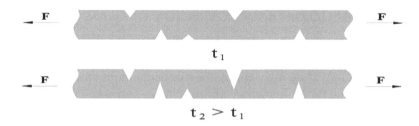

FIGURE 12.1 *The water attack on pre-existing surface flaws produces many small cracks along the fiber. The static force F opens the cracks and exposes fresh material to further attack. Strain-corrosion is the only known process that grows cracks under static loads.*

cracks under static loads. To clarify this, we consider the following two thought experiments.

In the first experiment, we subject bare (not resin-coated) glass filaments to the same tensile force F (a) in air, (b) immersed in water and (c) immersed in acid solution. The times to rupture of the bare glass filaments under the static force F depend on the environment.

- In air, the water presence is small and the time to fiber rupture is long.
- In acid solution, the presence of H^+ ions is high and time to rupture is short.
- In water, the time to rupture falls in between.

In the second thought experiment, we use resin-coated glass filaments under the same force F. The rupture times of the resin-coated fibers are as follows

- The time to rupture of the coated fibers in air is the same in both experiments, as expected.
- The time to rupture of the coated fibers in water is also the same. The resin coating provides no protection against water
- In acid solution, however, the coated filaments take a longer time to rupture than the bare fibers. The resin coat delays the acid penetration and, more important, reduces the acid concentration at the fibers surface.

This thought experiment explains the twofold effect of the matrix in protecting the fibers from chemical attack. First, the resin coat delays the penetration time and second – more important – it lowers the chemical concentration at the fiber surface. The reduction in chemical concentration at the fiber surface plays a fundamental role in the strain-corrosion durability of sewer pipes. As explained in Chapter 14, the low acid concentration at the fiber surface suggest the existence of a "strain-corrosion threshold" in sewer pipes. This is a topic for Chapter 14. The discussions in this chapter center on fiber strain-corrosion in the presence of water.

12.3   STATIC LOADINGS

In this section, we discuss the ply rupture under pure static loads. We begin with the study of single fibers, and subsequently extend the concepts to the rupture of multiple fibers in UD plies.

We begin with two statements. First, the manufacturing process introduces residual tensile strains on the outer surface of the glass fibers. Second, the apparently smooth fibers have small surface flaws. The surface flaws seed the development and growth of tiny cracks under the combined action of the residual strains and any reactive chemical present at the fiber surface.

Figure 12.2 shows the spontaneous surface cracking of a glass fiber exposed to 5.0% HCl at 60C for 2 hours. The spontaneous surface cracking ceases as soon as the self-limiting driving force (the residual strain) is dissipated. This situation changes, however, in the presence of non-self-limiting external forces that sustain the strain-corrosion process. The non-self-limiting external forces interact with the chemical at the fiber surface and the crack growth proceeds until rupture. The rate of crack growth on the glass fibers derives from the Paris static equation

$$\frac{da}{dt} = Y\left(\varepsilon \times \sqrt{\pi \times a}\right)^Z \tag{12.1}$$

Where

"a" is the crack size on the fiber.
"t" is the time
"ε" is the static tensile strain.
"Y" is a geometrical constant.
"Z" is a constant related to the rate of fiber deterioration.

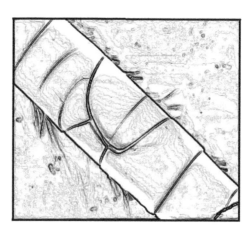

FIGURE 12.2 *Spontaneous surface cracking of glass fibers immersed in acid with no external load. The open cracks suggest the presence of residual surface strains. The spontaneous surface fiber cracking is similar to the environmental stress cracking observed in resin-rich liners. (Drawing by Gabriel Gonzalez)*

The static Paris law described in Equation (12.1) suggests a growth rate controlled by the crack size "a" and the fiber strain "ϵ". The exponent Z varies with the corrosive environment, which is water in the present case. The time to fiber rupture under constant strain comes from integration of Equation (12.1) from an initial crack size "a_0" to the critical size "a". After a few clever mathematical manipulations Equation (12.1) becomes

$$\log a = A_S - G_S \times \log t \tag{12.1A}$$

Where "a" is the critical crack size and "t" is the time to fiber rupture. The subscript "s" in the parameters As and Gs reflect the static character of the loading. Equation (12.1A) gives the time to rupture of a single glass filament subjected to constant static strains. Note that the critical crack size "a" defines the time to rupture of a single fiber. Now we apply this knowledge to multiple fibers.

The failure of UD plies statically loaded in the fiber direction involves an enormous number of fiber ruptures. Our task is to extend Equation (12.1A), derived for a single fiber, to many fibers. The first problem we face is the enormous variety of initial crack sizes "a_0". All initial cracks larger than a certain size will fracture immediately on application of the static force. This produces an enormous quantity of initial rupture points. The cracks that remain are small and do not rupture immediately. Instead, they grow and fail in response to a steady strain-corrosion process. The crack sizes and the density of ruptured points steadily increase with time. Eventually the density of ruptured points reach a critical value and the ply fails. The density of ruptured points, not the crack sizes, controls the time to ply rupture.

There is a fundamental connection between the density of ruptured fiber points and the critical crack size a. This connection is easy to see. The density of ruptured points increases in direct relation to the attainment of the critical size "a". All cracks reaching the critical size "a" fail and produce a ruptured point. The number of cracks that attain the critical crack size "a" is equal to the number of ruptured points. The density of ruptured points defines the residual ply strength. This density is equal to the number of cracks reaching the size "a" at the time "t". Therefore, we can replace the crack size "a" in Equation (12.1A) with the residual ply strength ε. From this reasoning, the regression equation to predict the rupture of UD plies loaded statically in the fiber direction is:

$$\log \varepsilon = A_S - G_S \log t \tag{12.2}$$

In the above equation, "ϵ" is the applied static strain in the fiber direction of the UD ply. This applied strain ruptures the ply at the time "t" and by definition is the ply strength at the time "t". The coefficients "As" and "Gs" are determined experimentally. The regression line (12.2) predicts the times to rupture of multi-fiber plies submitted to known applied static strains, and vice-versa. Equation (12.2) holds also for chopped fibers.

Equation (12.2) is the same Equation (8.3) of Chapter 8. As explained there, the regression parameter As derives from the short-term UD ply strength.

$$\log(T_r) = A_s$$

Where Tr is the ply rupture threshold (short-term rupture strength).

12.4   PURE CYCLIC LOADING OF UD PLIES

We next discuss the rupture of UD plies under pure cyclic loadings. As before, we start with the analysis of a single fiber. The mechanism of crack growth in single fibers under pure cyclic loads is different from the strain-corrosion of the pure static case. The previous section described the pure static strain-corrosion crack growth as a continuous process. The rate of static deterioration depends on the applied strain, the temperature, the aggressive environment and the glass composition. Under cyclic loads the rate of crack growth proceeds in a stepwise fashion, one little bit at a time, in response to the kinetic energy delivered in each load cycle. Furthermore, the rate of cyclic crack growth depends only on the peak strain regardless of the environment, temperature and glass composition.

The cyclic rate of crack growth proceeds in the same way as a hammer drives a nail on a piece of wood, one bit at a time, reflecting the kinetic energy of the blow. The rate of crack growth is a discontinuous stepwise process described by the Paris cyclic equation.

$$\frac{da}{dN} = Y\left(\varepsilon\sqrt{\pi \times a}\right)^Z \tag{12.3}$$

Where

   ε is the peak cyclic strain
   a is half the crack size
   N is the number of cycles
   Y and Z are parameters.

The rate of crack growth in Equation (12.3) is cycle-dependent, not time-dependent. The reader will note the number of cycles N in Equation (12.3) in place of the time variable t in Equation (12.1). Furthermore, the exponent "Z" in the cyclic Equation (12.3) does not depend on the operating temperature, the aggressive chemical and the glass composition. Temperature, moisture and glass composition have no effect on the rate of cyclic crack growth. Again, we are interested in the number of cycles leading to a single fiber rupture under constant peak strains. The number of cycles to rupture derives from a simple integration of Equation (12.3) from the initial crack size "a_0" to the critical size "a". The equation to predict the rupture of a single fiber is, as in the static case

$$\log a = A_C - G_C \times \log N \tag{12.3A}$$

Where "a" is the critical crack size and "N" is the number of cycles to rupture. The parameters Ac and Gc refer to cyclic loading. As discussed in the static case, we obtain the ply rupture equation by substituting the peak cyclic strength for the equivalent critical crack size in Equation (12.3A)

$$\log \varepsilon = A_C - G_C \times \log N \tag{12.4}$$

Where the "ε" is the peak cyclic strain in the fiber direction and N is the number of cycles to rupture the UD ply.

Example 12.1

This example defines the R ratio as well as the peak and the static strain components in complex loadings. Figure 12.3 illustrates the static and cyclic strain components.

$$\left(\varepsilon\right)_{static} = \varepsilon_{min} \quad \left(Static\ strain\ component\right)$$

$$\left(\varepsilon\right)_{cyclic} = \varepsilon_{max} - \varepsilon_{min} \quad \left(Cyclic\ strain\ component\right)$$

In the above, ε_{max} and ε_{min} are the extreme strain values. The R ratio is

$$R = \frac{\varepsilon_{min}}{\varepsilon_{max}}$$

The static and cyclic strain components expressed in terms of the R ratio and the peak strain ε_{max} are

$$\varepsilon_s = R\varepsilon_{max} = R\varepsilon \quad \left(Static\ strain\ component\right)$$

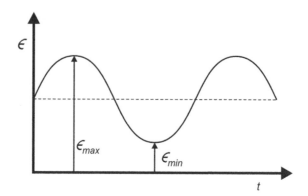

FIGURE 12.3 *The static strain component is the minimum strain ε_{min} = $R\varepsilon_{max}$. The cyclic strain component is $\varepsilon_{max} - \varepsilon_{min} = (1 - R)\varepsilon_{max}$.*

$$\varepsilon_c = \left(1 - R\right)\varepsilon_{max} = \left(1 - R\right)\varepsilon \quad \left(Cyclic\ strain\ component\right)$$

To facilitate the notation we have dropped the "max" subscript from the peak strain. The cyclic and static strain components shown above are direct inputs to the unified equation that compute the ply durability under simultaneous static and cyclic loadings.

12.5   THE WORK OF MARK GREENWOOD

Mark Greenwood creep-tested pultruded UD rods immersed in water. The generated regression line reflects the basic long-term hydrolytic stability of embedded glass fibers. The tested rods had the following specification:

- Diameter 6.0 mm
- Glass loading 75%
- High reactivity polyester resin
- Standard E glass or boron-free glass

The glass loading, rod diameter and type of resin are not relevant in the test. The important variable in this study is the glass composition, which reflects the strain-corrosion stability of the fiber. Segments of the pultruded rods were tensile loaded until rupture while immersed in water at room temperature. The tensile loads were constant throughout the tests.

As expected, the rods subjected to higher strains ruptured first. A statistical analysis of the failure data produced a straight line, as in Equation (12.2). The slope of the regression line depends on the glass resistance to water attack. Figure 12.4 shows the lines for E glass and boron-free glass. The boron-free fibers are less susceptible to strain-corrosion and holds up better in the presence of water than the traditional E glass. The failure strains extrapolated to 50 years indicate a long-term static strength Ss = 0.92% for the boron-free glass versus Ss = 0.41% for the traditional E glass. The interpretation is like follows.

1. The water damage lowers the long-term static strength of both rods.
2. The boron-free glass fibers cannot sustain elongations higher than 0.92% for 50 years in the presence of water.
3. The E glass fibers cannot sustain elongations higher than 0.41% for 50 years in the presence of water.
4. The boron-free glass has better strain-corrosion resistance than E glass.

A back extrapolation of Figure 12.4 indicates essentially the same short-term strength for the two glass compositions. The difference between E glass and boron-free glass is in the long-term, not in the short-term. Mark Greenwood never published the regression equations for his data. However, inspection of Figure 12.4 indicates the following

For boron-free glass:

$$\log(\varepsilon) = 0.400 - 0.077 \log(\text{hours}) \qquad (12.5A)$$

For E glass:

$$\log(\varepsilon) = 0.347 - 0.130 \log(\text{hours}) \qquad (12.5B)$$

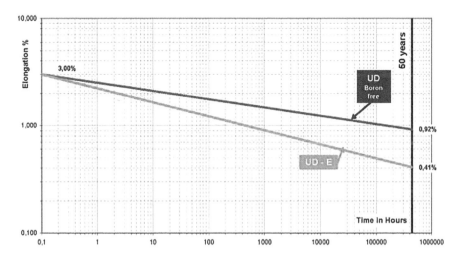

FIGURE 12.4 *Regression lines adapted from the work by Mark Greenwood on tensile loaded UD plies.*

Example 12.2

Compute the long-term and the short-term static strengths of UD plies of boron-free and E glass. Take 50 years (438 000 hours) as long-term and 0.1 hour as short-term.

Equation (12.5A) gives the short-term strength of UD plies of boron-free glass as

$$\log(\varepsilon) = 0.400 - 0.077 \log(\text{hours})$$

$$\log(\varepsilon) = 0.400 - 0.077 \times -1 = 0.477$$

$$S_S = 3.0\%$$

Equation (12.5B) gives the short-term strength of UD plies of E glass as

$$\log(\varepsilon) = 0.347 - 0.130 \log(\text{hours})$$

$$\log(\varepsilon) = 0.347 - 0.130 \times -1 = 0.477$$

$$S_S = 3.0\%$$

The two glass compositions have the same short-term strength, as expected.

The long-term tensile strength of UD plies of boron-free glass in 50 years is

$$\log S_S = 0.400 - 0.077 \log(438000)$$

$$\log S_S = 0.400 - 0.077 \times 5.641 = -0.034$$

$$S_S = 0.92\%$$

For UD plies of E glass the long-term 50 years strength is

$$\log S_S = 0.347 - 0.130 \log(438000)$$

$$\log S_S = 0.347 - 0.130 \times 5.641 = -0.386$$

$$S_S = 0.41\%$$

The meaning of the long-term strengths derive directly from the regression Equation (12.2). The long-term strengths are the residual strength of the deteriorated or damaged plies. The long-term strength is a metric of the accumulated damage.

12.6 THE WORK OF GUANGXU WEI

As we have just seen, Mark Greenwood determined the pure static (R = 1.0) regression lines for both E and boron-free UD plies loaded in the fiber direction. Guangxu Wei determined the pure cyclic (R = 0.0) regression lines of UD plies in the fiber direction and in the direction transverse to the fibers. See ref. 25. The pure cyclic (R = 0.0) UD equations reported by Guangxu Wei are:

$$\log(\varepsilon) = -0.602 - 0.040 \log(N) \quad \left(\text{Transverse direction}\right) \qquad (12.6\text{A})$$

$$\log(\varepsilon) = 0.519 - 0.089 \log(N) \quad \left(\text{Fiber direction}\right) \qquad (12.6\text{B})$$

Equation (12.6A) holds for pure cyclic loadings (R = 0.0) of transverse loaded UD plies. It computes the number of cycles to first rupture the transverse loaded UD ply under pure cyclic loadings. As discussed in Chapters 8 and 9, the first crack to appear in transverse loaded UD plies indicates the infiltration failure. Therefore, Equation (12.6A) is an infiltration – not rupture – equation. See Chapters 8 and 9.

Equation (12.6B) describes the pure cyclic ruptures (R = 0.0) of UD plies loaded in the fiber direction. Equation (12.6B) holds for both E and boron-free glasses.

The reader perhaps has noted that the strains in all previous equations were either *in the fiber direction or in the transverse direction to the fiber.* Both Mark Greenwood and Guangxu Wei developed their data in those directions. Mark's data were developed on UD rods tensile tested in the fiber direction. Guangxu Wei tested UD plies loaded in the fiber and in the transverse directions.

12.7 THE WORK OF THE BRITISH PLASTICS FEDERATION

The British Plastics Federation creep-tested polyester chopped plies under pure static loads, in the same way that Mark Greenwood did for UD rods. The static regression to describe the long-term pure static rupture of polyester chopped plies is

$$log\epsilon = \log(1.50) - 0.045 \times \log(hours) \tag{12.7}$$

Where 1.50% is the short-term rupture strength, or rupture threshold, of polyester chopped plies.

The long-term pure cyclic testing of chopped plies produced a similar equation.

$$log\epsilon = \log(1.50) - 0.075 \times \log(N) \tag{12.8}$$

In Equation (12.8), ϵ is the peak strain and 1.50% is, again, the short-term ply strength.

The pure cyclic and the pure static regression equations discussed in this chapter play a fundamental role in the unified equation to predict the ply (or laminate) durability under complex loadings.

12.8   THE UNIFIED EQUATION

Equations (12.2) and (12.4) are applicable to any ply subjected to pure static or pure cyclic loads. They are not valid if both loads act simultaneously. The unified equation, discussed in Chapter 16, computes the effect of the static and cyclic loads acting together The full and simplified versions of the unified equation are shown below, borrowed from Chapter 16.

$$\left(\frac{R \times \varepsilon \times SF}{S_S}\right)^{\frac{1}{G_S}} + \left(\frac{(1-R) \times \varepsilon \times SF}{S_C}\right)^{\frac{1}{G_C}} \tag{12.9A}$$

$$+ \left(\frac{R \times (1-R) \times \varepsilon^2 \times SF^2}{S_S \times S_C}\right)^{\frac{1}{G_{SC}}} = 1.00$$

$$\left(\frac{(1-R) \times \varepsilon \times SF}{S_C}\right)^{\frac{1}{G_C}} + \left(\frac{R \times (1-R) \times \varepsilon^2 \times SF^2}{S_S \times S_C}\right)^{\frac{1}{G_{SC}}} = 1.00 \tag{12.9B}$$

In the above equations, Gs and Gc are the pure static and the pure cyclic regression slopes mentioned earlier in this chapter. The long-term static and cyclic ply strengths Ss and Sc derive directly from the pure regression equations. The full Equation (12.9A) predicts the time-dependent long-term laminate rupture involving fiber strain-corrosion. The simplified version,

Equation (12.9B), predicts the time-independent infiltration, weep and stiffness failures. For details on the unified equation, see Chapter 16.

Example 12.3

Suppose an oil pipeline used in the transmission of pressurized water. Which failure occurs first, loss of stiffness, infiltration, weep or rupture?

Infiltration will probably occur ahead of the others, but is not important in this case, since water is not harmful to laminates. Our attention is therefore focused on the loss of stiffness, weep and rupture failures.

- Equation (12.7A) computes the long-term rupture from fiber degradation leading to long-term rupture. The safety factor SF, expressing the rupture durability, depends on the load ratio R, the peak strain, the number of cycles and the time duration.
- Equation (12.7B) computes the long-term failure from weeping and loss of stiffness caused by loads in the transverse direction of the UD plies. The safety factor SF in this case is fully cycle-dependent and not affected by the time duration.

The answer to the posed question will depend on the load ratio R, the number of cycles and the load duration. In some situations, weeping occurs first. In others, rupture may come first. Topics like this are easy to solve with the unified equation.

APPENDIX 12.1

SUDDEN DEATH

This appendix addresses the interesting composites phenomenon known as "sudden-death". We start with the events leading to the rupture of UD plies loaded in the fiber direction. To facilitate the presentation, the discussion is in terms of stresses, instead of strains. Figure 12.5 shows the cross section A – A of an UD ply tensile loaded by a force F in the fiber direction. The average stress in the fiber direction is

$$\sigma_0 = \frac{N}{A} \times a \times \sigma_{fiber} + \frac{(A - N \times a)}{A} \times \sigma_{resin} \tag{12.10}$$

Where

N – Is the number of fibers in the section A – A
A – Is the total area of the section A – A
a – Is the cross sectional area of the individual fibers
σ_{fiber} – is the fiber stress
σ_{resin} – is the resin stress

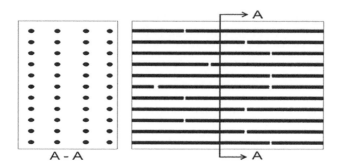

FIGURE 12.5 *The strained fibers display multiple ruptures at the weak points. These local fiber failures have negligible effect on the ply stiffness. Increments in the tensile force increase the number of fiber broken points. The ply rupture occurs when the density of ruptured points grow to a critical level. The initial ply modulus and strength remain relatively unaffected throughout this process. The ply dies a sudden death.*

Ignoring the small resin contribution, we have

$$\sigma_0 = \frac{N}{A} \times a \times \sigma_{fiber}$$

In accordance with the theory of fracture mechanics, the above stress (or strain) would immediately fail all fibers at the points where their crack sizes are larger than a certain critical value. These points of immediate rupture occur at random along the fiber length, as shown in Figures 12.5 and 12.6.

The loads not carried at the broken points transfer to the adjacent fibers and to the rest of the same fiber by shear at the glass-resin interface. See Figure 12.6. The broken fibers lose load carrying capacity in a small local segment "δ" at the ruptured points, known as "ineffective length" for obvious reasons. The rest of the fiber, that has not been broken, retains its load carrying capacity and maintains the strength and stiffness of the ply. We denote by "n" the number of ruptured fiber points that fall on a section A – A of width "δ". Equation (12.11) gives the strength of the lamina in this condition.

$$\sigma_0 = \frac{(N-n)}{A} \times a \times \sigma_{fiber} \tag{12.11}$$

In the above, "n" is the number of ruptured fiber points within the section A – A of width "δ". Equation (12.11) is the same Equation (12.10), except for the discount of the load no longer carried by the "n" broken fiber points on

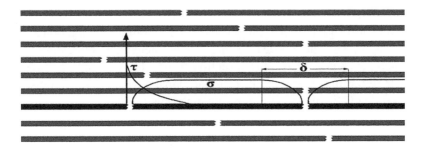

FIGURE 12.6 *The stresses not carried at the ruptured points are shear transferred to adjacent fibers and to the unbroken parts of the same fiber. See the ineffective fiber length "δ". The ply retains its original stiffness and strength even if a large number of fibers have failed. The sudden rupture takes place at a critical damage level.*

section $A - A$. The reader will agree that while the number of ruptured points on any fiber may be very large, the number of ruptured points falling exactly on the section $A - A$ is very small.

$$n \ll N$$

Entering the above in (12.11), we see that the ply strength is insensitive to the number of broken fibers. This is an interesting conclusion that goes against common sense. Common sense would lead us to expect the ply strength to fall as the fibers break but, instead, we see that it remains unchanged.

The strain-corrosion process grows the fiber crack sizes and increases the number of broken points at a constant tensile force F. The loads no longer carried by the broken points transfer to adjacent fibers. The probability of fiber rupture is higher in the neighborhood of the broken ends, because of the local load transfer. Eventually, the density of ruptured fiber points in the neighborhood of section $A - A$ becomes too high and the remaining fibers can no longer sustain the force F. At this point, the ply suddenly breaks. This description of the events is consistent with the "sudden-death" phenomenon reported by many researchers.

The foregoing description, derived for static strain-corrosion, is valid for cyclic fatigue as well. In both cases, the myriad of broken fiber points lead to the same "sudden-death" effect.

The foregoing discussion indicates that the ply global strength and stiffness remain essentially unchanged to the moment of rupture. Substantial reductions in stiffness occur locally on the section A – A, but the rest of the ply retains its original global properties such as modulus, elongation at break and strength. The sudden-death rupture is not detectable by testing the ply anywhere, even near the ruptured section A – A. The ply ruptures suddenly, while maintaining essentially its original global stiffness and strength. This "sudden-death" is typical of composite materials.

Often times we are tempted to estimate the residual life of aged equipment by measuring their strength retention. This protocol is valid for equipment in chemical environments that penetrate and destroy laminates. For details, see the Appendix 11.3 in Chapter 11. However, for equipment operating in mild and friendly environments, or those that have their corrosion barriers regularly replaced, this is a futile effort. The comparison of aged and pristine strength is not a valid method to estimate the residual life in such cases. The reason is, as explained above, the essentially unaltered global strength throughout the equipment life.

The measurement of the aged global strength may indicate a healthy laminate that nevertheless ruptures the next day. This uncertainty is true also in living organisms. A check-up of global health parameters is no guarantee against sudden death. The detection of the sudden-death phenomenon requires pinpoint accuracy and the ability to do measurements at the critical and random section A – A.

As a closing note, the resin contribution to the ply strength resides not on the small load that it carries, but on its ability to transfer stresses between broken fibers along the ineffective length "δ". Small ineffective lengths "δ" reduce the width of section A – A and increase the ply strength. We conclude that UD plies of tougher resins and better glass adhesion would have slightly higher rupture strengths in the fiber direction.

13 Infiltration, Weep and Stiffness Failures

13.1 INTRODUCTION

The preceding chapter discussed the long-term pure static and pure cyclic rupture failures of chopped and UD plies. The regression equations introduced to compute the ply durability under pure loads (static or cyclic) are simple and straightforward.

- Mark Greenwood developed the pure static long-term rupture equation for UD plies loaded in the fiber direction.
- Guangxu Wei developed the pure cyclic long-term rupture equation for UD plies loaded in the fiber direction
- Guangxu Wei also developed the pure cyclic long-term infiltration equation for transverse loaded polyester UD plies.
- The British Plastics Federation developed the pure cyclic and the pure static long-term rupture equations for polyester chopped plies.

The above equations predict the ply failure by rupture. The exception is the cyclic equation by Guangxu Wei on transverse loaded UD plies, which predict the infiltration failure. This chapter deals with the remaining load-dependent modes of ply failure – infiltration, weep and stiffness, and completes the load-dependent durability analysis of plies under pure static or pure cyclic loads. The regression equations introduced in this chapter, together with those in the previous chapter, provide the necessary inputs to the unified equation and solve the laminate durability problem under complex loadings. The unified equation – Chapter 16 – gives a complete analysis of the long-term ply – and laminate – failure under the simultaneous action of static and cyclic load components.

The plies submitted to cyclic loads develop many micro-cracks that grow in size until they meet fibers crossing their path, which arrest their growth. As a result, composite laminates never develop large cracks, except between plies (delamination) and alongside UD fibers. The delamination between plies and the macro-cracks in UD plies do not meet fibers in their path and grow to large sizes. See Appendix 10.1 in Chapter 10. Except for these two few and exceptional "macro-cracks", the composite laminates as a rule develop many small micro-cracks that grow in number, but not in size. The micro-crack density starts small in the early life of the composite, and steadily grows to a critical level where failure occurs by crack coalescence. The composite long-term load-dependent failures – weeping, infiltration, stiffness and rupture – result from the coalescence of a large number of accumulated micro-cracks, not from the growth of a single large crack. The critical accumulated damage – crack density – defines the ply long-term strength.

This chapter deals with the infiltration, weep and stiffness modes of long-term failure. The discussion focus mostly on the weep threshold. The concepts, however, hold for the infiltration and stiffness modes of failure as well. In all cases, the long-term strength of the critical ply defines the laminate durability.

The plies of chopped fibers have a fundamental role in our discussions, since they are critical to the long-term infiltration and weep modes of failure in commercial laminates.

13.2   THE CLASSICAL HDB

The regression weep lines of commercial pipes derive from the ASTM D 2992 B test protocol, which monitors the onset of weeping in pressurized water-filled pipes. The constant test pressures are higher than the weep threshold, otherwise the pipes would never weep. The cracks that develop in the pressurized specimens are stationary, and the times to weep are a function of the applied pressure, or strain. The failure points linking the applied strain and the times to weep generate regression lines of the form.

$$\log \varepsilon = A_S - G_S \log t \tag{13.1}$$

Equation (13.1) is a pipe equation, not a ply equation. It ignores the pipe wall thickness "e". The complete weep regression equation, including the pipe wall thickness, would be

$$\log \varepsilon = A_S + B_S \log e - G_S \log t \tag{13.1A}$$

In the above "ε" is the static, constant hoop strain that weeps the pipe of thickness "e" at the time "t". The weep time "t" measure the travel-time of the pressurized water in traversing the cracked wall of thickness "e". The intercept As in Equation (13.1) relates to the pipe's short-term weep strength, which is a pipe property not be confused with the weep threshold. The distinction between the pipe's short-term weep strength and the weep threshold is essential. The pipe's weep regression slope Gs varies with the resin toughness, wall thickness and the test temperature. For details, see Examples 10.3 and 10.6 in Chapter 10.

As we see, the weep times of pressurized pipes increase with the wall thickness. The ASTM D 2992B test protocol disregards the wall thickness and proposes the simplified Equation (13.1). In the above equations, "ε" is the pipe's hoop strain. If the applied strain is high, the cracks are many and the travel time is short. Higher strains produce shorter weep times. Contrary to widespread belief, the weep time "t" does not reflect any crack growth or progressive deterioration of the pipe. Rather, it reflects the time taken by the pressurized water to traverse the pathway of stationary cracks that form in the pipe wall. The weep times computed by the commercial regression Equation (13.1) are simply travel times, and have nothing to do with pipe deterioration or crack growth.

The pipe's long-term weep strength derives from the regression Equation (13.1) extrapolated to 20 years (oil pipes) or 50 years (sanitation pipes). The extrapolated hoop strain is the well-known long-term HDB – Hydrostatic Design Basis – the most widely quoted and the most meaningless of all composite pipe parameters. The common practice in the oil and sanitation industries defines the pipe's allowable hoop strain against long-term weep failure as the HDB – extrapolated to 20 or 50 years – divided by the safety factor SF = 1.8.

$$\left[allowable\ weep\ strain \right] = \frac{HDB}{1.8}$$

The HDB is the maximum static, constant, hoop strain that the pipe can sustain without weeping in the long-term. From the classical point of view, the HDB is the most important parameter in designing composite pipes. In our view, it is the most meaningless. Later in this chapter, we will argue against the HDB and propose its replacement by the weep threshold. For now, however, we proceed with our discussion of the classical HDB.

Figure 13.1 shows the experimental weep lines of oil and sanitation pipes. The oil pipes have no weep barrier and their regression line plots below that of sanitation pipes, which have a weep barrier of chopped glass.

Example 13.1

Compare the weep and rupture durability of pipes under (a) monotonically increasing pressure, (b) constant pressure, (c) constant peak cyclic pressure and (d) a combination of cyclic and static pressures.

a. Suppose a water-filled pipe submitted to a monotonically increasing pressure. The crack density grows in response to the increasing pressure. At first, the low pressure is not large enough to link up the cracks. As the pressure increases, however, the growing crack density links up and the pipe starts the weep process. The applied pressure may increase past the weep point, until the pipe bursts. The short-term weep failure usually occurs before the short-term burst rupture. However, if the rate of pressurization is too fast, or the wall thickness too large, the pipe may burst before weeping.

b. Consider now the water-filled pipe subjected to a non-increasing and static pressure. The static pressure does not cause crack growth. If the strain is less than the weep threshold, the pipe never weeps. In this scenario, the pipe eventually ruptures from long-term fiber strain-corrosion and never weeps.

c. Under cyclic pressure, which grows the crack density, the pipe may weep before rupture.

d. The discussion of combined pressures (cyclic + static) requires the unified equation introduced in Chapter 16.

It is instructive to ponder this example. The arguments presented are indeed powerful and show beyond a reasonable doubt that:

1. Static strains above the weep threshold always weep the pipe.
2. Static strains below the weep threshold fail the pipe by long-term rupture, not by weeping.

In the presence of combined static and cyclic loads, no such sweeping generalization is possible. The solution to the durability issue in such cases requires the unified equation discussed in Chapter 16.

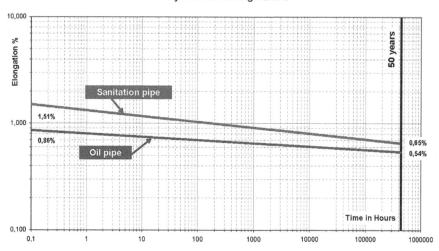

FIGURE 13.1 *Typical weep regression lines of sanitation and oil pipes. Notice the flatter line of oil pipes. The UD plies control the weeping of oil pipes. In sanitation pipes, the weep barrier of chopped glass has the control.*

Example 13.2

The weep regression line of sanitation pipes in Figure 13.1 indicates an extrapolated long-term (50 years) hoop strength of 0.65%. What is the 50 years HDB? What is the allowable hoop strain for a durability of 50 years?

The 50 years HDB is, by definition, 0.65%. The allowable 50 years hoop strain is

$$\left[allowable\ weep\ strain \right] = \frac{0.65\%}{1.8} = 0.36\%$$

Example 13.3

This example explains the differences between the weep lines of oil and sanitation pipes. From Figure 13.1 the differences requiring explanation are: (a) the slopes and (b) the relative positions of the two lines.

We begin with the relative position, which depend on the crack openings and the lengths of the pathways they form. We recall that the sanitation pipes are pressure tested under 2:1 loadings, which favor the crack openings. The sanitation pipes, by contrast, are pressure tested to 1:0 loadings and see only hoop – no axial – strains. Furthermore, for any given hoop strain, the weep barrier of sanitation pipes develops many small and short cracks, in contrast with (see Figure 13.2) the long and easy to traverse macro-cracks that arise in the UD plies of oil pipes. For any given hoop strain, the travel time of water in oil pipes is less than in sanitation pipes of the same thickness. These arguments explain why the weep lines of sanitation pipes plot above those of oil pipes.

Next, we address the different slopes. The few long UD macro-cracks in Figure 13.2 do not open much in response to increasing strains. It takes large strain increments to produce small increases in the crack opening. As a result, the weep times of oil

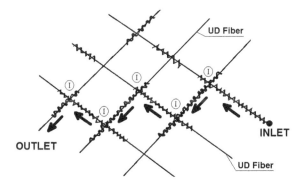

FIGURE 13.2 *The arrows indicate the pathway followed by the weeping water along the cracks depicted as wiggly lines. The advancing water moves from ply to ply at the crack intersection points "I".*

pipes respond poorly to pressure increments. By contrast, the total cross sectional area of the many cracks in the weep barrier of chopped strands responds better to pressure increments. The slope issue relates to crack cross sectional changes in response to pressure increments. In a nutshell…. the many cracks in the weep barrier of sanitation pipes display high cross sectional increases in response to strain increments. The opposite is true of the fewer cracks in oil pipes. That explains the difference in slope.

13.3 THE WEEP THRESHOLD

So far, we have discussed the old idea involving the classical HDB. We move now to the new ideas that support the concept of weep threshold. We start the discussion describing the mechanism of crack formation in the critical chopped ply of sanitation pipes under monotonic increasing pressures. To facilitate the presentation, we use a series of drawings labeled as Figure 13.3. Figure 13.3a shows a crack-free ply under low pressure. The next Figure 13.3b shows the onset of cracking in response to pressure increments. The subsequent figures describe the crack development. The captions explain the sequence of events. Figures 13.3a through 13.3d are applicable to chopped plies.

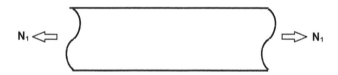

FIGURE 13.3A   *Under low pressures, below the infiltration threshold, the strains are too small and there is no significant cracking. The absence of cracks slows down the rate of chemical penetration. In this condition, the corrosive chemicals penetrate the laminate by the slow process of diffusion. The delayed penetration of corrosive chemicals below the infiltration threshold substantially increases the pipe's chemical durability.*

FIGURE 13.3B   *At higher pressures, the fiber-resin debonds initiate the first cracks. At this point, the chemical infiltration becomes significant, and has a substantial effect on the chemical life of the laminate. The onset of cracking marks the infiltration threshold. The infiltration threshold is a very important parameter in the design of composite equipment in corrosive environments. The drawing in this figure is applicable to chopped plies.*

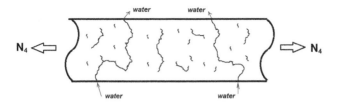

FIGURE 13.3C   *Further pressure increments increase the crack density. The reader will remember that stationary cracks do not grow under constant static loads. The cracks grow only if the pressure – or strain – is increased. The UD plies (not shown here) grow large macro-cracks parallel to fibers. The randomly dispersed chopped fibers in the weep barrier (shown here) keep the cracks small in the entire process. The larger cracks shown in the figure result from coalescence of smaller cracks. This is applicable to chopped plies.*

FIGURE 13.3D   *Eventually the many small cracks in the weep barrier of chopped fibers reach a critical density and coalesce. Further pressure increments will increase the number and opening of the already coalesced cracks. The cracks in one ply link up with similar cracks in neighbor plies. This allows the water to move from ply to ply and eventually weep. The critical strain – crack density – initiating the crack coalescence is the weep threshold. If the strain is less than the weep threshold, the pathway does not form and the pipe will not weep. If higher than the threshold, the pathway forms and the pipe will certainly weep. The time to weep depends on the number, opening and length of the cracks. It depends also on the pipe wall thickness. The higher the pressure, or the strain, the shorter is the weep time. Again, these concepts relate to stationary cracks that do not grow under static pressures. Again, this drawing describes the situation in chopped plies.*

The discussions in Figure 13.3 leave no doubt as to the crack density growth in response to increasing tensile strain. The crack density grows also in response to cyclic strains. Under constant pressures, the crack densities are stationary and the pipes strained below the weep threshold never weep. The concept of weep threshold is too recent and still not recognized. Recognition should come soon. The HDB concept described in ASTM D 2992 B protocol is currently the only recognized method to predict the long-term weep durability of pressurized pipes.

The above description of the weep failure in chopped plies is easy to extend to UD plies. Furthermore, a similar discussion is applicable to the resin dominated infiltration and stiffness failures.

13.4 LINER EFFECT ON THE WEEP THRESHOLD

This is the appropriate moment to correct a widespread misconception about the liner toughness. We start with the obvious statement that pipes with uncracked liners do not weep, which suggests that increments in the liner resiliency improve the pipe's weep resistance. This appealing argument is false. The well-bonded and thin liners – no matter how tough – cannot stop the growth of cracks emanating from the underlying laminate. The liners are simply too weak to resist the high energy released in the growth of large laminate cracks.

The liner breaks from cracks originating in the substrate, not on the liner itself. Contrary to widespread belief, the liner is not the critical ply in the weeping process. The critical weep ply in oil pipes is the UD ply in direct contact with the liner. In sanitation pipes, the critical ply is the weep barrier of chopped glass, also in direct contact with the liner. In both cases, the liner cracks emerge from the substrate plies. We will make extensive use of this fundamental fact in our further discussions. See the Appendix 11.2 in Chapter 11.

Many authors have observed and reported the above events. For a quick review, see references 5 and 9. The "in-situ rupture strains" of embedded liners are significantly lower than the lab values measured in isolated sheets. As we have just seen, the in-situ liner cracks originate in the substrate, not in the liner itself. The aluminum foil used in impermeable pipes provides an interesting illustration to this discussion. The extremely resilient foil has a gigantic elongation at break of 30%, at least 10 times higher than the short-term rupture strength of chopped plies. A quick consideration would expect the pipes with such a tough foil to be weep-proof. Not true. The tough aluminum foil certainly improves the pipe's weep threshold but do not make them weep-proof. The cracking of the substrate ply governs the in-situ foil rupture, which takes place at a global strain substantially less than the lab-measured value. When pressurized above the weep threshold, the impermeable pipes weep just like any other. The explanation for this "unusual" and "unexpected" result is straightforward and simple. For details, see Chapter 20.

In conclusion, the reader should make an effort to remember the counter-intuitive fact that it is the substrate ply bonded to the liner, and not the liner itself, that controls the weep process. The use of tougher liners has no appreciable effect on the pipe's weep threshold. Few people realize this important concept. Substantial improvements to the pipe's weep threshold come from the use of tough resin matrices in the critical ply and, of course, the use of aluminum foil.

13.5 INADEQUACY OF THE CLASSICAL HDB

The classical HDB derives from a false assumption combined with a meaningless variable. The false assumption is the growth of stationary cracks. The meaningless variable is the weep time, which is nothing more than the travel time of water in the cracked pipe wall. The pipe's weep lines predict the weep time at high static pressures, above the weep threshold, and outside the range of commercial service. The commercial pipes operate under low strains, below the weep threshold, and never weep under constant pressures. So, why bother with the pipe's weep line and the weep times?

Furthermore, the weep lines depend on the pipe's wall thickness and the extrapolated HDBs derived from them are not fundamental material properties. This background suffices to reject the classical HDBs as meaningless. The first author to reject the classical HDB was Frank Pickering (reference 8) who, in 1983, proposed its replacement for the hoop strain at the onset of nonlinear pipe response. That was a good start. This book takes one further step and proposes its replacement by the weep threshold of the critical ply. The weep threshold of the critical ply is a better indicator than the onset of pipe linearity. As we have said before, the weep threshold is a ply property, not a pipe property. We will have more to say about this in the following section. For now, it suffices to say that the classical HDB is a flawed predictor of long-term weep failure.

Figure 13.4 shows the horizontal line corresponding to the weep threshold $Tw = 0.80\%$ superimposed on the classical weep line of sanitation pipes. The figure shows the short-term inclined part of the classical regression line, valid for strains above the weep threshold. The extrapolated long-term HDB predicted by the classical line (0.65%) falls below the weep threshold of 0.80%.

13.6 MEASURING THE FAILURE THRESHOLDS

This book is about the durability of composites in industrial service, not about the measurement of ply properties. However, we believe a few words may be necessary to illustrate the methods to measure the still unfamiliar infiltration, stiffness and weep thresholds. We start by reminding the connection between the ply strength and the damage densities. This connection allows the use of simple ply strength tests to measure complex damage densities. The short-term critical damage densities – the thresholds – derive from simple measurements of the short-term ply strengths. The controversy in this otherwise simple protocol is the identification of the "failure point" of each failure mode. The failure points are special transitions observed in plies subjected to short-term quasi-static loads. The indicators of such transitions are:

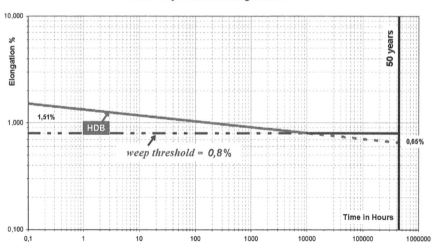

FIGURE 13.4 *Weep regression line of sanitation pipes superimposed on the weep threshold. The suggested weep threshold Tw = 0.80% is reasonable for chopped plies of polyester matrices with 3.0% elongation. The weep threshold Tw = 0.80% is substantially higher than the 50 years HDB = 0.65%. Looking at the figure, we see a weep line composed of two straight parts. The inclined part corresponds to strains above the weep threshold. The horizontal part corresponds to strains less than the weep threshold. The pipes operating at static strains less than the weep threshold never weep.*

- Sudden and sharp increase in acoustic emissions
- Sudden and sharp drop in light transmission
- Sudden and sharp change from linear to non-linear behavior

The ASME pressure code section X proposes an interesting test method – based on acoustic emission – that serves well to measure the ply thresholds. For details, see the ASME code section X, article RT – 8, titled "Test methods for determining damaged-based design criterion". The appendix in this chapter also describes an interesting method to measure the infiltration threshold.

Tables 8.3, 8.4 and 8.5 of Chapter 8 list all thresholds of commercial interest. The Tables 13.1 and 13.2 below are copies of Tables 8.4 and 8.5, repeated here to facilitate the discussion.

Let us pause for a moment and take a detailed look at Tables 13.1 and 13.2. The following points need consideration.

- The failure thresholds are short-term ply strengths valid for any combination of cyclic and static loading, short-term or long-term.

TABLE 13.1

Failure thresholds of transverse loaded UD plies. This is the same as Table 8.4 of Chapter 8.

| | Thresholds of transverse-loaded UD plies | |
	Polyester	Vinyl Ester
Infiltration	Ti = 0.20%	Ti = 0.30%
Weep	Tw = 0.25%	Tw = 0.40%
Stiffness	Ts = 0.40%	Ts = 0.60%
Rupture	Tr = 0.40%	Tr = 0.60%

TABLE 13.2

Failure thresholds of chopped plies. This is the same as Table 8.5 of Chapter 8.

| | Thresholds of chopped plies | |
	Polyester	Vinyl ester
Infiltration	Ti = 0.30%	Ti = 0.50%
Weep	Tw = 0.80%	Tw = 1.00%
Stiffness	------	------
Rupture	Tr = 1.50%	Tr = 2.50%

- The weep barrier of sanitation pipes is made of tough polyester with a 0.80% weep threshold. The classical 50 years HDB of sanitation pipes is 0.65%. Assuming a safety factor SF = 1.8, the allowable weep hoop strain would be 0.8/1.8 = 0.45% and 0.65/1.8 = 0.35% respectively for the weep threshold and the classical HDB concepts.
- The corrosion barriers of vinyl ester resins have infiltration threshold of 0.50%. Again, assuming a safety factor SF = 1.8, the allowable strain for corrosion barriers of vinyl ester resins would be 0.50/1.8 = 0.28%. This allowable strain is significantly above the 0.20% mentioned in the literature.
- The corrosion barriers of brittle bisphenol A resins have infiltration threshold of 0.30% and allowable strain of 0.30/1.8 = 0.17%. This allowable strain is considerably higher than the 0.10% mentioned in product standards for chemical service, such as NBS PS 15 69 and ASME RTP1.
- The vinyl ester resins are the best choice for oil pipes that have no weep barrier.
- The vinyl ester resins are the best choice for stiffness controlled wind blades.

The long-term safety factor SF for all load-dependent failures (infiltration, weep, stiffness and rupture) derives from the unified equation, and takes into account the combined (cyclic + static) loading. The SF = 1.8 listed in the commercial standards is a short-term safety factor.

The failure thresholds listed in Tables 13.1 and 13.2 are ply properties listed in the UD local ply directions 1 and 2. The structural analysis, however, computes the laminate strains in the global directions x and y. To make comparisons, we need to rotate the computed global strains to the local frame of each UD ply. Equation (13.2) shows the rotation matrix.

$$
\begin{bmatrix} \varepsilon_1 \\ \varepsilon_2 \\ \tfrac{1}{2}\gamma_{12} \end{bmatrix} = \begin{bmatrix} \cos^2\alpha & sen^2\alpha & 2(\cos\alpha)(sen\alpha) \\ sen^2\alpha & \cos^2\alpha & -2(\cos\alpha)(sen\alpha) \\ -(\cos\alpha)(sen\alpha) & (\cos\alpha)(sen\alpha) & \cos^2\alpha - sen^2\alpha \end{bmatrix} \times \begin{bmatrix} \varepsilon_x \\ \varepsilon_y \\ 0 \end{bmatrix}
$$

(13.2)

Where:

   α is the angle between the UD fibers and the longitudinal (axial) direction.
   ε_1 is the strain in the fiber direction
   ε_2 is the strain transverse to the fiber
   ε_x is the global strain in the axial direction
   ε_y is the global strain in the hoop direction
   γ_{12} is the local shear strain on the UD ply

Expanding Equation (13.2), we obtain the three strain components on the local reference frame 1–2 of the UD ply.

$$\varepsilon_1 = \varepsilon_x \cos^2\alpha + \varepsilon_y sen^2\alpha \tag{13.2A}$$

$$\varepsilon_2 = \varepsilon_x sen^2\alpha + \varepsilon_y \cos^2\alpha \tag{13.2B}$$

$$\gamma_{12} = 2sen\alpha\cos\alpha\left(\varepsilon_y - \varepsilon_x\right) \tag{13.2C}$$

Equation (13.2A) computes the UD strain component in the fiber direction. This strain controls the rupture durability of the UD ply discussed in the previous chapter.

Equation (13.2B) computes the UD strain component in the direction trans-verse to the fibers. This strain controls the weep and stiffness failures of oil pipes and wind blades.

Equation (13.2C) computes the UD shear strain, which is relevant in the esti-mation of the long-term laminate stiffness.

The chopped plies are isotropic and do not require strain rotation.

Example 13.4

There are two approaches to predict the weep failure of composite pipes under static pressures.

- The classical approach assumes the growth of initially small crack den-sities until the pipe weeps.
- The weep threshold approach proposes a threshold strain below which weeping never occurs.

A conclusive experiment to validate either approach requires a long testing time. In 2009, Hogni Johnson reported the results of a comprehensive test on deflected pipes subjected to strain-corrosion for a period of 30 years. The data from Hogni is the best experimental evidence available in favor of the weep threshold. For a brief discussion of this interesting experiment, see Chapter 14. Figure 13.5 shows the flattened regression line reported for the deflected pipes, exactly as the weep threshold proposed in this chapter. This is the best evidence available at this time in support of the weep threshold.

Some people may argue that a strain-corrosion test on deflected pipes is not a valid indicator of weep failure. A rebuttal of this argument is presented in Chapter 14 with the explanation that strain-corrosion of deflected pipes bears a very close relationship to weep failure. The results reported in Figure 13.5 are valid proof of the weep threshold.

Example 13.5

An oil pipe manufacturer reported the 20 years HDB as $\varepsilon_y = 0.40\%$. How would this value compare with the weep thresholds of UD plies in Table 13.1?

We recall that oil pipes do not have weep barriers and therefore the ± 55 UD is the critical weep ply. In addition, a 2:1 pressure loading applies when testing for the HDB of oil pipes. Referring to Chapter 4 we recall the relation between the axial and hoop strains for $\alpha = \pm 55$ angle-ply pipes subjected to a 2:1 loading.

$$\varepsilon_y = 4 \times \varepsilon_x$$

From the above we compute the axial strain on the UD ply corresponding to the reported hoop HDB = $\epsilon_y = 0.40\%$

$$\varepsilon_x = \frac{0.40\%}{4} = 0.10\%$$

From Equation (13.2B) the transverse long-term pipe strength corresponding to the reported HDB is

$$\varepsilon_2 = \varepsilon_x sen^2\alpha + \varepsilon_y \cos^2\alpha \qquad\qquad (13.2B)$$

$$\varepsilon_2 = 0.10 \times sen^2 55 + 0.40 \times \cos^2 55 = 0.20\%$$

Table 13.1 indicates the above transverse long-term strength as too low for pipes of tough vinyl ester or epoxy matrices. A more representative value would be $\varepsilon_2 = 0.40\%$. Assuming a conservative $\varepsilon_x = 0.30\%$ as the correct value and back calculating we obtain the actual hoop weep threshold as $\varepsilon_y = 0.60\%$, which is 50% higher than the reported HDB of 0.40%. This confirms our claim that the extrapolated HDB is significantly lower than the weep threshold.

This completes our discussion of the classical HDB versus the weep threshold. The substitution of the classical HDB for the weep threshold would bring the following benefits:

- The concept of weep threshold is consistent from a materials point of view.
- The weep threshold is easy to measure.
- The weep threshold is higher than the classical HDB.

The next section addresses the cyclic loadings.

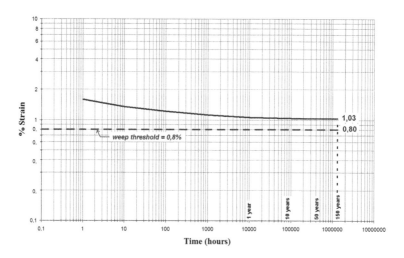

FIGURE 13.5 *The long-term strain-corrosion tests of sanitation pipes in 5.0% sulfuric acid shows a flattened regression line levelling off at a hoop strain of 1.03%. The graph also shows the proposed weep threshold of 0.8% for sanitation pipes.*

13.7 CYCLIC LOADINGS

The cyclic and static load actions differ in one important point, repeated many times in this book. The static loads do not grow cracks. The cyclic loads causing crack growth are the real game changers in the study of infiltration, weep and stiffness durability. Given enough time, the cyclic loads will grow the initially small defects present in pristine plies into the infiltration, stiffness and weep critical damages.

The laminate durability is the number of cycles required to grow the initial small defects into the long-term critical damage on the critical ply. In the general case of cyclic + static loadings, this computation requires the use of the unified equation. In the simple cases of pure cyclic loadings, the ply durability derives from the pure regression cyclic regression lines. This reasoning leads to the following statements:

The infiltration life of the critical ply is

$$\left[\textit{infiltration life} \right] = \left[\textit{cycles to reach the infiltration critical damage} \right]$$

The weep life is

$$\left[\textit{weep life} \right] = \left[\textit{cycles to reach the weep critical damage} \right]$$

The stiffness life is

$$\left[\textit{stiffness life} \right] = \left[\textit{cycles to reach the stiffness critical damage} \right]$$

The pure cyclic equations are available for all plies of commercial interest. The above lives are easy to compute in cases involving pure cyclic loadings See Table 9.1 in Chapter 9. The unified equation is required in cases involving complex loadings.

APPENDIX 13.1

MEASURING THE INFILTRATION THRESHOLD

The infiltration threshold is a property of corrosion barriers made of premium resins and chopped fibers. Although never explicitly recognized, the infiltration threshold is the intuitive basis for the allowable strains of laminates in chemical service. An interesting paper confirming this was published in 1993 at the 9th International Conference on Composite Materials.

See reference 12. The paper describes the measurement of the infiltration threshold of chopped plies, while not explicitly acknowledging its existence. The following excerpt from the original work was slightly adapted to comply with the nomenclature of this book.

"The mechanics of damage initiation in a number of tensile loaded plies of chopped fibers have been studied. Damage was characterized by undertaking quasi-static tensile tests in which the load is progressively increased until failure. By measuring the reduction in Young's modulus, the damage initiation strain has been determined. The technique can also be used to develop the regression equation for long-term fatigue loading. It is believed that this simple test provides a useful method for characterizing the strain at damage initiation in corrosion barriers".

The "strain at damage initiation" mentioned above is obviously the infiltration threshold defined in Chapter 8. We continue with the quote of the original paper.

"The stress-strain curve was determined for a chopped ply of Derakane 411 resin. The originally linear response became progressively non-linear with increasing stress. A closer examination of the specimens indicated the onset of linearity coincident with the beginning of acoustic emissions and loss of translucency. The onset and subsequent development of damage were then examined in more detail by a series of quasi-static load tests followed by measurement of the Young's Modulus. Figure 13.6 shows a typical damage development graph for this material. The onset of damage – ε_d – was determined by extrapolating the data back to a zero value as shown in Figure 13.6. The resulting ε_d for this particular test is 0.50%, a typical value for Derakane 411. The failure process was studied in more detail by stopping certain tests before fracture, then sectioning and polishing the cut edges before examination under a microscope. It was found that initial damage took the form of fiber–resin debonding in bundles oriented perpendicular to the applied load. With increasing load, the debonding became more intense and extended as resin cracks."

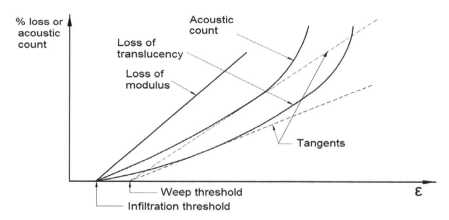

FIGURE 13.6 *The onset of fiber-resin debonds triggers the acoustic emissions and marks the infiltration threshold. The weep threshold occur at higher emissions. Like all ply property, the failure thresholds fall within a range of values. The infiltration and weep thresholds are arbitrarily set within the failure zones.*

TABLE 13.3

Infiltration thresholds for the corrosion barriers of several vinyl ester resins.

Resin matrix	Elongation at break	Infiltration threshold and glass loading		
		15%	30%	60%
DK 470	3.0%	------	0.32%	------
DK 411	5.0%	0.65%	0.50%	0.40%
DK 8084	15.0%	------	0.75%	------

The paper continues with a discussion of the effects of resin toughness and glass loading on the infiltration threshold of corrosion barriers. The reported data are in Table 13.3.

14 Laminate Strain-Corrosion

14.1 INTRODUCTION

There are two types of long-term failure from strain-corrosion in composites, both involving an accelerated chemical attack in the presence of strains. The first type, known as fiber strain-corrosion, has a global presence and deteriorates all fibers tensioned in the presence of water. We have already had a brief exposure to this type of strain-corrosion in Chapter 12, when discussing the process of long-term fiber rupture. The second type, known as laminate strain-corrosion, results from a local concentrated attack on small parts of the laminate bent in the presence of chemicals.

- The fiber strain-corrosion is a global process present in the entire laminate. As implied in the name, this type of deterioration is limited to the glass fibers. The resin matrix remains intact. The long-term laminate failure results from the accumulation and coalescence of ruptured fiber points, with no single visible crack.
- The laminate strain-corrosion – or simply strain-corrosion – is a local process present on small areas of the laminate subjected to bending strains in the presence of chemicals. The long-term failure results from the growth of a long and clearly identifiable laminate crack.

Both types of strain-corrosion deterioration lead to long-term laminate rupture. For details on fiber strain-corrosion, see Chapter 12. This chapter deals with the process of laminate strain-corrosion. It works like this:

14.2 THE MECHANISM OF LAMINATE STRAIN-CORROSION

Suppose a small ply defect near the surface of a laminate exposed to a corrosive chemical. The defect need not be at the surface. It suffices that it is near the exposed surface. Suppose also an external load straining the laminate. The ply defect amplifies the strain field and locally accelerates the rate of chemical attack. Remember we are discussing the case of strained laminates immersed in corrosive products. The combination of magnified strain and chemical attack produce many micro-cracks emanating from the ply defect and transverse to the strain direction. Soon a single growing dominant crack emerges which, the larger it becomes the faster it grows. The seeding defect postulated here lie in a ply near the laminate surface and within reach of the aggressive chemical. The corrosive attack occurs on the overstrained fibers near the ply defect – as described in Chapter 12 – except that in this case it comes from a chemical product and not necessarily from water. The deterioration process is stronger on the fibers lying in the strain direction. The fibers perpendicular to the strain direction play no part in the process. The deteriorated fibers – those lying in the strain direction – hold the crack growth for a while, until they fully corrode and fail.

The fibers most tensioned, those aligned with the strain, are the ones that fail first, one by one, forming a crack that grows perpendicular to the laminate. The crack grows across the laminate, perpendicular to the strain direction forming two smooth surfaces that look as if cut by a sharp razor. Such smooth crack surfaces are in sharp contrast with the jagged, brushy surfaces observed in the rupture of laminates. The process continues as described, with a smooth crack growing across the fibers, cutting through the laminate, until the remaining thickness no longer sustains the load. At this point, the laminate ruptures.

The above is a fair description of the way laminates strain-corrode. The reader will note the same fiber degradation mechanism described in Chapter 12, except for the corrosive chemicals taking the place of water. The differences between fiber and laminate strain-corrosion are:

- The fiber strain-corrosion develops no visible cracks and involves the entire laminate. The laminate strain-corrosion develops a single large local crack.
- The fiber strain-corrosion comes from water. The laminate strain-corrosion comes from aggressive chemicals.
- The laminate strain-corrosion is restricted to bending loads.

The failures from laminate strain-corrosion are easy to identify. Their distinctive features are (a) a long and deep crack, (b) cutting across the laminate with (c) smooth surfaces and (d) no jagged or brushy fibers.

The strain-corrosion deterioration may involve the resin, the fibers or both. The resin deterioration is not relevant, since the fibers arrest the crack growth. As far as resins go, the strain-corrosion damage is limited to the shallow and harmless mud-cracks observed in liners exposed under tension to highly aggressive chemicals. Such environmental stress cracks exhaust all there is to say about resin strain-corrosion.

Things are different, however, when the chemical attack is on the fibers. The attacked fibers corrode and fail, producing a small laminate crack that opens under tension and exposes fresh material. The exposed fresh fibers hold the crack growth for a while until they too are strain-corroded.

In laminate strain-corrosion, the unchecked crack growth is perpendicular to the laminate, cutting through the fibers. This situation is in sharp contrast with the fundamental composite effect that denies the possibility of cracks growing across the fibers. What happens to the laminate outstanding toughness and fatigue strength when the composite effect is lost? Well... the outstanding toughness and fatigue resistance of the laminate simply vanish near the strain-corrosion crack. The following statement is close to reality.

All sudden in-service ruptures of composite structures have their origin in laminate strain-corrosion.

With the exception of gross operational errors and other accidents, all cases of premature in-service sudden laminate ruptures result from laminate strain-corrosion.

14.3   LAMINATE STRAIN-CORROSION AND BENDING LOADS

The preceding chapters highlighted the effects of chemicals, water and tensile loads on the durability of composites. The discussions there made no mention of bending loads. Now, as we discuss laminate strain-corrosion, the bending loads take a special significance. What makes bending loads so special in the process of laminate strain-corrosion?

As explained in Chapter 10, there are two classes of chemicals corrosive to composites.

- The non-reactive chemicals quickly pervade the entire laminate to reach all fibers in all plies. Water is the most notorious and important of these chemicals. In fact, as discussed in Chapters 9 and 12, the static long-term ply rupture comes from fiber strain-corrosion in the presence of water.
- The reactive chemicals slowly penetrate the laminate and for all practical purposes never reach the fibers in the structural plies.

The reader may wonder why linger on facts that have already been mentioned several times. Our purpose in doing so is to set the stage to explain the action of bending loads, which we proceed to do.

Figure 14.1a shows the penetrated depth on a tensile loaded laminate exposed to an aggressive chemical. The tensile strain is the same on all plies and the computation of the penetrated depth "Δe" is as explained in Chapter 11. Figure 14.1b shows the same laminate exposed to the same chemical under a bending load. Three interesting things happen to the strains in this new

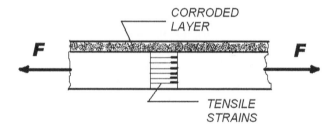

FIGURE 14.1A *The chemical presence causes a uniform degradation of the corrosion barrier in tensile loaded laminates. See Chapter 11.*

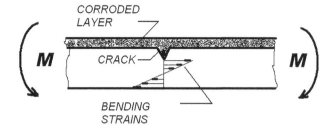

FIGURE 14.1B *The tensile strains in Figure 14.1a are uniform in the laminate thickness. By contrast, the bending strains in Figure 14.1b concentrate mostly on the outer fibers near the laminate surface. The large bending strains near the laminate surface start the strain-corrosion process.*

scenario. First, they are now bending, not tensile. Second, they are larger near the exposed laminate surface than near the center. Third, they become significantly larger as the chemical degradation reduces the local load bearing thickness of the laminate. These events favor crack initiation in the defects near the exposed surface. A dominant crack soon emerges that starts the laminate strain-corrosion process.

It can be shown mathematically (see any book on fracture mechanics) that bending loads give higher strain intensity factors than tensile loads. Have you ever tried to break a pencil? You would instinctively do it in bending, not in tension. The bending loads grow the initially small surface flaw into a large crack that produce a fast and localized rupture. This phenomenon is exclusive to bending loads in the presence of chemicals that attack the fibers. As a rule, it does not happen with tensile loads.

There is one rare instance of strain-corrosion caused by tensile loads. Such a rare situation occurs in exceptional cases of large pre-existing surface cracks running perpendicular to the strain direction. This unlikely scenario favors the crack growth by tensile strain-corrosion, as in the bending case discussed above. See Figure 14.2.

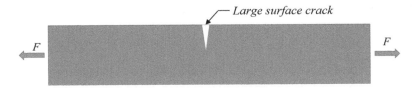

— Large surface crack

F   F

FIGURE 14.2   *Tensile strain-corrosion is rare and requires the pre-existence of a surface crack transverse to the strain direction.*

Example 14.1

Explain the differences between laminate tensile and bending strain-corrosion.

The small surface defects seeding the bending strain-corrosion usually hide beneath healthy liners and are difficult to spot. The bending strain-corrosion process develops underneath healthy liners, in the absence of visible surface cracks. The basic requirements to get the bending strain-corrosion process going are the presence of a ply defect near the laminate surface, a bending strain and contact with an aggressive chemical. The absence of visible surface flaws makes the early detection of bending strain-corrosion practically impossible. The best strategy to minimize the risk of strain-corrosion is to increase the laminate thickness in those areas subjected to bending strains. This is the procedure followed in product standards like ASME RTP1, which demand increased thickness in the knuckle transition of storage tanks.

The tensile strain-corrosion requires the presence of a large surface crack that are clearly visible to the inspector. The presence of surface cracks – like cracked liners – is inadmissible in aggressive chemicals, as they are a possible seed for tensile strain-corrosion. For details, see Appendix 11.2.

Summarizing this section, the laminate strain-corrosion process occurs on those parts of the laminate that are subjected to bending loads, like knuckle transitions in vertical tanks, pipeline saddle supports, pipe to flange transitions, etc. One important practical application involving laminate strain-corrosion is underground sanitation pipes. Most of the following discussion centers on the strain-corrosion of underground sanitation pipelines.

14.4   LAMINATE STRAIN-CORROSION CRACKS

The laminate strain-corrosion cracks start as small flaws that grow under bending loads in the presence of chemicals. The strain-corrosion cracks nucleated by the flaws are easy to identify and difficult to explain. They are usually long and deep, with smooth surfaces cutting across the fibers and perpendicular to the laminate. The strain-corrosion rupture develops in two stages. The initial stage grows the crack with smooth surfaces, and locally reduce the laminate thickness. In the final stage, the thin remaining laminate cannot hold the large bending strain and ruptures in the usual way, forming jagged and brushy surfaces. The initial stage results from strain-corrosion. The final stage is the normal rupture of the residual thin laminate, not strong enough to hold the bending load.

The strain-corroded fibers fail by chemical action, not by mechanical stresses. The aggressive chemical cuts the fibers like a sharp razor, forming smooth crack surfaces with no jagged, frayed and brushy fibers. Furthermore, in most cases the strain-corroded crack walls have a pristine look, indicating absence of chemical attack on the resin. The inspector sees in front of him a long and deep, smooth-walled crack surrounded by a perfectly healthy laminate. The pristine look of the resin shows no sign of chemical attack. Where did such a large crack come from? What caused it? This puzzling and mysterious crack is hard to explain. The non-initiate inspector may hack his brain and never figure out an explanation. See Figure 14.3a.

Example 14.2

Figure 14.3b shows a piece of pipe failed by strain-corrosion. The pipeline carried a solution of acidic chlorine in a pulp bleaching plant. The chlorine attacks the resin with little effect on the glass fibers. The acid, on the contrary, corrodes the fibers with little effect on the resin. The pipeline installation and operation followed strict and correct procedures. Still, one of the pipes locally failed forming an ugly large longitudinal crack.

This is one example of those rare cases of tensile strain-corrosion. What most likely happened was a large impact that cracked the liner. The liner crack magnified the tensile strain in the presence of acid, setting the stage for the strain-corrosion process. The rupture evolved like this.

1. The chlorine solution inside the pipe strain-corroded and mud-cracked the resin in the liner. This is a normal and expected liner situation, and has nothing to do with the problem at hand.
2. The liner crack magnified the tensile strain field from the internal pressure.
3. The strained fibers in the corrosion barrier had direct contact with the acid solution. This started the strain-corrosion process that grew the crack.
4. The crack grew from the inside of the pipe, moving to the outer surface.
5. The chlorine had no part on the strain-corrosion process. The acid is the sole culprit.
6. The outside pipe surface, not in contact with the acid, broke in the usual way showing typical jagged and brushy fibers.

The long and deep longitudinal crack has neat, smooth-walled faces that cut across the fibers. Such cracks are typical of all laminate strain-corrosion.

The chemicals in this case are not overly aggressive to the resin and take a long time to penetrate beyond the liner. If the strain-corrosion process had not developed, this failed pipe would have provided an extremely long service life.

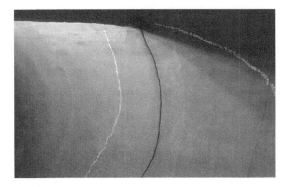

FIGURE 14.3A *Laminate strain-corrosion. The inspector sees a long, deep and smooth-walled crack surrounded by a perfectly healthy laminate. The pristine look of the resin shows no sign of chemical attack. Where did such a large crack come from?*

FIGURE 14.3B *Illustrating the strain- corrosion crack of the pipe discussed in Example 14.2. The crack has neat and smooth surfaces, instead of the jagged, splintered and brushy surfaces typical of normal ruptures. The chlorine attack on the corrosion barrier is barely perceptible. The strain-corrosion crack initiated and progressed from the inside of the pipe.*

14.5 PREDICTING THE TIME TO STRAIN-CORROSION RUPTURE

Predicting the time to strain-corrosion rupture is a complex task, affected by the magnitude of the bending strain, the chemical concentration and the corrosion strength of the glass fibers. To solve this problem, we should take into account the time taken by the chemical to (a) traverse the protective liner and (b) corrode the fibers. The following discussion assumes laminates with corrosion barriers of chopped fibers protected by resin-rich liners, like those found in chemical service and sanitation. The strain-corrosion process starts when the aggressive chemical traverses the liner and reaches the fibers in the corrosion barrier. The time to rupture is the sum of two partial times, corresponding to the liner penetration and to the fiber corrosion. The following equation applies.

$$[rupture\ time] = [penetration\ time] + [corrosion\ time] \tag{14.1}$$

The penetration time depends on the liner preservation in service.

1. Fully preserved liner slows down the chemical penetration.
2. Fully destroyed liners allow the quick access of the chemical. In such cases, the rupture time is equal to the corrosion time.

As we see, well-preserved liners slow down the access of chemicals to the underlying fibers. Since as a rule the aggressive chemicals quickly crack the liners, we propose to ignore them altogether. This is the same assumption made in Chapter 11, where we proposed to compute the service life of corrosion barriers ignoring the liner. With this simplification Equation (14.1) reduces to

$$[rupture\ time] = [corrosion\ time] \tag{14.1A}$$

The regression equation to predict the strain-corrosion time of unprotected – no liner – corrosion barriers under static bending forces is similar to the one developed in Chapter 11 for the service life of corrosion barriers.

$$\log(\epsilon) = A + B \times \log(c) + C \times \log(t) + \frac{D}{T} \tag{14.2}$$

Where
  ε is the static bending strain on the critical ply
  c is the chemical concentration in the solution
  t is the time to failure
  T is the absolute temperature
  A, B, C and D are the experimental strain-corrosion parameters

Assuming room temperature service and a fixed concentration of the corrosive solution, Equation (14.2) reduces to

$$\log(\epsilon) = A_{sc} - G_{sc} \times \log(t) \tag{14.2A}$$

In Equation (14.2A), Asc and Gsc are the regression strain-corrosion parameters experimentally measured at a constant temperature and a fixed chemical concentration. It is customary to represent the long-term strain-corrosion strength of composite plies as $\varepsilon = Sb$. From Equation (14.2A), the Sb value for a lifetime of 50 years is

$$\log(S_b) = A_{sc} - G_{sc} \times \log(50 \ years) \tag{14.3}$$

14.6 STRAIN-CORROSION IN UNDERGROUND SANITATION PIPES

The underground composite pipes operate in friendly and non-aggressive chemicals like urban sewage, water and a few mild industrial effluents. The really nasty and aggressive chemicals, like those in the chemical process industry and most industrial effluents, use above ground pipelines for fear of soil contamination. This is a wise and safe precaution, considering the possibility of laminate strain-corrosion in underground service.

The underground sanitation pipes operate in non-aggressive, stable and well-defined conditions. The transported fluid is either water or mild domestic sewage and the working temperature varies within a small range. The constant bending strains are easy to predict as a function of the pipe geometry and deflection. These stable and well-defined operating conditions facilitate the analysis of the strain-corrosion process in underground sanitation pipes. The laminate construction of sanitation pipes is in Figure 4.1, Chapter 4. The critical plies and the likely places of deterioration in underground pipes are marked in Figure 14.4.

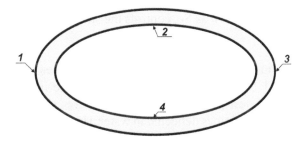

FIGURE 14.4 *The maximum bending strains in underground and deflected pipes occur at the locations shown by the numerals. The innermost weep barrier is tensile loaded at the points 2 and 4. The outermost UD plies are tensile loaded at the locations 1 and 3.*

The mild chemicals involved in sanitation service – water and urban sewage – do not harm the resin and fully preserve the liner. The intact liner shields and retards the penetration of the sewer-generated sulfuric acid into the weep barrier. However, from Figure 11.1 of Chapter 11, the presence of intact liners have no effect on the acid concentration at the fiber surface. While retarding the acid penetration, the intact liner provides no long-term protection against strain-corrosion. We repeat. The intact liner does not affect the long-term acid concentration in the corrosion barrier and therefore gives no long-term protection to the pipe. In keeping with everything done so far, the discussion of the strain-corrosion life of underground sanitation pipes ignores the liner presence.

The critical plies to strain-corrosion in underground sanitation pipes are:

- *Outer UD plies.* The outermost UD ply is strain-corroded by the moisture in the soil at the locations 1 and 3 of Figure 14.4. The long-term strain-corrosion strength of this ply is the same as the long-term tensile rupture strength Ss of UD plies immersed in water, discussed in Chapter 12. In both situations, the UD fibers are statically loaded in the fiber direction in the presence of water. For details, see the comments in Example 14.3 below. As explained in Equation (14.3), it is usual to refer to the long-term strength of the outer UD plies in underground pipes as Sb. Therefore, Sb = Ss.
- *Inner chopped fibers.* The innermost ply of chopped fibers is strain-corroded by acid (sewer service) or water at the locations 2 and 4 of Figure 14.4. In water service, the Sb of the inner weep barrier is higher than the Sb of the outer UD fibers and, for that reason, is not measured. The outer UD plies always fail first in water. In sewer service, however, the long-term strength of the inner weep barrier – usually called Sc – may be less than the Sb of the outer UD plies in water. In this situation, the long-term strength of the weep barrier in acid, known as Sc, controls the strain-corrosion durability of sewer pipes.

The critical strain-corrosion ply of underground pipes in water service is the outermost UD ply, which fails sooner than the inner weep barrier. The ASTM D 5365 test protocol allegedly measures the long-term strain-corrosion strength Sb of pipes deflected in water. However, as explained later, the ASTM protocol tests fully immersed pipe specimens and give results coming from both the outer UD and the inner chopped plies. The long-term strength Sb measured in ASTM D 5365 comprises a meaningless and useless mixture of outer UD and inner chopped fiber failures. The following example expands on this.

Example 14.3

Explain the significance of the two ASTM strain-corrosion test methods to measure the long-term strength of underground composite pipes.

The soil overburden deflects the underground pipes, giving rise to high bending strains. This condition makes the strain-corrosion failure a fundamental durability issue for underground pipes carrying water or urban sewage. There are two ASTM tests, both designed to determine the long-term bending strength of underground composite pipes in sanitation service.

The ASTM D 5365 measures the long-term bending strength of pipes deflected in water. The ASTM D 3681 does a similar evaluation for pipes deflected in a 5% solution of sulfuric acid. The regression lines generated in these tests allow the prediction of the long-term strength. Extrapolation of the water regression line gives the long-term bending strength of the pipe in water, known as Sb. Likewise the acid regression line gives the long-term bending strength in sewer service, known as Sc. Both methods measure the residual strength of pipe specimens at room temperature.

- The ASTM D 5365 test protocol evaluates the long-term bending strength Sb of deflected pipe specimens fully immersed in water. The full immersion implies water contact with the inner weep barrier as well as with the outer UD fibers. Therefore, the ruptures may come from either the outer UD fibers or the inner chopped fibers. The test method does not discriminate between these two obviously different failures. From a statistical point of view, the randomly dispersed inner chopped fibers should sustain higher strains than the aligned outer UD fibers. The observed ruptures should therefore come from strain-corrosion of the outer UD fibers. This, however, is not certain since the test results would depend on the laminate construction. The mixing of ruptures coming from UD and chopped fibers produce useless data. The ASTM D 5365 protocol needs modification to test separately and independently for the UD and chopped plies.
- The ASTM D 3681 test protocol measures the long-term bending strength Sc of pipes deflected in a 5% solution of sulfuric acid. The test protocol places the acid solution in direct contact with the liner, to simulate the acidic sewer. The problem here is similar to the one discussed in the measurement of the HDB. See Chapter 13. In both cases, the test protocol strains the specimens above the weep threshold. The overstrained pipe specimens allow the fast ingress of the acid solution and produces invalid data. The extrapolated Sc value from the ASTM D 3681 protocol is not representative of the long-term strain-corrosion strength of pipes in sewer service.

The following is an interesting discussion. The two ASTM test protocols just discussed strain the pipe specimens above the weep threshold. This highly cracked condition allows direct access of the test solution into the structural plies. This is of no consequence to the measurement of Sb, since the residual water in the pipe wall is enough to cause fiber strain-corrosion by itself, with no need of the "extra" test water. The same is not true of the acid solution.

The acid solution reaches the structural plies in full strength in the test specimens cracked above the weep threshold.

Below the weep threshold, the uncracked weep barrier keeps a low acid concentration in the structural plies. See Figure 11.1 in Chapter 11, as well as the findings of Regester in the Appendix 8.1 of Chapter 8. Above the weep threshold, however, the cracked weep barrier allows direct access of the full 5.0% acid concentration to the structural plies. The strain-corrosion rupture goes fast in such condition. The regression line developed by ASTM D 3681 in acid produces low Sc values that are not representative of real operating conditions. This is an undisputed fact.

Taking the above argument a bit further, we can imagine a sanitation sewer pipe with its weep barrier cracked from mishandling and impact. The acidic sewer would penetrate fast in such pipe and cause a short strain-corrosion life. The pipeline owner is always worried about damaging the weep barrier. The impermeable pipe technology described in Chapter 20 provides a simple and low cost solution to this concern.

14.7   STATIC STRAIN-CORROSION REGRESSION LINES OF SANITATION PIPES

This section discusses the static strain-corrosion regression equations available for underground sanitation pipes. The discussion of cyclic regression equations requires knowledge of the unified equation. The general form of the static regression strain-corrosion line is as in Equation (14.2A).

$$\log(\epsilon) = A_{sc} - G_{sc} \times \log(hours) \tag{14.2A}$$

The Greek letter "ϵ" denotes the sustained static hoop bending strain that ruptures the pipe by strain-corrosion at the time "t" usually expressed in hours. In other words, the ϵ symbol is the long-term bending strength of the pipe. The intercept Asc is related to the hoop short-term bending strength of the pipe, and the slope Gsc measures the rate of chemical attack on the glass fibers. Table 14.1 shows typical values of Asc and Gsc reported for sanitation pipes tested in water and in a 5% solution of sulfuric acid. The parameters in Table 14.1, derived from the faulty ASTM protocols discussed in the previous section, have no meaning. Their discussion here serves only an educational purpose. The values of Asc have been adjusted for strains in % and the time in hours.

The regression equations corresponding to the parameters in Table 14.1 are:

TABLE 14.1

The table shows the static strain-corrosion regression parameters and the 50 years bending strength of commercial sanitation pipes. The values of Asc have been adjusted to give the elongations in % and the failure time in hours.

	Asc	Gsc	Bending strength at 50 years
5% H₂SO₄ @ 25°C (ASTM D3681)	0.220	0.071	$S_c = 0.66\%$
Water @ 25°C (ASTM D5365)	0.334	0.039	$S_b = 1.30\%$

$$\log \varepsilon = 0.220 - 0.071 \times \log(hours) \; \left(\text{Pipes in 5\% } H_2SO_4\right) \qquad (14.4)$$

$$\log \varepsilon = 0.334 - 0.039 \times \log(hours) \; \left(\text{Pipes in water}\right) \qquad (14.5)$$

Figure 14.5 shows a plot of the above equations. As expected, the 5% sulfuric acid is far more aggressive than water. For a failure time of 50 years, the bending strength is Sb = 1.30% in water and just Sc = 0.66% in acid. Figure 14.5 shows identical short-term rupture strength in water and in acid, as expected. We will return to these numbers later.

Example 14.4

Describe the significance of the long-term pipe strengths Sc and Sb and their relations to the ply long-term and short-term strengths discussed in Chapter 8.

Strain corrosion of sanitation pipes

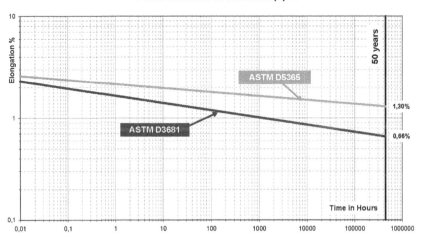

FIGURE 14.5 *Strain-corrosion lines of sanitation pipes in water (ASTM D 5365) and in 5% sulfuric acid (ASTM D 3681).*

The Sb strength is the long-term strain-corrosion bending strength of pipes deflected in water. There are two Sb values, one for the inner ply of chopped fibers and another for the outer ply of UD fibers. The Sb value of practical interest relates to the outer UD fibers. The Sb value for the outer UD ply is equal to the Ss value developed for UD plies tensile loaded in water in the fiber direction. Therefore, Sb = Ss.

The Sc strength has essentially the same meaning as the Sb, except the testing is in a 5% solution of sulfuric acid. Since these concepts are applicable to static loadings, we can conclude (see Example 14.6) that the long-term Sc strength in acid is equal to the weep threshold Tw. Therefore, Sc = Tw.

The concepts behind the long-term pipe strengths Sb and Sc are correct, even if their measurements by the ASTM test protocols are not. Both the Sb and the Sc are pipe properties defined by plies. We can say they are ply properties. The foregoing discussion allows the computation of the theoretical long-term Sb and Sc strengths of sanitation pipes. The values are in Table 14.2.

A close inspection of Table 14.2 reveals the following.

- The long-term bending strength Sb of the outer UD ply in water is the same long-term static strength Ss of UD plies discussed in Chapter 12. For boron-free glass, the 50 years value is Ss = Sb = 0.92%.
- Likewise, the long-term bending strength Sb of the inner chopped fibers in water is the same as the long-term static tensile strength Ss of chopped fibers in water. This Ss value comes from the static equation developed by the British Plastics Federation that gives Ss = Sb = 0.85% for brittle polyester resins. The value for the tough polyester used in sanitation pipes should be higher.
- The Sc strength of the outer UD plies – not in contact with sewage – is not applicable.
- The long-term bending strength Sc of the inner ply of chopped fibers in acid is equal to its short-term weep threshold, since this is a static situation. This assumption is justified by the low acid presence in the

---

TABLE 14.2

Suggested values for the long-term strain-corrosion strengths of underground pipes in water (Sb) and urban sewer (Sc). The listed values assume the UD plies in the hoop direction, as in hoop-chop pipes. Furthermore, since this is a static situation, the long-term inner ply strength in acid, Sc, is equal to the short-term weep strength, Tw.

	Outer UD ply in hoop-chop pipes	**Inner ply of chopped fibers**
Sb in water (50 years)	Equal to the long-term tensile strength of the UD ply (Sb = Ss = 0.92%)	Not available
Sc in sewer	Not applicable	Equal to the weep threshold of the chopped ply (Sc = Tw = 0.80%)

structural plies below the weep threshold. For details, see Figure 11.1 of Chapter 11 and the Appendix 8.1 in Chapter 8. See also Examples 14.5 and 14.6. See Section 14.8.

Example 14.5

The solubility of sulfuric acid in bisphenol A polyester @ 100C measured by Regester is very low. See Chapter 8. Describe the practical implication of such a low acid solubility.

The implication comes directly from the definition of solubility. The 5.0% acid sewer solution carried in the pipe gives a very low acid concentration in the corrosion/weep barrier. Such a low acid concentration would hardly harm the embedded chopped fibers and suggest a mild strain-corrosion process for strains below the weep threshold.

Example 14.6

Explain the protective mechanisms of the weep barrier and the liner in the strain-corrosion of sewer pipes.

The 5.0% acid solution present in urban sewage is not overly aggressive to polyester resins. However, like all acids, it has a considerable strain-corrosion effect on the fibers. The liner provides short-term protection by slowing down the acid penetration. The liner does not reduce the acid concentration in the weep barrier and gives no long-term protection.

The low solubility of sulfuric acid in polyesters assures a low acid concentration in the laminate, and little harm to the UD and chopped fibers. This situation holds as long as the weep barrier is below the weep threshold. Remember this is a static situation. However, should the pipe be strained above the weep threshold, the cracked weep barrier allows direct ingress of the full concentration acid into the underlying structural UD plies. The presence of high concentration acid in the structural plies cause quick strain-corrosion failure. We have the following situation:

- Above the weep threshold, the cracked weep barrier allows the fast ingress of the full strength acid into the structural plies. This promotes a fast strain-corrosion process.
- Below the weep threshold, the weep barrier keeps the low acid concentration in the structural fibers. This slows down the strain-corrosion process.

The strain-corrosion testing per ASTM D3681 deflects the pipe specimens above the weep threshold and produces regression equations that have no use in the estimation of the long-term strength Sc. In static service, the long-term strain-corrosion strength Sc of underground sewage pipes is equal to the weep threshold of the chopped plies. See Table 2.

Example 14.7

Compute the 50 years Sb strength of underground pipes of boron-free glass. Suppose the UD plies laminated with $\alpha = \pm 70$ degrees, $\alpha = \pm 55$ degrees and $\alpha = 90$ degrees. The long-term static strength of boron-free fibers at 50 years is Ss = 0.92%. See Chapter 12.

The long-term Sb strength defined in the pipe standards and recognized by the pipe manufacturers is the global hoop strain, not the strain in the fiber direction. The recognized Sb value is therefore equal to ϵ_y, not equal to ϵ_1. The strain rotation from the local fiber direction to the global hoop direction is according to Equation (12.2A) developed in Chapter 12.

$$\varepsilon_1 = \varepsilon_x \cos^2 \alpha + \varepsilon_y sen^2 \alpha$$

Since the global strain in the axial direction of underground pipes is zero, the above equation reduces to

$$\varepsilon_1 = \varepsilon_y sen^2 \alpha$$

From which we have

$$S_b = \epsilon_y = \frac{\epsilon_1}{sen^2 \alpha}$$

For $\alpha = \pm 55$ degrees

$$S_b = \frac{0.92\%}{sen^2 55} = 1.37\%$$

For $\alpha = \pm 70$ degrees

$$S_b = \frac{0.92\%}{sen^2 70} = 1.04\%$$

For $\alpha = \pm 90$ degrees (hoop-chop pipes)

$$S_b = \frac{0.92\%}{sen^2 90} = 0.92\%$$

It is interesting to compare the above Sb values with the 50 years value Sb = 1.30% reported by a pipe manufacturer in Table 14.1.

14.8   STRAIN-CORROSION THRESHOLD

As discussed in Chapters 8 and 13, the analysis of the resin-dominated pipe durability develops around three modes of failures defined by weep, infiltration and stiffness thresholds. This raises the question.... *Is there a strain-corrosion threshold as well?*

To answer this question we consider the extremely low concentration of sulfuric acid in the weep barrier of sewer pipes. There are two reasons for such a low concentration. First, the acid concentration in the sewer solution is low, not more than 5.0%. Second, the low solubility of sulfuric acid in polyester further reduces this already low concentration. These two facts explain the extremely low presence of sulfuric acid in the weep barrier of sewer pipes.

The weak acidic sewer causes little damage to underground composite pipes. The strain-corrosion process is barely perceptible. Hogni Jonsson has recently published a report strongly supporting this. His findings, shown in Figure 14.6, cover pipe specimens continuously deflected in a 5% solution of sulfuric acid for a period of 30 years. The data show the strain-corrosion regression line turning horizontal for bending strains near the weep threshold. Figure 14.6 certainly suggests the existence of a strain-corrosion threshold equal to the weep threshold of the weep barrier. This, however, is not strictly correct, since the diffusion process remains active and the acid concentration in the weep barrier, although low, will eventually weaken and rupture the fibers.

In conclusion, there is no absolute strain-corrosion threshold for underground pipes in sewer service. The low acid presence at the embedded fibers give such a false impression. However, the arguments just presented, supported by the regression line in Figure 14.6, are so compelling that, for

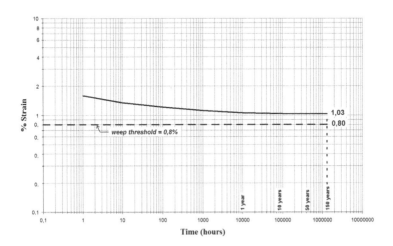

FIGURE 14.6 *The long-term strain-corrosion tests of sanitation pipes in sulfuric acid shows a flattening of the regression line at the weep threshold. The flattened line suggests a weep threshold of 1.03%, versus our conservative suggestion of Tw = 0.80%. (Drawing by Gabriel Gonzalez).*

practical purposes, we can safely assume the existence of a strain-corrosion threshold equal to the weep threshold of the chopped fibers ply. Table 14.2 recognizes this.

14.9 CYCLIC STRAIN-CORROSION

The preceding discussion is applicable to static loads. The strain-corrosion cases combining static and cyclic loads are in principle solved by the unified equation below, imported from Chapter 16.

$$\left(\frac{R\varepsilon \times SF}{S_s}\right)^{\frac{1}{G_s}} + \left(\frac{(1-R)\varepsilon \times SF}{S_c}\right)^{\frac{1}{G_c}} + \left(\frac{R(1-R)\times \varepsilon^2 \times SF^2}{S_s \times S_c}\right)^{\frac{1}{G_{sc}}} = 1.00$$

static	*cyclic*	*interactive*
contribution	*contribution*	*contribution*

The above is the full version of the unified equation, since we wish to account for the combined effects of the static fiber strain-corrosion with the cyclic damage, both present in this case. However, there is a problem…. The laminate strain-corrosion rupture discussed in this chapter is not the same as the fiber strain-corrosion leading to the long-term rupture described by the unified equation. The main points of discrepancy are:

- The fiber strain-corrosion comes from rupture of tensile loaded fibers in water.
- The laminate strain-corrosion comes from rupture of laminates strained in contact with chemicals, not necessarily in water
- The laminate strain-corrosion process involves the growth of a single crack, not the coalescence of many small fiber breaks.

The above considerations rule out the unified equation in applications involving laminate strain-corrosion. However, in view of our previous discussion, we may ignore the low acid presence in sanitation pipes and solve their cyclic durability problem assuming they carry water. The error would be small, given the low acid concentration at the embedded fibers. We have arrived at an unsuspected and amazing conclusion:

The long-term performance of composite pipes in sewer is essentially the same as in water.

Example 14.8

Prof. Mandell mentions an interesting case of strain-corrosion associated with cyclic fatigue. The application is high voltage UD rods insulators in electrical transmission lines. The rods operate subjected to static tensile strains as well as to wind induced vibrations. The wind vibrations have two effects. First, they generate the bending strains leading to the strain-corrosion process. Second, they promote crack growth by a fatigue mechanism. The corrosive chemical in this case is nitric acid generated by atmospheric nitrogen ionized by electric discharges. The stage is thus set for the strain-corrosion process. The wind vibrations provide the bending strains and the nitric acid provides the corrosive chemical. Once the surface cracks start, the tensile load also promotes strain-corrosion, as explained earlier in this chapter. The pultruded rods of UD fibers have no liner or corrosion barrier.

The rods of standard E glass have a short life in this application. This is due to (1) the absence of a protective corrosion barrier allowing the direct access of the acid to the strained UD fibers and (2) the bending cyclic vibrations. We quote from Prof. Mandell himself:

> *"Some failures occur at strains which appear to be less than 10% of the short term value (the quality of the UD rods appears to be very good and the strength near the cracks is close to the initial values). The aspect of these field failures which is most unusual is the mode of crack growth. Cracks propagate in a planar fashion perpendicular to the fibers, with no significant splitting or debonding along the fibers. The fracture surfaces are almost perfectly flat over most of the 2 cm rod diameter, with fracture surface features which can be traced back to crack origin, as with many homogeneous materials. Along with the main failure crack, there are often several small cracks which have grown a short distance in from the surface."*

The above is a good description of the strain-corrosion rupture. Should the rods have been subjected to pure tensile strains, with no wind vibration, the strain-corrosion rupture would probably not occur. The induced vibration provides the bending loads required to initiate and sustain the strain-corrosion process. The rupture times are probably short in this case, since the cyclic-strain-corrosion interaction should be particularly strong.

15 Abrasion Life

15.1 INTRODUCTION

The remarkable performance of composite pipes in aggressive and abrasive service derives mostly from their inherent corrosion resistance. This is a distinctive advantage of composites pipes over their metal counterparts protected by thin oxide layers. The protective oxide layers of the metallic pipes wear away quickly in abrasive service, exposing unprotected material to a fast and aggressive combination of erosion and corrosion. The absence of the dreadful combination erosion-corrosion, together with recent developments in the technology of abrasion resistant barriers, explain the excellent durability and acceptance of composite pipelines in abrasive service.

There is a striking similarity in the analysis of the durability of the corrosion and abrasion barriers. The lab tests performed by the pipe manufacturers have established the best material formulations for enhanced abrasion resistance. The abrasion barriers consisting of hard ceramic granular fillers like silicon and tungsten carbide, in combination with resilient resin matrices, offer the best solution. However, knowing the best materials combination is hardly enough to estimate the abrasion life. Knowing the best formulation of resin and ceramic filler that would last longer in abrasive service is not enough.

In analogy with the chemical life discussed in Chapter 11, the abrasion barrier is the critical ply in control of the pipe durability. As in the chemical case, the composite pipeline is unfit for service when the protective abrasion barrier is eroded away, exposing the structural plies. This is in perfect analogy with the chemical life discussed in Chapter 11. This chapter introduces a quantitative method to predict the durability of abrasion barriers.

15.2 THE ERODED DEPTH Δe

The abrasive slurry carried in the pipeline erodes away the protective inner barrier while leaving the structural plies intact, as in Figure 15.1. It takes a long time for the erosive action to advance through the protective barrier and reach the structural plies. The rate of thickness erosion measures the performance of the protective barrier. The analogy with the chemical penetration discussed in Chapter 11 is obvious. Furthermore, the residual abrasion life is

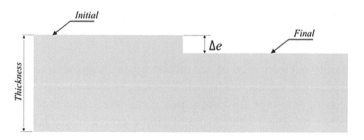

FIGURE 15.1 *The worn out depth "Δe" measures the extent of erosion. The eroded depth is determined directly by measuring the reduction in wall thickness.*

determined by measuring the residual barrier thickness after a certain period in service.

Given enough time, the abrasive slurry destroys the protective barrier and initiates a rapid erosion of the structural plies. In analogy with the pipes for chemical service, the structural plies designed for the best mechanical properties have poor abrasion resistance.

15.3   THE GOVERNING WEAR EQUATION

Having measured the eroded depth, we next derive a mathematical relationship for the wear rate. The derivation takes into consideration the many process and slurry variables known to affect the wear rate. The most important are:

- Pressure
- Temperature
- Particle size
- Hardness
- Slurry concentration
- Flow velocity

In analogy with the chemical case, we estimate the wear rate with a power function of the form

$$\frac{d(\Delta e)}{dt} = K \times H^{\alpha} \times V^{\beta} \times D^{\gamma} \times P^{\delta} \times C^{\epsilon} \times T^{\theta} \tag{15.1}$$

Where

  K is a parameter characteristic of the abrasive barrier
  P is the pressure
  T is the temperature
  D is the particle size
  H is the particle hardness
  C is the slurry concentration
  V is the flow velocity
  α, β, γ, δ, ε and θ are experimental parameters

Equation (15.1) has many variables and is too complex. The experimental determination of the exponents would involve a series of expensive, laborious and impractical lab tests. The practical approach in such cases consists in assuming constant average values for the variables of lesser relevance in practical service. By doing so, we drop the less relevant variables from Equation (15.1) and capture their effects in the constant K. This is the same approach that we have taken in Chapters 11 and 14 to estimate the service life and the strain-corrosion durability of composite laminates in chemical service. The simplified lab tests to measure the exponent parameters appearing in Equation (15.1) assigns typical, constant values, to each of the ignored variables. This simplification reduces the number of variables and has a significant impact on the cost of the experimental tests. Looking at Equation (15.1), we can assume constant and typical values for the pressure, temperature, slurry concentration and particle size. With this simplification, the only remaining variables are the particle hardness H and the flow velocity V. Equation (15.1) reduces to

$$\frac{d(\Delta e)}{dt} = K \times H^{\alpha} \times V^{\beta} \tag{15.2}$$

Integrating Equation (15.2) gives the eroded depth

$$\Delta e = K \times H^{\alpha} \times V^{\beta} \times t \tag{15.3}$$

Taking logarithms of Equation (15.3), we have

$$\log(\Delta e) = \log K + \alpha \times \log(H) + \beta \times \log(V) + \log(t)$$

Or

$$\log(\Delta e) = A + \alpha \times \log(H) + \beta \times \log(V) + \log(t) \tag{15.4}$$

Equation (15.4) is similar to the one proposed in Chapter 11 to compute the penetrated depth of aggressive chemicals in the corrosion barrier. As explained in Chapter 11, we make use of a multilinear regression analysis to determine the wear parameters α, β and A in Equation (15.4). Such parameters, of course, hold for pipelines lined with the abrasion barrier used in the test.

The assigned and constant temperature, pressure, particle size and slurry concentration are those commonly found in practice. The test variables that remain are the particle hardness H, the flow velocity V and the time t. The values of the eroded depth Δe are a function of these variables. The desired wear parameters A, α and β in Equation (15.4) come from a multilinear regression analysis of several eroded depths measured as a function of the variables V, H and t. The tested pipeline would ideally include typical fittings to allow the simultaneous determination of their own wear parameters

A simplified test like the one just described may still be too complex and expensive. Even ignoring all variables except three, we still face a sizeable amount of testing. A further simplification would involve the retention of just two variables, flow velocity and time, but the accuracy of Equation (15.4) in this case would perhaps be too low.

15.4   MEASURING THE WEAR PARAMETERS

There is no standardized test method at this time to measure the abrasion parameters. The current abrasion protocols proposed by the pipe man-ufactures are excellent to screen and compare formulations and materials performance, but are useless to compute the wear parameters. In this section, we discuss a protocol to measure the abrasion parameters of straight pipe sections.

The protocol involves the rotation of straight pipe sections partially filled with abrasive slurries of different hardness at different velocities. The eroded depths "Δe" come from direct measurement at several time intervals and flow velocities for different abrasive materials. The measured data points, corre-sponding to the eroded depth "Δe", exposure time "t", abrasive hardness "H" and flow velocity "V" are annotated and then processed by the usual least squares method to produce a statistical estimation of the wear parameters. This is a typical linear regression problem in three independent variables – velocity, hardness and time. The regression Equation (15.4) obtained in this way predicts the service life of the tested abrasion barrier.

The method just described may give reasonably good values for the wear parameters of straight pipe sections. The determination of the wear parameters for fittings would be more complex and require a more elaborate test set up. To eliminate this difficulty, we consider the fittings as expendable items discarded and routinely replaced during the pipeline operation. The fittings construction use the same erosive barriers of the straight pipe sections. This simplification greatly facilitates the fabrication and testing, with a negligible impact on the overall performance of the pipeline.

The following examples illustrate the application of the regression Equation (15.4). We assume an abrasion barrier of vinyl ester resin loaded with tungsten carbide. We also assume the following wear parameters

$A = -2.53$, $\alpha = 1.03$ and $\beta = 0.80$.

The above values are valid for "Δe" in mm, flow velocity V in m/s, particle hardness H in mhos and exposure time t in months. The wear parameters enter Equation (4) to determine (a) the service life of a 2.0 mm abrasion barrier and (b) the required thickness of the abrasion barrier for a projected service life of 25 years.

Example 15.1

Estimate the service life of a 2.0 mm abrasion barrier. To increase the service life the pipeline is rotated by 180 degrees when the eroded depth reaches $\Delta e = 2.0$ mm. The service life is estimated for calcite slurry (mhos hardness H = 3) at a flow velocity V = 5 m/s.

From Equation (15.4)

$$\log(\Delta e) = A + \alpha \times \log(H) + \beta \times \log(V) + \log(t)$$

$$\log 2.0 = -2.53 + 1.03 \times \log 3.0 + 0.80 \times \log 5.0 + \log(t)$$

Solving for the time "t" we obtain $t \approx 60$ months (5 years).

The pipeline is rotated by 180 degrees after 5 years in service, to last another 5 years. The expected life is, then, 10 years.

Example 15.2

Estimate the required thickness of the abrasion barrier for a service life of 20 years. Considering rotation of the pipes, we design the barrier thickness to last 10 years. Per Equation (15.4) the required thickness is

$$\log \Delta e = -2.53 + 1.03 \times \log 3.0 + 0.80 \times \log 5.0 + \log(10 \times 12)$$

$\Delta e = 4.0$ mm.

16 The Unified Equation

16.1 INTRODUCTION

Suppose a generic laminate operating in any chemical environment at any temperature and subjected to any combination of cyclic and static loadings. This comprehensive assumption covers all possible applications of industrial composites. The goal of this book is the estimation of the durability of any laminate operating in any imaginable situation for all modes of failure. At first glance, this is an impossible task. How could anyone estimate the durability of any laminate for all possible modes of failure, all chemicals, and all loadings? The current state of the art is capable at best to give partial answers to the rupture durability of, perhaps, a few laminates subjected to specific loadings. To move from such a limited number of available solutions to all possible solutions seems too bold, far-fetched, and frightening. Still, this is the proposed goal of this book.

Our approach to this daunting problem is deceptively simple. It consists essentially in breaking down this complex task into several easily solvable parts governed by few well-known deterioration mechanisms. The first major step toward this simplification concerns the recognition of the eight modes of long-term failure and their driving mechanism (see Chapter 8). The second step deals with the solution of the durability problem in each isolated case. The following is a recap of what we have done in the preceding chapters:

- The durability related to the three **load-independent** modes of failure—chemical, anomalous, and abrasion—have simple well-known causes and are easy to solve. Their solution is available in several chapters of this book. See Chapter 11 for the chemical durability. The durability in abrasive service is in Chapter 15. The anomalous failure durability is in Chapter 20.
- The durability related to the four **load-dependent** modes of failure—infiltration, weeping, loss of stiffness and rupture—although having simple known causes, are not so easy to compute. The laminate deterioration in such cases come from the accumulated damages caused by strongly interacting static and cyclic load components. The durability solution to such a complex interaction problem requires the use of the unified equation introduced in this chapter.
- The durability related to the combined **load + chemical** mode of failure—laminate strain-corrosion—requires specific regression

equations that are not available at this time. In Chapter 14, we present a special solution to the highly complex strain-corrosion problem of sanitation pipes.

This chapter and Chapters 17–19 deal with the unified equation and its application to the remaining durability issues involving the four load-dependent modes of deterioration. The unified equation completes our goal of solving all durability issues of any laminate in any possible operating condition.

16.2 PURE REGRESSION LINES

In Chapter 9, we introduced the pure regression lines (static and cyclic) to estimate the rupture failure of chopped and UD plies. Those lines play a central role in the development of the unified equation. We begin this chapter with a brief recap of the pure regression equations discussed in Chapter 9. Let us open the discussion with a recap of the process of fiber long-term rupture.

There are two causes of long-term fiber rupture, the first involving strain-corrosion from pure static loads, and the second involving pure cyclic fatigue (see Chapters 9 and 12). The difference between these modes of failure lies in the driving mechanism of fiber damage, which is fatigue in the cyclic case and strain-corrosion in the static case. The Paris law governs the crack growth rates in both cases:

$$\frac{da}{dt} = Y\left(\varepsilon_S \times \sqrt{\pi \times a}\right)^Z \quad \left(\text{Pure static strain}\right)$$

$$\frac{da}{dN} = Y\left(\varepsilon_C \times \sqrt{\pi \times a}\right)^Z \quad \left(\text{Pure cyclic strain}\right)$$

Where

  a is the crack length.
  t   is the time variables.
  N is the number of cycles.
  ε_S is the pure static strain.
  ε_C is the peak of the pure cyclic strain.
  Y is a geometry-dependent parameter.
  Z is a material-dependent parameter.

The growth rates indicated in the Paris laws depend on the crack size a and the strains ε_S or ε_C. The exponent Z accounts for the nature of the corrosive environment, which in this case is water. The Paris laws are valid for homogeneous materials, like metals, resins and glass fibers that grow self-similar cracks. They do not apply to composites that fail by the coalescence of many small cracks. This characteristic behavior of composite materials explains why their regression equations use strains, instead of crack sizes. Integration of the Paris laws gives the regression equations to predict the fiber rupture under pure static or pure cyclic loads. They are:

$$\log \varepsilon_S = A_S - G_S \log(t) \tag{16.1}$$

and

$$\log \varepsilon_C = A_C - G_C \log(N) \tag{16.2}$$

The above equations, expressed in strains instead of crack densities, compute the pure static and pure cyclic long-term ply strengths at any given time t or number of cycles N. Equations (16.1) and (16.2) are pure, since the applied loads act independently, with no interaction. As explained, the regression parameters A and G come from statistical methods and take different subscripts s or c to denote the static or cyclic cases.

Next we rearrange the pure regression Equations (16.1) and (16.2) in a suitable form for the derivation of the unified equation. Suppose a ply under the simultaneous action of static and cyclic load components. The ply lifetime is X years and Y cycles. Remember… such a ply sees the simultaneous action of cyclic and static load in its lifetime. We illustrate the derivation for the static load component. The derivation for the cyclic component is similar. For a lifetime of X years, the long-term static ply strength is determined from Equation (16.1) as

$$\log S_S = A_S - G_S \log(X) \tag{16.1A}$$

Where Ss is the long-term pure static strength of the ply, corresponding to a lifetime of X years. Combining Equations (16.1) and (16.1A), we have

$$\frac{X}{t} = \left(\frac{\varepsilon_S}{S_S} \right)^{\frac{1}{G_S}} \tag{16.1B}$$

In Equation (16.1B), the static strain ε_S and the time t describe any pair of rupture points. For any given time t there is a corresponding static strain ε_S

that fails the ply. In a similar manner, the long-term strength S_S corresponds to the failure time X.

The analog equation for the pure cyclic case is:

$$\frac{Y}{N} = \left(\frac{\varepsilon_C}{S_C}\right)^{\frac{1}{G_C}}$$

(16.2A)

Where Sc is the long-term pure cyclic strength of the ply, corresponding to the number of cycles Y. The value of Sc comes from Equation (16.2) with N = Y:

$$\log S_C = A_C - G_C \log(Y)$$

Equations (16.1B) and (16.2A) are applicable to pure static and pure cyclic loads acting alone. Our objective is to combine them into a unified equation in which both loads act at the same time. To do so, the number of cycles and the time duration must have a common basis or—as explained in Chapters 17 and 18—fall on the same load path. In other words, the number of cycles Y must correspond exactly to the lifetime X years. For example, if the target lifetime is X = 50 years, then Y must be the number of cycles in 50 years. The following definitions apply:

Ss is the long-term static strength in X years.
Sc is the long-term cyclic strength in Y cycles, corresponding to X
    years.
X is the ply lifetime in years.
Y is the number of cycles occurring in X years.
ε_S is the long-term static strength for any time t.
ε_C is the long-term cyclic strength for any number of cycles N.
N is the number of cycles that fail the ply under ε_C.
t is the time that fail the ply under ε_S.

16.3   DERIVING THE UNIFIED EQUATION

The left hand side of Equations (16.1B) and (16.2A) gives the fractions of the ply lifetime consumed by the static or the cyclic loads acting alone during X years and Y cycles. The reader should pause for a moment to understand this. Assume we have a ply failing at X = 50 years under static and cyclic load components. The number of cycles in 50 years is Y = 300 000. The pure regression Equations (16.1) and (16.2) give the number of cycles N and the time t that would fail the ply if the pure static and the pure cyclic loads acted alone.

- Suppose we compute N = 1 000 000 cycles to failure when the pure cyclic load $\mathcal{E}c$ acts alone. The actual number of cycles that fail the ply under the combined loading in X = 50 years is Y = 300 000, which is less than the N = 1 000 000 cycles required by the pure loading. The measured life of X = 50 years "consumes" 300/1000 = 0.30 = 30% of the total pure cyclic lifetime.
- Suppose also we obtain t = 200 years to failure if the static load $\mathcal{E}_S$ acted alone. By the same reasoning, the measured lifetime X = 50 years "consumes" 50/200 = 0.40 = 40% of the total pure static lifetime.

From the above considerations, we see that neither the pure static nor the pure cyclic loads fail the ply in X = 50 years when acting alone. The question is ... Would the loads fail the ply in X = 50 years and Y = 300 000 cycles if they acted simultaneously? The two loads acting together would consume 40% + 30% = 70% of the total ply lifetime in 50 years and 300 000 cycles. Therefore, after 50 years of uninterrupted and simultaneous loading, the ply retains 100% − 70% = 30% of its life. The reader will note that *the ply retains 30% of its life, not 30% of its strength*. The unified equation is about life, not about residual strength.

Based on these arguments, the combined loading that fails the ply in X years and Y cycles is obtained when the partial lives consumed by the pure static and the pure cyclic load components add up to 100%, or

$$\left(\frac{X}{t}\right)_{STATIC} + \left(\frac{Y}{N}\right)_{CYCLIC} = 1.00$$

This, in view of Equations (16.1B) and (16.2A), is the same as

$$\left(\frac{\varepsilon_S}{S_S}\right)^{\frac{1}{G_S}} + \left(\frac{\varepsilon_C}{S_C}\right)^{\frac{1}{G_C}} = 1.00 \tag{16.3}$$

In Equation (16.3) Gs and Gc are the known pure regression slopes, Ss is the long-term pure static strength at X years, and Sc is the long-term pure cyclic strength at Y cycles. In real life, the applied strains $\mathcal{E}_S$ and $\mathcal{E}_C$ (both known) are small and do not fail the ply in X years and Y cycles. To cause failure, they are multiplied by a long-term factor SF. Entering this SF factor in Equation (16.3), we obtain

$$\left(\frac{\varepsilon_S \times SF}{S_S}\right)^{\frac{1}{G_S}} + \left(\frac{\varepsilon_C \times SF}{S_C}\right)^{\frac{1}{G_C}} = 1.00 \tag{16.3A}$$

Equation (16.3A) computes the safety factor SF that multiplies the applied loading to fail the ply in the apprised time. SF is the long-term safety factor. Equation (16.3A) ignores the interaction between the cyclic and the static loads. The complete unified equation, including the interaction, is

$$
\underbrace{\left(\frac{\varepsilon_s \times SF}{S_S}\right)^{\frac{1}{G_S}}}_{\substack{static \\ damage}} + \underbrace{\left(\frac{\varepsilon_c \times SF}{S_C}\right)^{\frac{1}{G_C}}}_{\substack{cyclic \\ damage}} + \underbrace{\left(\frac{\varepsilon_s \times \varepsilon_c \times SF^2}{S_S \times S_C}\right)^{\frac{1}{G_{SC}}}}_{\substack{interactive \\ damage}} = 1.00
\tag{16.4}
$$

Equation (16.4) is the final form of the unified equation. The interaction parameter Gsc is determined experimentally as shown in Chapter 17. The only unknown in Equation (16.4) is the long-term safety factor SF. The unified equation computes the long-term safety factor SF for any given combination of static and cyclic load components acting simultaneously for any period of time X and number of cycles Y. The unified equation is not concerned with residual ply strengths. Instead, it gives the long-term safety factor SF for continuous loadings over the lifetime of X years and Y cycles. The long-term safety factor SF is, I believe, of more interest to the engineer than the residual ply strength.

The unified equation is a neat, symmetrical, and elegant mathematical expression that, although derived for tensile rupture, is equally applicable to all long-term modes of failure. When applied to the fiber direction of UD plies, the unified equation predicts the long-term SF to rupture. The analysis performed in the transverse direction of UD plies predicts the long-term SF to weeping, stiffness or infiltration failures, as the case may be. Similar arguments apply to chopped fibers.

The unified equation solves all load-dependent durability issues of any laminate. Its application involves the following steps:

- The choice of the critical ply.
- The computation of the total strains (mechanical + residual) on the critical ply. The unified equation holds for plies, not laminates.
- The decomposition of the complex loading into its pure static and pure cyclic components.
- The computation of the partial life loss from each pure loading component acting alone.
- The actual life loss is the sum of the partial losses from each pure load component, plus the interaction loss.
- The laminate durability is simultaneous with the failure of the critical ply.

These steps allow the computation of the load-dependent durability of any laminate submitted to any loading. The focus is on the critical ply, not the laminate. The inputs required by the unified equation are:

- The total static and cyclic strain components on the ply frame
- The slope Gs of the pure static regression line
- The slope Gc of the pure cyclic regression line
- The interaction parameter Gsc
- The long-term pure static strength Ss for the lifetime X years
- The long-term pure cyclic strength Sc for Y cycles occurring in X years

The total static and cyclic strain components entering the unified equation include the mechanical and residual contribution computed for the critical ply as shown in the Part I of this book. The pure slopes Gs and Gc come from the pure regression equations listed in Chapter 9. The long-term pure strengths Ss and Sc derive from the same pure equations. The remaining unknowns in Equation (16.4) are the long-term safety factor SF and the interaction parameter Gsc. Chapters 17 and 18 discuss the determination of the interaction parameter Gsc.

The unified equation computes the infiltration, stiffness, weep, and rupture durability of any ply. The critical ply and the mode of failure depend on the intended service. The designer knows the mode of failure and the critical ply in each case. The modes of failure could be the long-term rupture or the weep failure of ± 55UD plies in oil pipes. They could also be the long-term weep failure of chopped fibers in sanitation pipes, or the long-term stiffness failure of transverse loaded UD plies in wind blades.

The classical idea of durability is commonly associated with the time or the number of cycles to failure. The unified equation works in reverse and computes the long-term safety factor SF from pre-defined lifetimes X and number of cycles Y. The pre-defined lifetime X and the corresponding number of cycles Y are not explicit in the unified equation. They are implicit in the pure long-term ply strengths Ss and Sc.

As a closing note, we recognize that it is easier to work with the unified equation expressed in terms of the R ratio and the operating peak strain, instead of the static and cyclic strain components. Introducing this new notation, the unified Equation (16.4) becomes

$$\left(\frac{R \times \varepsilon_{\max} \times SF}{S_S}\right)^{\frac{1}{G_S}} + \left(\frac{(1-R) \times \varepsilon_{\max} \times SF}{S_C}\right)^{\frac{1}{G_C}} + \left(\frac{R \times (1-R) \times \varepsilon_{\max}^2 \times SF^2}{S_S \times S_C}\right)^{\frac{1}{G_{SC}}} = 1.00$$

Dropping the subscript "max" from the above equation, we have

$$\left(\frac{R \times \varepsilon \times SF}{S_S}\right)^{\frac{1}{G_S}} + \left(\frac{(1-R) \times \varepsilon \times SF}{S_C}\right)^{\frac{1}{G_C}} + \left(\frac{R \times (1-R) \times \varepsilon^2 \times SF^2}{S_S \times S_C}\right)^{\frac{1}{G_{SC}}} = 1.00 \quad (16.5)$$

In Equation (16.5), R is the load ratio and ε is the peak strain. This notation is in full agreement with the classical authors.

16.4   APPLYING THE UNIFIED EQUATION

The unified equation computes the long-term safety factor SF of any ply embedded in any laminate subjected to any loading. It covers all modes of load-dependent failures such as infiltration, loss of stiffness, weeping, or rupture. Its application starts with the definition of the applied loading and the desired lifetime. It goes like this:

- The user informs the operating conditions.
- The user informs the expected number of cycles Y and the target lifetime X.
- The designer identifies the mode of failure and the critical ply.
- The designer computes the total peak strain and the R ratio on the critical ply.
- The user supplied durability X and Y define the load path (see Chapter 17).
- The designer picks the pure regression lines applicable to the critical ply and the mode of failure. The pure regression lines are in Chapter 9.
- The parameters Gs and Gc come from the above pure regression lines.
- The long-term pure strengths Ss and Sc derive from the pure regression equations and the expected durability informed by the user.
- The interaction parameter Gsc come from the tables in Chapters 17 and 18.
- The only unknown in the unified equation is the long-term safety factor SF.

The unified equation computes the laminate long-term safety factor SF from user-defined loadings and desired durability. The designer computes the total ply strains in accordance with the protocol described in the Part I of this book. The long-term ply strengths come from the pure regression equations tabulated in Chapter 9. The reader will note that the unified equation does not compute the lifetime. Rather, it computes the long-term safety factor SF for the user-specified loading and lifetime. The computations hold for any of the four long-term load-dependent modes of failure.

Example 16.1

This numerical example computes the rupture durability of underground hoop-chop pipes used in water transmission. The operating conditions are as follows:

Hoop static tensile strain component	$\epsilon_s = 0.25\%$
Hoop static bending strain component	$\epsilon_b = 0.15\%$
Hoop cyclic tensile strain component	$\epsilon_c = 0.10\%$
Target lifetime	X = 50 years = 438 000 hours
Number of cycles in 50 years	Y = 158 000 000 cycles

There is no regression equation for this particular loading. The unified equation is the only way to solve this problem. The critical UD ply controls the long-term rupture. We solve the problem for two scenarios:

- Pipes of E glass
- Pipes of boron-free glass

Since the UD fibers in hoop-chop laminates align in the hoop direction, no strain rotation is required.

The minimum hoop strain comes from adding the known tensile and bending static components:

$$\epsilon_{min} = 0.25 + 0.15 = 0.40\%$$

The maximum hoop strain comes in the same way, including the cyclic component.

$$\epsilon_{max} = 0.25 + 0.15 + 0.10 = 0.50\%$$

The R ratio is:

$$R = \frac{0.40}{0.50} = 0.80$$

The load path (loading frequency) is:

$$Frequency = \frac{158,000,000}{50 \times 365 \times 24 \times 60 \times 60} = 0.10 \, Hertz$$

TABLE 16.1

Inputs in the unified equation for a target lifetime of X = 50 years and Y = 158 000 000 cycles.

Parameter	Value	Source
Gs	0.077	Mark Greenwood, boron-free glass (Chapter12)
	0.130	Mark Greenwood, E glass (Chapter 12)
Gc	0.089	Guangxu Wei, fiber direction (Chapter 12)
	0.040	Guangxu Wei, transverse direction (Chapter 13)
Gsc	NA	Gsc varies with the lifetime, the number of cycles and the strain ratio R. (Chapters 17 and 18)
Ss	0.92%	Mark Greenwood for boron-free glass and a target lifetime of X = 50 years (Chapter 12)
	0.41%	Mark Greenwood for E glass and a target lifetime of X = 50 years (Chapter 12)
Sc	0.62%	Guangxu Wei for Y = 158 000 000 cycles in the fiber direction (Chapter 13)

First, as a curiosity, we determine the short-term safety factor against rupture. The short-term SF is the ratio between the short-term rupture strength and the total strain in the fiber direction:

$$SF = \frac{3.0\%}{0.50\%} = 6$$

The short-term safety factor is independent of the glass composition.

Let us now move to the computation of the long-term safety factor. The input data needed in the unified equation for a target lifetime of X = 50 years and Y = 158 000 000 cycles are in Table 16.1. The long-term static strengths Ss of the UD plies come from the pure static equations developed in Chapter 12.

For boron-free glass in 50 years:

$$\log \varepsilon_S = 0.400 - 0.077 \log\left(hours\right)$$

$$\log S_S = 0.400 - 0.077 \times \log(438000)$$

$$\log S_S = 0.400 - 0.077 \times 5.641 = -0.034$$

$$S_S = 0.92\%$$

For E glass in 50 years:

$$\log \varepsilon_S = 0.347 - 0.130 \log \left(hours \right)$$

$$\log S_S = 0.347 - 0.130 \times 5.641 = -0.386$$

$$S_S = 0.41\%$$

The long-term cyclic strength Sc of UD plies loaded in the fiber direction comes from the pure cyclic Equation (16.6):

$$\log S_C = 0.519 - 0.089 \log Y \qquad\qquad (16.6)$$

Where Y = 158 000 000 cycles is the number of cycles expected in the target lifetime X = 50 years. Equation (16.6) is valid for pure cyclic loads in the fiber direction.

$$\log S_C = 0.519 - 0.089 \log 158 \times 10^6$$

Which gives $S_C = 0.62\%$.

We now proceed to apply the unified equation. We address the boron-free pipes first, by entering the appropriate inputs in the unified Equation (16.5):

$$\left(\frac{R \times \varepsilon \times SF}{S_S} \right)^{\frac{1}{G_S}} + \left(\frac{(1-R) \times \varepsilon \times SF}{S_C} \right)^{\frac{1}{G_C}} + \left(\frac{R(1-R) \times \varepsilon^2 \times SF^2}{S_S \times S_C} \right)^{\frac{1}{G_{SC}}} = 1.00$$

$$\left(\frac{0.80 \times 0.50 \times SF}{0.92} \right)^{\frac{1}{0.077}} + \left(\frac{(1-0.80) \times 0.50 \times SF}{0.62} \right)^{\frac{1}{0.089}}$$

$$+ \left(\frac{0.80(1-0.80) \times 0.50^2 \times SF^2}{0.92 \times 0.62} \right)^{\frac{1}{G_{SC}}} = 1.00$$

The interaction parameter Gsc is not known for the load path 0.10 Hz and R = 0.80. However, for illustration purposes, we assume Gsc = 2. Hence,

$$\left(\frac{0.40 \times SF}{0.92} \right)^{\frac{1}{0.077}} + \left(\frac{0.10 \times SF}{0.62} \right)^{\frac{1}{0.089}} + \left(\frac{0.40 \times 0.10 \times SF^2}{0.92 \times 0.62} \right)^{\frac{1}{2}} = 1.00$$

Solving the above equation, we obtain SF = 1.8. The meaning of this safety factor is as follows. The applied loading by itself is not enough to fail the pipe in 50 years.

For rupture to occur in 50 years, the applied loading should be 1.8 times higher. Therefore, the long-term safety factor is SF = 1.8.

Setting SF = 1.8 in the above equation, we have an idea of the partial life losses from each loading

$$\left(\frac{0.40\times1.8}{0.92}\right)^{\frac{1}{0.077}} + \left(\frac{0.15\times1.8}{0.62}\right)^{\frac{1}{0.089}} + \left(\frac{0.40\times0.15\times1.8^2}{0.92\times0.62}\right)^{\frac{1}{2}} = 1.00$$

The above computation indicates that the pure static and the pure cyclic loading components per se have little effect on the durability. The real damage comes from the interaction. This conclusion, however, is purely illustrative, since the correct value of Gsc is not available for this loading.

A similar analysis is possible for pipes of E glass. The long-term pure cyclic strength Sc for E glass is 0.62%, the same as that for boron-free glass. The long-term static strength for E glass is (Table 16.1) Ss = 0.41%. Entering these values in Equation (16.5),

$$\left(\frac{0.40\times SF}{0.41}\right)^{\frac{1}{0.130}} + \left(\frac{0.10\times SF}{0.62}\right)^{\frac{1}{0.089}} + \left(\frac{0.40\times0.10\times SF^2}{0.41\times0.62}\right)^{\frac{1}{2}} = 1.00$$

which gives SF = 0.73, meaning the pipes of E glass under the stated loading rupture before completing 50 years in service.

This example illustrates the superiority of boron-free glass vis-à-vis the traditional E formulation. The higher hydrolytic stability of boron-free glass produces pipes of higher resistance to long-term rupture.

Example 16.2

Let us use the unified Equation (16.5) to analyze the 50 years rupture of the pipes described in Example 5.2. The operating conditions are:

- The HDT of the vinyl ester resin matrix is HDT = 105°C.
- The peak post-cure temperature is PCT = 100°C.
- The pipes carry water at room temperature OT = 25°C.
- The operating pressure is P = 10 bar under a 2:1 loading.
- The pipe diameter is D = 500 mm.
- The water pickup of vinyl ester resins is Δm = 1.0%.

The mode of failure is rupture. The total strain in the fiber direction is ε_1 = 0.0012 = 0.12%, as computed in Example 5.4. There is no cyclic load component. See Example 5.2. First, as a curiosity, we determine the short-term safety factor against rupture, defined as the ratio between the short-term rupture strength and the total strain in the fiber direction:

$$SF = \frac{3.0\%}{0.12\%} = 25$$

This healthy short-term safety factor is not dependent on the glass formulation. We now proceed to the computation of the long-term safety factor. We assume pipes of boron-free glass with the following static regression equation.

$$\log \varepsilon_S = 0.400 - 0.077 \log(hours)$$

$$\log S_S = 0.400 - 0.077 \log\left(438000\right)$$

$$S_s = 0.92\%$$

The unified equation is:

$$\left(\frac{R \times \varepsilon \times SF}{S_S}\right)^{\frac{1}{G_S}} + \left(\frac{(1-R) \times \varepsilon \times SF}{S_C}\right)^{\frac{1}{G_C}} + \left(\frac{R(1-R) \times \varepsilon^2 \times SF^2}{S_S \times S_C}\right)^{\frac{1}{G_{SC}}} = 1.00$$

In the absence of cyclic loads, R = 1.0 and the unified equation reduces to

$$\left(\frac{\varepsilon \times SF}{S_S}\right)^{\frac{1}{G_S}} = 1.00$$

$$\left(\frac{0.12 \times SF}{0.92}\right)^{\frac{1}{0.077}} = 1.00$$

Which gives the long-term safety factor SF = 7.6.

This example indicates a 50 years long-term rupture safety factor SF = 7.6, against the short-term safety factor SF = 25.

Example 16.3

We continue with the analysis of the pipe described in Example 5.2, assuming the presence of a cyclic load component of amplitude $\epsilon_c = 0.10\%$ in the fiber direction. The expected number of cycles in 50 years is Y = 10 000 000. The analysis is for long-term rupture and the focus is on the fiber direction.

The long-term pure cyclic strength in the fiber direction is computed from Equation (16.6) for Y = 10 000 000 cycles is

$$\log S_C = 0.519 - 0.089 \log Y$$

$$\log S_C = 0.519 - 0.089 \log 10^7$$

$$Sc = 0.79\%$$

The minimum strain in the fiber direction, as computed in Example 5.4, is $\mathcal{E}_{min}$ = 0.12%. The maximum strain in the fiber direction is $\mathcal{E}_{max}$ = 0.12% + 0.10% = 0.22%. The R ratio in the fiber direction is:

$$R = \frac{0.12}{0.12 + 0.10} = 0.55$$

The loading frequency is:

$$Frequency = \frac{10000000}{50 \times 365 \times 24 \times 60 \times 60} = 0.01 \; cycles/s$$

Again, we do not have the Gsc parameter for the load path of 0.01 Hertz and the load ratio R = 0.55. Assuming, for illustration purposes, Gsc = 5, we obtain

$$\left(\frac{R \times \varepsilon \times SF}{S_s}\right)^{\frac{1}{G_s}} + \left(\frac{(1-R) \times \varepsilon \times SF}{S_c}\right)^{\frac{1}{G_c}} + \left(\frac{R(1-R) \times \varepsilon^2 \times SF^2}{S_s \times S_c}\right)^{\frac{1}{G_{sc}}} = 1.00$$

$$\left(\frac{0.55 \times 0.22 \times SF}{0.92}\right)^{\frac{1}{0.077}} + \left(\frac{(1-0.55) \times 0.22 \times SF}{0.79}\right)^{\frac{1}{0.089}}$$

$$+ \left(\frac{0.55(1-0.55) \times 0.22^2 \times SF^2}{0.92 \times 0.79}\right)^{\frac{1}{5}} = 1.00$$

Which gives SF = 5.7

The reader will note the arbitrarily chosen Gsc = 5 in this example, different from the Gsc = 2 in Example 16.1. The actual values of Gsc have not yet been measured and are not known at this time. The protocol to measure the interaction parameter Gsc is discussed in Chapters 17 and 18.

16.5 TIME-INDEPENDENT FAILURES

So far, we have been dealing with long-term rupture failures from strains in the fiber direction. The fiber-dominated failures involve the time variable and requires the complete form of the unified equation. The resin-dominated failures, on the other hand, do not involve the time variable and use a simplified version of the unified equation. Let us derive the simplified unified equation. There are two simplifications.

- The time (static) contribution is set equal to zero, since the resin damages do not come from strain-corrosion.
- For the same reason, the long-term pure static strength Ss is constant and equal to the short-term value. In other words, Ss is equal to the infiltration, weep, or stiffness threshold, as the case may be.

The simplified—time-independent—version of the unified equation is

$$\left(\frac{(1-R)\times \varepsilon \times SF}{S_C}\right)^{\frac{1}{G_C}} + \left(\frac{R(1-R)\times \varepsilon^2 \times SF^2}{S_S \times S_C}\right)^{\frac{1}{G_{SC}}} = 1.00 \qquad (16.7)$$

In the above equation, the long-term pure static strength Ss is set equal to the applicable failure threshold.

$S_S = Ti$ (Equal to the infiltration threshold)
$S_S = Tw$ (Equal to the weep threshold)
$S_S = T_S$ (Equal to the stiffness threshold)

The pure long-term cyclic strength Sc derives from the applicable pure cyclic regression equation tabulated in Chapter 9 and the user-informed number of cycles Y. The simplified Equation (16.7) applies to transversely loaded UD plies and to chopped plies.

Example 16.4

We expand the analysis of the vinyl ester pipe discussed in Example 5.2 by computing its weep durability. The long-term weep failure of oil pipes comes from transverse cracks on the UD plies. The focus is now on the weep failure and we shift the analysis from the fiber direction to the transverse direction of the UD plies.

The pure weep regression equation of transverse loaded vinyl ester UD plies isin Table 9.3.

$$\log(\varepsilon) = \log(0.40) - 0.051 \times \log(N)$$

The long-term pure cyclic weep strength is

$$\log S_C = -0.398 - 0.051 \log Y$$

Where Y = 10 000 000 is the number of cycles informed by the user.

$$\log S_C = -0.398 - 0.051 \log 10^7$$

$$Sc = 0.18\%$$

The long-term static strength in the transverse direction of the vinyl ester UD ply is equal to the weep threshold Ss = Tw = 0.50%. The cyclic transverse strain component is $\epsilon_C = 0.065\%$. The static transverse strain component is $\epsilon_S = 0.15\%$. The peak transverse strain is $0.15 + 0.065 = 0.22\%$. The transverse R ratio is R = 0.15/0.22 = 0.68. The loading frequency is 0.01 Hz.

The interaction parameter Gsc depends on R and Y and is not available. For illustration purposes we assume Gsc = 2 500.

This is a time-independent problem requiring the simplified Equation (16.7)

$$\left(\frac{(1-R) \times \epsilon \times SF}{S_C} \right)^{\frac{1}{G_C}} + \left(\frac{R(1-R) \times \epsilon^2 \times SF^2}{S_S \times S_C} \right)^{\frac{1}{G_{SC}}} = 1.00$$

Entering the data in the above equation, we have

$$\left(\frac{(1-0.68) \times 0.22 \times SF}{0.18} \right)^{\frac{1}{0.051}} + \left(\frac{0.68(1-0.68) \times 0.22^2 \times SF^2}{0.50 \times 0.18} \right)^{\frac{1}{2500}} = 1.00$$

From which we obtain the long-term safety factor SF = 1.8. The unified equation predicts a weep long-term safety factor SF = 1.8 for the vinyl ester pipe after 50 years of continuous service.

The short-term safety factor in this case, as a matter of curiosity, is

$$SF = \frac{0.50\%}{0.22\%} = 2.3$$

Example 16.5

Repeat the analysis of Example 16.4 for pipes of brittle polyester.

The pure weep regression line of transverse loaded polyester UD plies is in Table 9.3.

$$\log(\varepsilon) = \log(0.25) - 0.039 \times \log(N)$$

The long-term pure cyclic weep strength of transversely loaded polyester UD plies is:

$$\log S_C = -0.602 - 0.039 \log Y$$

$$\log S_C = -0.602 - 0.039 \log 10^7$$

$$Sc = 0.13\%$$

We apply the simplified version of the unified equation

$$\left(\frac{(1-R) \times \varepsilon \times SF}{S_C} \right)^{\frac{1}{G_C}} + \left(\frac{R(1-R) \times \varepsilon^2 \times SF^2}{S_S \times S_C} \right)^{\frac{1}{G_{SC}}} = 1.00$$

$$\left(\frac{(1-0.68) \times 0.22 \times SF}{0.13} \right)^{\frac{1}{0.039}} + \left(\frac{0.68(1-0.68) \times 0.22^2 \times SF^2}{0.25 \times 0.13} \right)^{\frac{1}{2500}} = 1.00$$

Solving the above equation, we obtain SF = 1.3.

The unified equation predicts that pipes of brittle polyester barely pass the weep requirement after 50 years of continuous use.

Again, as a curiosity, the short-term SF for the brittle polyester pipe is:

$$SF = \frac{0.25\%}{0.22\%} = 1.1$$

This example brings out an interesting point. The reader will note the long-term safety factor of 1.3 higher than the short-term value of 1.1. There is no contradiction in this curious fact, since the two safety factors are not related. The long-term safety factors can be—and often times are—larger than the short-term safety factor.

Example 16.6

Compute the weep long-term SF of sanitation pipes carrying pressurized water. Assume the following operating conditions:

Static hoop strain component = 0.35%
Cyclic hoop strain component = 0.15%

Target lifetime X = 50 years
Expected number of cycles Y = 10^8 cycles

The pure weep regression equation of polyester chopped plies is in Table 9.3.

$$\log(\varepsilon) = \log(0.80) - 0.061 \times \log(N)$$

The pure long-term weep strength is:

$$\log S_C = \log(0.80) - 0.061 \log 10^8$$

$Sc = 0.26\%$

The long-term static strength is:

$Ss = Tw = 0.80\%$

The operating peak hoop strain is

$\varepsilon = 0.35 + 0.15 = 0.50\%$

The minimum hoop strain is 0.35%

The R ratio is 0.35/0.50 = 0.70

The interaction parameter Gsc for R = 0.70 is not available. We assume Gsc = 10 for illustration purposes. The long-term weep safety factor SF comes from the simplified version of the unified equation

$$\left(\frac{(1-R) \times \varepsilon \times SF}{S_C} \right)^{\frac{1}{G_C}} + \left(\frac{R(1-R) \times \varepsilon^2 \times SF^2}{S_S \times S_C} \right)^{\frac{1}{G_{SC}}} = 1.00$$

$$\left(\frac{(1-0.70) \times 0.50 \times SF}{0.26} \right)^{\frac{1}{0.061}} + \left(\frac{0.70(1-0.70) \times 0.50^2 \times SF^2}{0.80 \times 0.26} \right)^{\frac{1}{10}} = 1.00$$

Solving the above equation, we obtain SF = 1.4.

The short-term SF is:

$$SF = \frac{0.80}{0.50} = 1.6$$

Example 16.7

Explain why the long-term static strengths Ss of time-independent failures—infiltration, weep and stiffness—are equal to the thresholds.

The thresholds are short-term strengths. By definition of time independence, the long-term and short-term strengths have the same value. Therefore, the long-term strengths of time-independent failures are equal to the thresholds.

16.6   BLOCK LOADING

The unified equation assumes the same loading applied throughout the entire lifetime of the ply. We know this never happens. In a lifetime of X years, the laminate may see a variety of loadings and load paths, which invalidate the computations based on a single loading and a single load path.

This apparent serious limitation is easy to solve. The solution consists in grouping the anticipated complex loading in several simple load blocks of similar R ratios and load paths. We end up with several load blocks of known durations, R ratios and load paths. The long-term safety factor for the actual loading derives from the partial safety factors of each individual load block. The only difficulty in this method is, perhaps, the accurate grouping of the actual loading into representative blocks. This solution makes two assumptions:

- The sequence of the load blocks is not relevant.
- The sequential load blocks operate on pristine, undamaged laminates

Example 16.8

Consider the design of a pipeline submitted to a variable loading distributed over 45 years of continuous service. A careful analysis of the expected loading comes up with the approximate simple load blocks:

1. 10 years peak pressure of 20 bar, minimum pressure of 10 bar at 0.5 Hz
2. 15 years peak pressure of 20 bar, minimum pressure of 10 bar at 0.1 Hz
3. 5 years peak pressure of 18 bar, minimum pressure of 12 bar at 1.0 Hz
4. 10 years peak pressure of 30 bar, minimum pressure 20 bar at 0.01 Hz
5. 5 years at constant pressure of 30 bar. No cyclic loading.

The above load blocks represent a careful estimation of the expected operating conditions of the pipeline. The unified equation computes the long-term safety factors SF for each individual load block.

The safety factor SF_1, for the load block 1 is computed for a lifetime of 10 years, $R = 0.50$, and a load path of 0.5 Hz. The unified equation readily gives the safety factor SF_1 for this loading.

The safety factor SF_2 for the load block 2 comes from a lifetime of 15 years, $R = 0.50$, and a load path of 0.1 Hz.

The safety factor SF_3 for the load block 3 comes from a lifetime of 5 years, $R = 0.67$, and a load path of 1.0 Hz.

The safety factor SF_4 for the load block 4 comes from a lifetime of 10 years, $R = 0.67$, and a load path of 0.01 Hz.

The safety factor SF_5 for the load block 5 comes from a lifetime of 5 years under a static load of 30 bars.

Each load block produces a long-term safety factor SF_i. We proceed next to compute the overall long-term safety factor SF. The life lost in each load block is:

$$\left(lost\ life\ per\ block\right)_i = \frac{Total\ life}{SF_i}$$

Where SF_i is the partial long-term safety factor for the i_{th} load block. The total life loss accumulated in all blocks is

$$life\ loss = \sum \left(lost\ life\ per\ block\right)_i = \sum \frac{total\ life}{SF_i} \qquad (16.8)$$

The life loss from the actual loading is:

$$life\ loss = \frac{total\ life}{SF} \qquad (16.9)$$

The overall long-term safety factor SF comes from the equality of Equations (16.8) and (16.9)

$$\frac{1}{SF} = \sum \frac{1}{SF_i} \qquad (16.10)$$

In expanded form, the overall long-term SF is:

$$\frac{1}{SF} = \frac{1}{SF_1} + \frac{1}{SF_2} + \frac{1}{SF_3} + \frac{1}{SF_4} + \frac{1}{SF_5}$$

The above derivation assumes the individual load blocks acting on pristine, non-aged laminates. Equation (16.10) is valid for all four long-term modes of failure. The overall long-term safety factor SF is a multiplier of the actual loading that produces ply failure in the prescribed lifetime.

Example 16.9

The computation of the load-dependent durability starts from pristine laminates with intact plies displaying their full stiffness. The total ply strains, computed on pristine laminates, enter the unified equation as indicated in this chapter. This situation changes when the critical ply is the last – not the first – to fail. The strain in the critical ply increases along the laminate lifetime, as the weaker plies fail and lose stiffness. The new peak strain on the critical ply corresponds to a new loading. The accurate analysis of the durability corresponding to the last ply to fail, therefore, should discount the failed weaker plies and compute the overall long-term SF using the load block concept discussed above.

Example 16.10

Explain the difference between the short-term and the long-term safety factors.

The short-term safety factors are load multipliers that bring the peak applied strain to the short-term strengths, or failure thresholds.

$$SF = \frac{failure\ threshold}{peak\ strain} \quad (Short\text{-}term)$$

The four failure thresholds are known ply properties. The computation of the total peak strain is as explained in the Part I of this book. The short-term safety factors SF do not involve the time variable, or the number of cycles. The classical literature deals exclusively with the short-term safety factor.

The long-term safety factors come from the unified equation and represent load multipliers that bring the applied peak strain to values that fail the ply at the time X and number of cycles Y. The long-term SF takes into consideration the loading parameters, like R ratio, load path, lifetime X, number of cycles Y, and the interaction between the static and cyclic load components.

$$\left(\frac{R \times \varepsilon \times SF}{S_s}\right)^{\frac{1}{G_S}} + \left(\frac{(1-R) \times \varepsilon \times SF}{S_c}\right)^{\frac{1}{G_c}} + \left(\frac{R(1-R) \times \varepsilon^2 \times SF^2}{S_S \times S_C}\right)^{\frac{1}{G_{SC}}} = 1.00 \left(Long\text{-}term\right)$$

This book deals with the unified equation and the long-term safety factors. The computation of the short-term safety factors is too obvious and receives just a passing mention in a few cases.

The short-term and long-term safety factors are not related. They apply to different situations. It is common to find safety factors larger for the long term than for the short term. The prudent engineer determines both values. There are loadings that give satisfactory long-term safety factors and yet fail the ply in the first cycle.

APPENDIX 16.1

THE UNIFIED EQUATION AND THE GOODMAN DIAGRAMS

The classical constant life diagrams, or Goodman diagrams, are condensed representations of the unified equation. The Goodman diagrams for laminates have limited and specific applicability and, for these reasons, will not be discussed here. The ply diagrams, however, are direct competitors of the unified equation in the solution of long-term load-dependent failures.

Figure 16.1 shows a typical Goodman diagram of a ply. The static and cyclic strain components are marked respectively on the horizontal and the vertical axis. The radial lines emanating from the origin contain the experimental failure points corresponding to the load ratios R. The failure points corresponding to constant cycles to failure fall on the marked triangular-shaped lines. The Goodman diagrams condense the entire failure

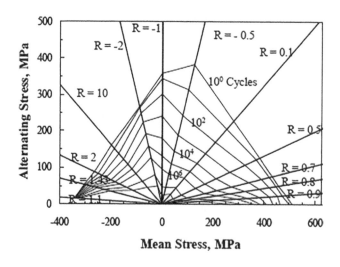

FIGURE 16.1 *The Goodman diagrams are direct competitors of the regression equations of isolated plies. They give the same results as the unified equation for time-independent failures like weeping, loss of stiffness, and infiltration. They ignore the time variable and are not valid for time-dependent ruptures.*

space of the ply in a single graph, covering all possible failure points from different R ratios, number of cycles, and peak strains. They are an excellent tool to make quick computations of time-independent failures.

The Goodman diagrams ignore the time variable and, for that reason, cannot describe the time-dependent long-term ruptures involving fiber strain-corrosion.

The basic difference between the ply Goodman diagrams and the unified equation is in the handling of the time variable. The Goodman diagrams ignore the time variable, which is a corner stone in the unified equation. This limitation excludes Goodman diagrams from the analysis of long-term ply ruptures involving fiber strain-corrosion. The long-term rupture durability, so easily captured by the unified equation, is beyond the reach of the Goodman diagrams.

The unified equation and the Goodman diagrams have equal competence and accuracy in the analysis of time-independent failures—infiltration, stiffness and weeping. This is easy to see, since they derive from the same regression equations. The two methods give the same answer for the long-term safety factor SF of time-independent failures.

The Goodman diagrams applied to time-independent, fully cycle-dependent, long-term failures of plies is a graphical representation of the unified equation.

APPENDIX 16.2

PLY THRESHOLDS TRANSFERRED TO PIPES

Figure 16.2 shows the infiltration and weep thresholds of the critical ply of two identical pipes of different resin toughness. The difference between the two pipes is in the tough or brittle nature of the resin used in their manufacture. The failure thresholds are ply properties that directly transfer to the host pipes. The brittle pipes (plies) have lower infiltration and weep thresholds than the tougher pipes.

Figure 16.2 also shows the short-term weep strengths that are pipe, not ply, properties. Compared to brittle pipes, the tough pipes have lower short-term weep strengths and higher thresholds. For further details, see Figure 10.3 of Chapter 10. All strengths in Figure 16.2 are short-term.

Let us now see what happens in the long-term. We start with static loadings and the time effect. The time variable is marked on the horizontal axis of Figure 16.3. The loading is static.

The infiltration and weep thresholds are ply properties directly transferred to the pipes. They are constant properties that plot horizontally as a function of time. The pipe's weep regression lines plot as discussed in Chapter 13.

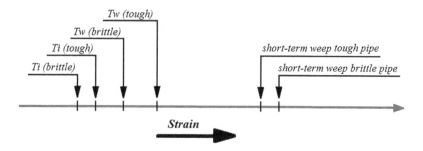

FIGURE 16.2 *The failure thresholds on the left are pipe properties borrowed from the embedded critical plies. The short-term weep strengths on the right are pipe and not ply properties.*

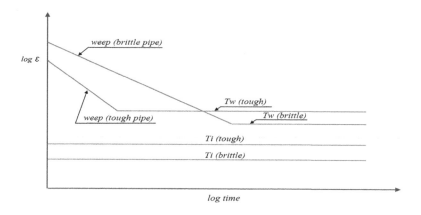

FIGURE 16.3 *The short-term pipe weep strengths of Figure 16.2 are marked on the vertical axis. The time variable is marked on the horizontal axis. Note the lower short-term weep strength of the tough pipe. The infiltration and weep thresholds plot horizontally.*

The weep regression lines are pipe features unrelated to the weep thresholds. Note the pipe weep lines turning horizontal on meeting the weep thresholds.

The points of interest in Figure 16.3 are:

- The infiltration and weep thresholds are ply properties borrowed by the pipes.
- The weep lines are pipe features determined according to ASTM D 2992B. They turn horizontal on meeting the weep thresholds.
- The tougher pipes have steeper weep regression lines.
- The short-term weep strain is lower for tougher pipes.

The weep regression lines are pipe features defined by the wall thickness and the applied pressure, both unrelated to the thresholds. The tough pipe grows few long cracks and has a steeper weep line than the brittle pipe. The pipes operating below the infiltration and weep thresholds under static loadings have infinite infiltration and weep durability. They simply never fail in these conditions. It takes cyclic fatigue loads to fail such pipes.

Next, we discuss the effect of cyclic loads. The cyclic load components grow the crack densities in the critical ply and eventually fail the pipe. The unified equation computes the long-term SF for a given number of cycles to ply failure—infiltration, weeping, stiffness and rupture. The reader will remember that the unified equation is about plies, not about laminates and pipes.

APPENDIX 16.3

LAMINATE REGRESSION EQUATIONS

The central topic in the determination of the load-dependent durability of laminates is how to handle the complex loadings involving the simultaneous action of cyclic and static components. As discussed in Chapter 9, the classical solution to this problem consists in developing regression equations at constant R ratios and load paths for every laminate of interest. The failure points produce the following regression equation:

$$log\varepsilon = A - BlogN$$

In the above equation N is the number of cycles that fail the laminate subjected to a cyclic peak strain ε at a specific load ratio R and load path. The static load component is implicit in the R ratio. The loading frequency is implicit in the load path. The laminate regression slope B captures the time variable. The laminate regression equations are specific to the tested laminate and loading. Any change in the laminate construction, materials, R ratio, and load path (loading frequency) would require new and expensive fatigue tests to develop a new regression equation.

The laminate regression equations are difficult to obtain and not plentiful in the literature. They are available for a few laminates of high commercial interest, subjected to standard loadings. This is the case of wind blades and some aircraft parts.

The regression equations are not available, except in a few cases, for plies. This is a shame, since the interaction parameter Gsc needed in the unified equation derives from the regression equations of plies. It is unfortunate that the enormous effort spent in the development of laminate regression equations did not contemplate plies. The load-dependent durability problem of composites would be completely solved today, had the work on laminates been directed to plies.

17 The Interaction Parameter Gsc

17.1 INTRODUCTION

The unified equation introduced in Chapter 16 computes the long-term safety factors SF of plies—and laminates—submitted to the simultaneous action of cyclic and static loads. The long-term SF is a load multiplier that fails the critical ply in the time X and number of cycles Y specified by the designer. The unified equation is applicable to all laminates, all loadings and all modes of failure.

The complete form of the unified equation is

$$\left(\frac{R\varepsilon \times SF}{S_S} \right)^{\frac{1}{G_S}} + \left(\frac{(1-R)\varepsilon \times SF}{S_C} \right)^{\frac{1}{G_C}} + \left(\frac{R(1-R)\varepsilon^2 \times SF^2}{S_S \times S_C} \right)^{\frac{1}{G_{SC}}} = 1.00 \qquad (17.1)$$

The inputs required in the unified equation are:

- The load ratio R
- The total peak strain ε on the critical ply
- The long-term ply pure static strength Ss for the pre-defined lifetime X
- The long-term ply pure cyclic strength Sc for the pre-defined Y cycles
- The pure static slope, Gs
- The pure cyclic slope, Gc
- The interaction parameter, Gsc

The pre-defined lifetime X and number of cycles Y enter the unified equation indirectly through the ply long-term pure strengths Ss and Sc. The slopes of the pure static Gs and pure cyclic Gc regression lines are available, for all plies, loadings and modes of failure (see Chapter 9). For a description of the computation protocol leading to these pure regression parameters, see Chapter 9. The only unknowns in Equation (17.1) are the long-term safety factor SF and the interaction parameter Gsc. This chapter is about the computation of the interaction parameter Gsc from experimental data.

17.2   THE LOAD PATHS

The interaction parameter Gsc is computed for any ply and any loading by solving the unified Equation (17.1) fed with known experimental failure points. The known failure points derive from experimental regression equations determined for a few chosen R ratios and loading frequencies (load paths). The computation protocol is straightforward.

- The experimental failure data come from regression equations developed for fixed load paths and load ratios R.
- The failure data enter the unified equation. The safety factor SF for the failure points is, of course, SF = 1.0. That leaves Gsc as the only unknown in Equation (17.1).
- The Gsc values are computed by solving Equation (17.1).

The Gsc values computed in this way assure perfect agreement of the unified equation with the experimental data. It is unfortunate that experimental regression equations for UD and chopped plies at R ratios other than the pure cases R = 1.0 and R = 0.0 are not available. To my knowledge, the only ply equations available at load ratios other than R = 0.0 and R = 1.0, are those developed by Guangxu Wei for UD plies at R = 0.50.

Guangxu Wei developed his cyclic UD equations at the loading frequency of 20 Hz. Such a high frequency is desirable to shorten the test duration, but does not capture the time effect very well. The early works on composite load-dependent durability assumed the fiber-dominated fatigue rupture to be time independent. Such an erroneous belief is prevalent to this day. The classical regression equations to compute the fatigue rupture of composite laminates makes no mention of the time variable. This simplification may be justifiable for short-duration laboratory tests, but is highly suspicious for long lifetimes measured in decades.

A possible justification for the early workers failing to capture the time effect may be the short duration of their tests. The number of cycles N relates to the test duration by the following relationship:

$$N = \left(frequency \right)\left(time \right) \tag{17.2}$$

Taking the Guangxu Wei tests conducted at 20 Hertz as an example, we have

$$N = \left(20 \right)\left(seconds \right) \tag{17.3}$$

From Equation (17.3), the tests by Guangxu Wei to N = 10^8 cycles at 20 Hz were completed in just 1400 hours. Such a short test duration, less than 60 days, is hardly long enough to capture the time-dependent fiber strain-corrosion. No wonder, then, the effects of time and loading frequency escaped detection by the early workers.

Let us proceed with this discussion. Taking logarithms of both members of Equation (17.3), we have

$$logN = 1.30 + log(time) \qquad (17.4)$$

In Equation (17.4), the time variable enters in seconds. Figure 17.1 is a plot of Equation (17.4) illustrating the linear load path of the tests by Guangxu Wei at 20 Hz. This is the only load path available at this time for UD plies. In fact, the only load path for any ply. Obviously, we need additional tests, conducted at different load paths. We will have more to say about this as we move on. The pure static load path is not cycle-dependent and plots along the vertical axis in Figure 17.1. Likewise, the pure cyclic load path is not time dependent and plots along the horizontal axis.

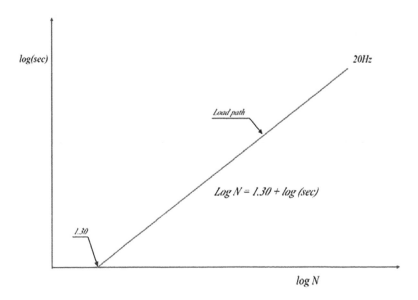

FIGURE 17.1   *The line shows the load path for the test frequency 20 Hz. This is the only load path reported in the literature. The pure static load path coincides with the time axis. The pure cyclic load path coincides with the cycle axis.*

17.3 COMPUTING THE GSC

This section describes the protocol to compute the interaction parameter Gsc from experimental data. The experimental failure data come from ply—not laminate—regression equations developed at constant test frequencies and R ratios. The test frequency defines the load path, as shown in Figure 17.1.

To compute the interaction parameter Gsc, we need the ply pure regression equations and experimental failure points. The pure regression equations are available in Chapter 9. The experimental failure points are not available, but should be. The pure regression equations gives the exponents Gs and Gc in the unified equation. They also give the ply long-term strengths Ss and Sc. The failure points come from regression equations developed at fixed R ratios and loading frequencies. These data points are fed into the unified equation with the safety factor SF = 1.0. This allows the computation of the interaction parameter Gsc, for the load ratio R and the load path of the experimental regression equation.

As we see, the protocol to compute the interaction Gsc values uses failure data from experimental regression lines developed at constant R ratios and test frequencies. To cover a wide range of situations, the computation requires many regression lines, at several R ratios and load paths. For example, the experimental ply regression lines should be developed for R = 0.10, R = 0.30, R = 0.50, R = 0.70, and R = 0.9. For each of these R ratios, the test frequencies could be 10 Hz, 5 Hz, 1 Hz, and 0.1 Hz.

The loading frequencies would produce load paths similar to the one described in Equation (17.4) and Figure 17.1. For example, a test frequency of 0.1 Hz would produce the load path

$$N = 0.1 \times seconds$$

$$logN = -1 + log(seconds)$$

The load path for a test frequency of 10 Hz is:

$$N = 10 \times seconds$$

$$logN = 1 + log(seconds)$$

Figure 17.2 plots several load paths for a fixed load ratio R. The load paths provide the failure data points necessary to feed Equation (17.1) and compute the Gsc values. There is a Gsc value for every data point in the load path (see Figure 17.2). The value G_{sc}^1 corresponds to the failure point t_1 and N_1 on the load path 1. The value G_{sc}^2 corresponds to the failure point t_2 and N_2 on the load path 2.

The following protocol computes the Gsc values along any load paths:

- Define the failure points by arbitrarily choosing the time t and the number of cycles N.
- Compute the peak failure strains for each pair of t and N using the ply regression equation.
- The computed peak strains and the corresponding N are, of course, known experimental failure points.
- The static and cyclic pure slopes Gs and Gc come from the pure regression equations.
- The long-term ply strengths Ss and Sc come from the pure regression equations (known) and each failure point t and N.
- The above data enter the unified equation.
- The Gsc value is computed by setting the safety factor SF = 1.0.
- The Gsc values are tabulated for every load path as a function of N and t.
- The Gsc values for any combination of N and t are obtained by interpolation between the tabulated load paths.

Chapter 18 gives several numerical examples to clarify the computation protocol just outlined. The important point to have in mind at this time is the

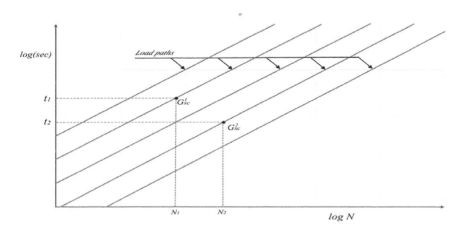

FIGURE 17.2 *Shows two Gsc values for two failure points. The Gsc values are computed from the unified equation for any time and number of cycles on known load paths. For failure points X and Y outside the load paths, the Gsc values are interpolated.*

recognition of a different Gsc value for every point in the load paths. The Gsc value to be used in the unified Equation (17.1) falls on the load path corresponding to the user-informed durability X and Y. As a rule, the informed X and Y never fall on a tabulated load path. Rather, they fall between known load paths as shown in Figure 17.2. The value of the interaction parameter Gsc to enter the unified equation is obtained by interpolation.

17.4 INTERPOLATING THE GSC VALUES

The R ratios and load paths in actual service are never the same as those that have been tested and tabulated. The correct Gsc value to be used in Equation (17.1) is not readily available and requires interpolation between the tabulated values. The interpolation follows two steps: The first step is between load paths (see Figure 17.3) and the second step interpolates between R ratios. It works like this.

The interpolation between load paths starts with the marking of the user-defined lifetime X and number of cycles Y in Figure 17.3. The value we are seeking interpolates between the four known neighbor Gsc values in Figure 17.3. There are two interpolation routes:

- Polynomial technique—This is the method used in finite element analysis.
- Simple averaging—This is probably the preferred way.

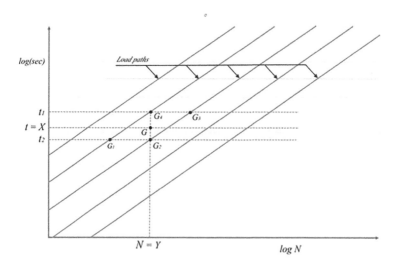

FIGURE 17.3 *The figure shows the load paths for a given R ratio. The parameter Gsc is known along the load paths. Outside the load paths, the Gsc values are obtained by interpolation. Good interpolations require many load paths.*

The interpolation between R ratios is similar.

As we see, the computation of the actual Gsc value to be used in the unified Equation (17.1) requires two interpolations: one between load paths and another between R ratios.

17.5 COMPUTING THE FIBER-DOMINATED DURABILITY

To compute the time-dependent rupture durability we need the complete form of the unified equation. The computation protocol is as follows:

- Compute the peak and minimum strains in the critical ply, as shown in the Part I of this book.
- Compute the R ratio.
- Compute the long-term pure static Ss and pure cyclic Sc strengths of the critical ply.
- The static and cyclic pure slopes Gs and Gc come from the pure rupture equations.
- Mark the target durability X and Y in Figure 17.3.
- Interpolate for Gsc between load paths in Figure 17.3.
- Interpolate for Gsc between R ratios.
- Enter the computed Gsc value in the unified equation
- Compute the long-term SF

The generation of good Gsc values requires many experimental regression equations developed at several R ratios and load frequencies. The ply equations to do this are not available at this time. The current lack of experimental regression equations limits the application of the unified equation in many practical situations involving fiber-dominated rupture.

The computation of the time-dependent rupture durability seems very complicated, and indeed it is. In the Appendix 18.1 of Chapter 18, the reader will find a simplified method to compute the long-term safety factor SF.

17.6 COMPUTING THE RESIN-DOMINATED DURABILITY

The time-independent resin-dominated failures use a simplified version of the unified equation. The computation process involves the following steps:

- Compute the peak and minimum strains on the critical ply, as shown in the Part I of this book.
- Compute the R ratio.
- Choose the applicable pure regression line (infiltration, weep or stiffness). See Chapter 9.

- The long-term static strength Ss is equal to the infiltration, stiffness or weep threshold, as the case may be.
- Compute the long-term pure cyclic ply strength Sc using the pure regression line.
- The pure slope Gc comes from the pure regression equation.
- Derive the actual Gsc by interpolating between tabulated values.
- Enter the interpolated Gsc in the simplified unified equation.
- Compute the long-term SF.

The time-independence and the absence of load paths facilitate the interpolation in these cycle-dependent durability cases.

17.7 A FEW Gsc VALUES

The numerical computation of the Gsc values from experimental regression equations is discussed in Chapter 18. The sample values in Tables 17.1 and 17.2 are from Chapter 18.

Example 17.1

Describe the use of the unified equation in oil pipes.

The oil pipes never see chemical service and require no check for infiltration. The failure of oil pipes comes from either weeping, loss of stiffness, or long-term rupture.

The computation of the rupture durability estimates the long-term safety factor SF for the lifetime X and cycles Y prescribed by the pipeline owner. The strains of interest are those in the fiber direction of the critical UD ply and, since the fiber strain-corrosion is time dependent, we use the complete form of the unified equation.

The computation of the stiffness and weep durability estimate the long-term safety factor SF using the stiffness or weep threshold as the long-term static strength Ss. The pure long-term cyclic strengths Sc is derived from the transverse pure cyclic regression equation and the number of cycles Y is prescribed by the pipeline owner. The strains of interest are those in the transverse direction of the critical UD ply and, since the resin is not strain-corroded, we use the simplified form of the unified equation. The Gsc values of transversely loaded UD plies are time-independent.

Example 17.2

Explain the use of the unified equation in sanitation pipes.

The computation of the time-dependent long-term rupture of sanitation pipes requires the complete unified equation. The inner barrier of chopped plies controls

TABLE 17.1

Rupture G_{sc} values for UD plies tensile loaded in the fiber direction at a load path of 20 Hz. The Gsc values in the fiber direction are resin independent.

Rupture Gsc in the fiber direction
Load path 20 Hz

Lifetime (seconds)	N	R				
		0.0	0.1	0.5	0.9	1.0
5×10	10^3	0.0	8.70	683.23	NA	0.0
5×10^2	10^4	0.0	8.76	348.48	NA	0.0
5×10^3	10^5	0.0	8.84	177.41	NA	0.0
5×10^4	10^6	0.0	8.91	90.44	NA	0.0
5×10^5	10^7	0.0	9.00	45.40	NA	0.0
5×10^6	10^8	0.0	9.07	22.78	NA	0.0
5×10^7	10^9	0.0	9.15	11.16	NA	0.0
5×10^8	10^{10}	0.0	9.22	6.08	NA	0.0
50 years	3.17×10^{10}	0.0	9.26	3.70	NA	0.0

TABLE 17.2

Weep G_{sc} values for transverse loaded vinyl ester UD plies.

Gsc of transverse loaded vinyl ester UD plies
The time variable and the loading frequency are not relevant

Lifetime (Seconds)	N	R				
		0.0	0.1	0.5	0.9	1.0
—	10^3	0.0	20.38	332 400	NA	0.0
—	10^4	0.0	21.24	159 300	NA	0.0
—	10^5	0.0	22.12	107 300	NA	0.0
—	10^6	0.0	22.96	71 500	NA	0.0
—	10^7	0.0	23.83	62 500	NA	0.0
—	10^8	0.0	24.72	40 250	NA	0.0
—	10^9	0.0	25.57	27 900	NA	0.0
—	10^{10}	0.0	26.42	14 430	NA	0.0
—	3.17×10^{10}	0.0	26.84	13 340	NA	0.0

the long-term weep failure and requires the time-independent, simplified version of the unified equation. The pipeline owner informs the expected durability X years and Y cycles.

The foregoing discussion highlights the difficulty of using the unified equation in the solution of long-term load-dependent failures. For a simplified approximate solution, see Appendix 18.1.

18 Numerical Computation of Gsc

18.1 INTRODUCTION

This chapter deals with the numerical computation of the interaction parameter Gsc. The computation protocol consists in setting the SF = 1.0 and solving the unified equation fed with known long-term failure points. The failure points come from experimental ply regression equations measured at a fixed R ratio and constant loading frequency. There are seven experimental ply equations available at this time:

- The pure static (R = 1.0) rupture line of UD plies loaded in the fiber direction.
- The two cyclic rupture equations (R = 0.1 and R = 0.5) of UD plies loaded in the fiber direction at 20 Hz.
- The two cyclic infiltration equations (R = 0.1 and R = 0.5) of transversely loaded UD plies at 20 Hz.
- The pure static (R = 1.0) rupture equation for chopped fibers.
- The cyclic (R = 0.1) rupture equation for chopped fibers.

The pure cyclic equations (R = 0.0) are not available due to technical difficulties. In their absence, we assume the equations derived at R = 0.1 as pure cyclic. There is a specific regression equation for every ply, every loading, and every load-dependent mode of failure. The cost and effort to measure these equations are considerable. This book suggests many important simplifications to alleviate this situation.

One such simplification is the assumption of the same long-term cyclic strength for all plies, the so-called fatigue threshold. The same fatigue threshold (defined in Chapter 9 as $T_0 = 0.05\%$) applies to all plies, regardless of the resin and the mode of failure. This is a reasonable assumption, considering that in the long-term, the accumulated damage is so high as to equalize all ply strengths. See Chapter 9.

The knowledge of the ply thresholds (short-term and long-term) allows the derivation of all ply regression equations from the experimental measurement of just one line, usually the rupture line. The knowledge of one regression equation, together with the short-term and long-term thresholds, suffices to compute all regression equations. This fundamental simplification allows the computation of all pure regression equations for UD and chopped plies for all modes of failure (see Chapter 9).

The ply regression equations capture all variables related to their load-dependent long-term failures. They include the peak strain, the time variable, the load path, the number of cycles, the load ratio R, and the number of cycles N. Needless to say, the Gsc values derived from regression equations also depend on the load ratio R, the time variable t, the load path, and the number of cycles N.

The Gsc values to use in the unified equation are interpolated between tabulated values. The goodness of the interpolation is contingent on the availability of many R ratios and load paths, which requires a large number of experimental equations.

The pure regression equations for UD and chopped plies are available in Chapter 9. The non-pure regression equations for R ratios other than R = 0.0 and R = 1.0 are very scarce. The only non-pure regression equations available at this time are those developed by Guangxu Wei for polyester UD plies at 20 Hz, R = 0.5 and R = 0.1 (see Ref.[25]). These equations are listed below.

$$log\epsilon = 0.519 - 0.065 \times logN \quad \left(\text{UD ply, fiber direction, rupture, R = 0.5, 20 Hz}\right)$$

$$log\epsilon = 0.477 - 0.089 \times logN \quad \left(\text{UD ply, fiber direction, rupture, R = 0.1, 20 Hz}\right)$$

$$log\epsilon = -0.699 - 0.027 \times logN \quad \left(\text{UD ply, transverse loaded, infiltration, R = 0.5, 20 Hz}\right)$$

$$log\epsilon = -0.699 - 0.034 logN \quad \left(\text{UD ply, transverse loaded, infiltration, R = 0.1, 20 Hz}\right)$$

These are the only non-pure ply equations available in the published literature. The scarcity of such equations is a serious impediment to the practical use of the unified equation. A special effort to develop a comprehensive set of non-pure equations, covering several load ratios R and load-paths, is essential at this time. The above four equations, developed for R = 0.10, R = 0.50 and 20 Hz, are clearly not enough to provide a comprehensive set of Gsc values.

18.2 COMPUTING THE TIME-DEPENDENT GSC

The computation of the Gsc values from non-pure regression equations is very simple. The protocol uses the full version of the unified equation, with SF = 1.

$$\left(\frac{R\varepsilon}{S_S}\right)^{\frac{1}{G_S}}+\left(\frac{(1-R)\varepsilon}{S_C}\right)^{\frac{1}{G_C}}+\left(\frac{R(1-R)\varepsilon^2}{S_S\times S_C}\right)^{\frac{1}{G_{SC}}}=1.00$$

Solving for Gsc, we have

$$G_{SC}=\frac{log\left[\dfrac{R\left(1-R\right)\times\epsilon^2}{S_S\times S_C}\right]}{log\left[1-\left(\dfrac{R\times\epsilon}{S_S}\right)^{\frac{1}{G_s}}-\left(\dfrac{\left(1-R\right)\times\epsilon}{S_C}\right)^{\frac{1}{G_C}}\right]}$$

The above equation computes the time-dependent rupture Gsc of chopped plies and UD plies loaded in the fiber direction. The peak strain ε comes from the non-pure rupture regression equation. The cyclic S_C and static S_S strengths come from the pure rupture equations in Chapter 9. The slopes G_S and G_C of the pure static and pure cyclic rupture equations are also available in Chapter 9. The pure regression equations are available. The problem is the lack of non-pure equations.

The only non-pure regression equations available are:

$$log\epsilon = 0.477-0.065\times logN \quad \left(\text{UD ply, fiber direction, rupture, R}=0.5,\ 20\ \text{Hz}\right)$$

$$log\epsilon = 0.477-0.089\times log\,N \quad \left(\text{UD ply, fiber direction, rupture, R}=0.1,\ 20\ \text{Hz}\right)$$

We illustrate the computation of Gsc for $N = 10^9$ cycles, load path 20Hz and load ratio $R = 0.5$. The test duration in hours, corresponding to $N = 10^9$ cycles and a load path of 20 Hz is

$$hours = \frac{10^9}{20\times 60\times 60} = 14\ 000\ hours$$

The peak strain comes from the rupture regression equation for $R = 0.50$.

$$log\epsilon = 0.477-0.065\times logN$$

In the above equation, ε is the peak strain in the fiber direction and N is the number of cycles to failure. This equation is valid for the load path of 20 Hz and R = 0.5. The peak strain that fails the ply in N = 10⁹ cycles is:

$$log\varepsilon = 0.477 - 0.065 log\,(10^9)$$

$$\epsilon = 0.780\%$$

We have just computed the peak strain that fails the ply in the fiber direction at R = 0.5, N = 10⁹ cycles, and t = 14 000 hours. This peak strain derives from an experimental rupture equation and is accurate. Next, we compute the ply long-term pure static and pure cyclic strengths. The long-term pure static strength comes from the pure static equation listed in Chapter 9 for UD plies loaded in the fiber direction.

$$log\,\varepsilon = 0.400 - 0.077 \log(hours) \quad \left(\text{Pure static}\right)$$

The time variable corresponding to N = 10⁹ cycles is 14 000 hours. The long-term static strength is:

$$\log S_S = 0.400 - 0.077 \log(14000)$$

$$Ss = 1.204\%$$

The long-term pure cyclic strength comes from the pure cyclic equation of UD plies loaded in the fiber direction. This equation is in Chapter 9.

$$log\,\varepsilon = 0.477 - 0.089 \times \log N \quad \left(\text{Pure cyclic}\right)$$

$$\log Sc = 0.477 - 0.089 \times \log(10^9)$$

$$Sc = 0.474\%$$

Entering the above values in the equation to compute Gsc, we have

$$G_{SC} = \frac{log\left[\dfrac{R\left(1-R\right)\times \epsilon^2}{S_S \times S_C}\right]}{log\left[1-\left(\dfrac{R\times \epsilon}{S_S}\right)^{1/G_s} - \left(\dfrac{\left(1-R\right)\times \epsilon}{S_C}\right)^{1/G_C}\right]}$$

$$G_{SC} = \cfrac{\log\left[\cfrac{0.5(1-0.5)\times 0.780^2}{1.204 \times 0.474}\right]}{\log\left[1 - \left(\cfrac{0.5 \times 0.780}{1.204}\right)^{\frac{1}{0.077}} - \left(\cfrac{(1-0.5)\times 0.780}{0.474}\right)^{\frac{1}{0.089}}\right]}$$

From which we obtain Gsc = 11.16 for N = 10^9 cycles and a loading frequency of 20 Hz. The same procedure is applicable to other values of N and t along the load path of 20 Hz and R = 0.5. The results are in Table 18.1. The Gsc values computed in this way are exact and produce the same results as the non-pure regression equations that originated them.

The Gsc values in the fiber direction for R = 0.1 and 20 Hz are computed in same way. The regression equation to use in this case is

$$\log\epsilon = 0.477 - 0.089 \times \log N \quad \text{(UD ply, fiber direction, rupture, R = 0.1, 20 Hz)}$$

The load ratio is R = 0.1 and the load path is 20 Hz. The computed Gsc values are in Table 18.2.

18.3   COMPUTING THE TIME-INDEPENDENT GSC

The weep, infiltration, and stiffness failures of chopped and transversely loaded UD plies are resin-dominated and time-independent. The Gsc values in these cases derive from the simplified version of the unified equation.

TABLE 18.1
Rupture Gsc values for UD plies loaded in the fiber direction at 20 Hz and R = 0.5. The rupture Gsc values in the fiber direction are resin-independent.

N	Seconds	ϵ(%)	R	Ss(%)	Sc(%)	Gsc
10^3	5×10	1.914	0.5	3.492	1.622	683.23
10^4	5×10^2	1.648	0.5	2.924	1.321	348.48
10^5	5×10^3	1.419	0.5	2.449	1.076	177.41
10^6	5×10^4	1.222	0.5	2.051	0.877	90.44
10^7	5×10^5	1.052	0.5	1.718	0.714	45.40
10^8	5×10^6	0.906	0.5	1.439	0.582	22.78
10^9	5×10^7	0.780	0.5	1.205	0.474	11.16
10^{10}	5×10^8	0.671	0.5	1.009	0.386	6.08
3.17×10^{10}	50 years	0.623	0.5	0.924	0.349	3.70

TABLE 18.2

Rupture Gsc values for UD plies loaded in the fiber direction at 20 Hz and R = 0.1. The peak strain ε and the long-term cyclic strength S_C come from the same regression equation and have equal values. This is because we have assumed the pure equation (R = 0.0) equal to the equation for R = 0.1. Such an approximation is required in the absence of a regression equation for R = 0.0. The above values are fiber-dominated and resin-independent.

N	Seconds	ε(%)	R	Ss(%)	Sc(%)	Gsc
10^3	5×10	1.622	0.1	3.492	1.622	8.70
10^4	5×10^2	1.321	0.1	2.924	1.321	8.76
10^5	5×10^3	1.076	0.1	2.449	1.076	8.84
10^6	5×10^4	0.877	0.1	2.051	0.877	8.91
10^7	5×10^5	0.714	0.1	1.718	0.714	9.00
10^8	5×10^6	0.582	0.1	1.439	0.582	9.07
10^9	5×10^7	0.474	0.1	1.205	0.474	9.15
10^{10}	5×10^8	0.386	0.1	1.009	0.386	9.22
3.17×10^{10}	50 years	0.349	0.1	0.924	0.349	9.26

$$\left(\frac{(1-R) \times \varepsilon \times SF}{S_C} \right)^{\frac{1}{G_C}} + \left(\frac{R(1-R) \times \varepsilon^2 \times SF^2}{S_S \times S_C} \right)^{\frac{1}{G_{SC}}} = 1.00$$

Setting SF = 1.0 and solving for Gsc, we have

$$G_{SC} = \frac{\log\left[\dfrac{R\left(1-R\right) \times \epsilon^2}{S_S \times S_C} \right]}{\log\left[1 - \left(\dfrac{\left(1-R\right) \times \epsilon}{S_C} \right)^{1/G_C} \right]}$$

The long-term static strength S_S in time-independent failures is equal to the short-term threshold corresponding to the mode of failure under investigation. For example, the long-term ply static strengths for weep and infiltration failures are as follows:

Ss = Ti = 0.20% (Infiltration threshold, polyester UD plies)
Ss = Ti = 0.30% (Infiltration threshold, vinyl ester UD plies)
Ss = Ti = 0.20% (Infiltration threshold, polyester chopped fibers)
Ss = Ti = 0.50% (Infiltration threshold, vinyl ester chopped plies)

Ss = Tw = 0.25% (Weep threshold, polyester UD plies)
Ss = Tw = 0.40% (Weep threshold, vinyl ester UD plies)
Ss = Tw = 0.80% (Weep threshold, polyester chopped plies)
Ss = Tw = 1.00% (Weep threshold, vinyl ester chopped plies)

The resin-dominance removes the time variable from the unified equation, making the stiffness, weep, and infiltration failures fully cycle-dependent. The cycle-dependent interaction parameters Gsc are a function of N alone. This simplification allows the use of the Goodman diagrams to estimate the ply stiffness, weep, and infiltration durability on an equal standing with the unified equation. The reader will understand that the Goodman diagrams coincide with the unified equation only in the solutions of time-independent cases involving infiltration, weep, and stiffness failures. They differ from the unified equation in the solution of time-dependent long-term ruptures associated with fiber strain-corrosion.

Example 18.1

Compute the time-independent weep Gsc value of transverse loaded polyester UD plies at $N = 10^9$ cycles and $R = 0.5$. This is a time-independent situation in which the load path (loading frequency) is not relevant.

The infiltration line of transverse loaded polyester UD plies at $R = 0.5$ is

$$log\epsilon = log(0.20) - 0.027 \times logN$$

To compute the weep Gsc at $R = 0.5$ we need the weep regression equation of transverse loaded UD plies at $R = 0.5$. Let us derive this equation from the above infiltration line. The weep line we are looking for passes through the short-term weep threshold Tw = 0.25% and intercepts the known infiltration equation at the fatigue threshold $T_0 = 0.05\%$. From simple geometric considerations of Figure 9.2, the weep equation at $R = 0.5$ is

$$log\epsilon = log(0.25) - \frac{log(0.25/0.05)}{log(0.20/0.05)} \times 0.027 \times logN$$

$$log\epsilon = -0.602 - 0.031 \times logN$$

Where Tw = 0.25% is the weep threshold, Ti = 0.20% is the infiltration threshold and $T_0 = 0.05\%$ is the fatigue threshold of transverse loaded polyester UD plies. The G = 0.027 in the above equation is the experimentally measured infiltration slope at $R = 0.5$.

The peak strain that weeps the ply at R = 0.5 and N = 10⁹ is:

$$log\epsilon = -0.602 - 0.031 \times log10^9$$

$$\epsilon = 0.132\%$$

The long-term pure static strength of transverse loaded polyester UD plies in this case is equal to the weep threshold,

$$Ss = Tw = 0.25\%$$

The long-term pure weep cyclic strength of transversely loaded polyester UD plies comes from the weep regression equation listed in Chapter 9.

$$log\epsilon = log(0.25) - 0.039 \times logN$$

Taking N = 10⁹, we have

$$log\,S_C = log(0.25) - 0.039 \times log10^9$$

$$Sc = 0.111\%$$

Entering the above values in the simplified equation, we have

$$G_{SC} = \frac{log\left[\dfrac{R(1-R)\times\epsilon^2}{S_S \times S_C}\right]}{log\left[1-\left(\dfrac{(1-R)\times\epsilon}{S_C}\right)^{1/G_C}\right]}$$

$$G_{SC} = \frac{log\left[\dfrac{0.5(1-0.5)\times0.132^2}{0.25\times0.111}\right]}{log\left[1-\left(\dfrac{(1-0.5)\times0.132}{0.111}\right)^{1/0.039}\right]} = 1\,140\,000$$

Where Gc = 0.039 is the slope of the pure weep cyclic line of transverse loaded polyester UD plies. See Chapter 9.

Table 18.3 lists the computed Gsc values as a function of the number of cycles N.

TABLE 18.3

Weep Gsc values for transverse loaded polyester UD plies at R = 0.5. The time variable and the loading frequency are not relevant. The computed Gsc values are surprisingly high, indicating a large interaction at the loading ratio R = 0.5.

N	Seconds	ϵ	R	Ss	Sc	Gsc
10^3	—	0.202	0.5	0.25	0.191	19,200,000
10^4	—	0.188	0.5	0.25	0.175	13,300,000
10^5	—	0.175	0.5	0.25	0.160	8,700,000
10^6	—	0.163	0.5	0.25	0.144	3,700,000
10^7	—	0.152	0.5	0.25	0.133	3,000,000
10^8	—	0.141	0.5	0.25	0.122	2,300,000
10^9	—	0.132	0.5	0.25	0.111	1,140,000
10^{10}	—	0.122	0.5	0.25	0.102	1,000,000
3.17×10^{10}	—	0.118	0.5	0.25	0.097	670,000

TABLE 18.4

Weep Gsc values for transverse loaded polyester UD plies at R = 0.1. The time variable and the loading frequency are not relevant. The peak strain ϵ and the long-term cyclic strength S_C derive from the same regression equation. Such an approximation is required in the absence of the pure regression equation at R = 0.0.

N	Seconds	ϵ	R	Ss	Sc	Gsc
10^3	—	0.191	0.1	0.25	0.191	38.54
10^4	—	0.175	0.1	0.25	0.175	39.80
10^5	—	0.160	0.1	0.25	0.160	41.09
10^6	—	0.144	0.1	0.25	0.144	42.61
10^7	—	0.133	0.1	0.25	0.133	43.75
10^8	—	0.122	0.1	0.25	0.122	45.00
10^9	—	0.111	0.1	0.25	0.111	46.36
10^{10}	—	0.102	0.1	0.25	0.102	47.57
3.17×10^{10}	—	0.097	0.1	0.25	0.097	48.30

Table 18.4 lists the weep Gsc values of transverse loaded polyester UD plies at R = 0.1. In deriving these values we have used the weep regression equation at R = 0.1 as a substitute for the pure weep cyclic equation. This is so, because the "true" pure weep line at R = 0.0 is not available.

The pure weep cyclic equation of transverse loaded polyester UD plies is in Chapter 9.

$$\log \epsilon = \log(0.25) - 0.039 \times \log N$$

The long-term pure cyclic weep strengths Sc and the peak strains ϵ derive from the above equation. The pure static weep transverse strength is constant at Ss = Tw = 0.25%. The load ratio is R = 0.1. The pure weep slope is Gc = 0.039. To compute the interaction parameter Gsc at R = 0.1 we simply enter these inputs into the simplified version of the unified equation

$$G_{SC} = \frac{\log\left[\dfrac{R(1-R) \times \epsilon^2}{S_S \times S_C}\right]}{\log\left[1 - \left(\dfrac{(1-R) \times \epsilon}{S_C}\right)^{1/G_C}\right]}$$

Tables 18.5 and 18.6 list the weep Gsc values of transverse loaded vinyl ester UD plies at R = 0.1 and R = 0.5. The weep regression equations of vinyl ester UD plies derive from those of polyester, as explained in Chapter 9. The following computations is a recap of the discussions of Chapter 9.

The weep slopes of the transverse loaded vinyl ester UD ply are

TABLE 18.5
Weep Gsc values of transverse loaded vinyl ester UD plies at R = 0.5. The time variable and the loading frequency are not relevant.

N	Seconds	ϵ	R	Ss	Sc	Gsc
10^3	—	0.301	0.5	0.40	0.281	332,400
10^4	—	0.274	0.5	0.40	0.250	159,300
10^5		0.249	0.5	0.40	0.222	107,300
10^6	—	0.227	0.5	0.40	0.198	71,500
10^7	—	0.207	0.5	0.40	0.176	62,500
10^8	—	0.188	0.5	0.40	0.156	40,250
10^9	—	0.171	0.5	0.40	0.139	27,900
10^{10}	—	0.156	0.5	0.40	0.124	14,430
3.17×10^{10}	—	0.148	0.5	0.40	0.117	13,340

$$G_C = \frac{\log(0.40/0.05)}{\log(0.20/0.05)} \times 0.027 = 0.041 \quad \left(\text{Vinyl ester, R} = 0.5\right)$$

$$G_C = \frac{\log(0.40/0.05)}{\log(0.20/0.05)} \times 0.034 = 0.051 \quad \left(\text{Vinyl ester, R} = 0.1\right)$$

From the above slopes, the weep regression equations of transverse loaded vinyl ester UD plies are:

$$\log \epsilon = \log 0.40 - 0.051 \log N \quad \left(\text{Weep, vinyl ester, UD, R} = 0.1\right)$$

$$\log \epsilon = \log 0.40 - 0.041 \log N \quad \left(\text{Weep, vinyl ester, UD, R} = 0.5\right)$$

The above equations, together with the pure cyclic weep equation listed in Chapter 9, allow the computation of the Gsc values listed in Tables 18.5 and 18.6. For technical reasons, the pure cyclic equation is taken equal to the equation at R = 0.1.

The Gsc values of UD plies loaded in the fiber direction are in Table 18.7. The time-dependent values in the fiber direction are valid for rupture failure and a load path of 20 Hz. The Gsc values in the fiber direction are resin-independent.

TABLE 18.6

Weep Gsc values of transverse loaded vinyl ester UD plies at R = 0.1. The time variable and the loading frequency are not relevant. The peak strain ε and the long-term cyclic strength S_C come from the same regression equation at R = 0.1.

N	Seconds	ε	R	Ss	Sc	Gsc
10^3	—	0.281	0.1	0.40	0.281	20.38
10^4	—	0.250	0.1	0.40	0.250	21.24
10^5		0.222	0.1	0.40	0.222	22.12
10^6	—	0.198	0.1	0.40	0.198	22.96
10^7	—	0.176	0.1	0.40	0.176	23.83
10^8	—	0.156	0.1	0.40	0.156	24.72
10^9	—	0.139	0.1	0.40	0.139	25.57
10^{10}	—	0.124	0.1	0.40	0.124	26.42
3.17×10^{10}	—	0.117	0.1	0.40	0.117	26.84

TABLE 18.7

Rupture G_{sc} values of fiber loaded UD plies at several R ratios for a load path of 20 Hz. The above values are resin-independent.

Load-path 20 Hertz

Lifetime (seconds)	N	R				
		0.0	0.1	0.5	0.9	1.0
5×10	10^3	0.0	8.70	683.23	NA	0.0
5×10^2	10^4	0.0	8.76	348.48	NA	0.0
5×10^3	10^5	0.0	8.84	177.41	NA	0.0
5×10^4	10^6	0.0	8.91	90.44	NA	0.0
5×10^5	10^7	0.0	9.00	45.40	NA	0.0
5×10^6	10^8	0.0	9.07	22.78	NA	0.0
5×10^7	10^9	0.0	9.15	11.16	NA	0.0
5×10^8	10^{10}	0.0	9.22	6.08	NA	0.0
50 years	3.17×10^{10}	0.0	9.26	3.70	NA	0.0

TABLE 18.8

Weep G_{sc} values for transverse loaded polyester UD plies at several R ratios.

The time variable and the loading frequency are not relevant

Lifetime (Seconds)	N	R				
		0.0	0.1	0.5	0.9	1.0
—	10^3	0.0	38.54	19,200,000	NA	0.0
—	10^4	0.0	39.80	13,300,000	NA	0.0
—	10^5	0.0	41.09	8,700,000	NA	0.0
—	10^6	0.0	42.61	3,700,000	NA	0.0
—	10^7	0.0	43.75	3,000,000	NA	0.0
—	10^8	0.0	45.00	2,300,000	NA	0.0
—	10^9	0.0	46.36	1,140,000	NA	0.0
—	10^{10}	0.0	47.57	1,000,000	NA	0.0
—	3.17×10^{10}	0.0	48.30	670,000	NA	0.0

Table 18.8 lists the time-independent weep Gsc values of transverse loaded polyester UD plies at several R ratios.

The Gsc values at R = 1.0 and R = 0.0 are equal to zero, as they should, since there is no interaction in pure cyclic and in pure static loadings. In the neighborhood of R = 0.5, the transverse Gsc of UD plies increases substantially for low N and short time durations, reflecting a very strong interaction. Such an increased interaction suggests the existence of a singularity. To avoid this

TABLE 18.9

Weep G_{sc} values of transverse loaded vinyl ester UD plies at several R ratios. These values enter the unified equation to compute the weep durability of oil pipes.

The time variable and the loading frequency are not relevant

Lifetime (Seconds)	N	R				
		0.0	**0.1**	**0.5**	**0.9**	**1.0**
—	10^3	0.0	20.38	332,400	NA	0.0
—	10^4	0.0	21.24	159,300	NA	0.0
—	10^5	0.0	22.12	107,300	NA	0.0
—	10^6	0.0	22.96	71,500	NA	0.0
—	10^7	0.0	23.83	62,500	NA	0.0
—	10^8	0.0	24.72	40,250	NA	0.0
—	10^9	0.0	25.57	27,900	NA	0.0
—	10^{10}	0.0	26.42	14,430	NA	0.0
—	3.17×10^{10}	0.0	26.84	13,340	NA	0.0

singularity, the unified equation is not applicable to low values of N and short lifetimes. This limitation holds for the classical regression equations as well, as explained in Appendix 9.2.

Example 18.2

Suppose an oil pipeline operating as indicated in Table 18.10. It is required to determine its safety factor SF against rupture failure for a period of X = 20 years of continuous operation.

Let us compute the long-term rupture safety factor SF for the user-defined lifetime X = 20 years. We focus on the ply strain in the fiber direction. First, we add the static and cyclic strain components to compute the peak global hoop and axial strains.

TABLE 18.10

Operating conditions for the oil pipeline of Example 18.2.

Winding angle	$\alpha = \pm 55°$
Static hoop strain component	$\epsilon_y = 0.20\%$
Static axial strain component	$\epsilon_x = 0.15\%$
Cyclic hoop strain component	$\Delta\epsilon_y = 0.10\%$
Cyclic axial strain component	$\Delta\epsilon_x = 0.05\%$
Expected cycles in 20 years	$Y = 10^8$
Target lifetime	$X = 20$ years

$\epsilon_y = 0.20 + 0.10 = 0.30\%$ $\left(Peak\ hoop\right)$

$\epsilon_x = 0.15 + 0.05 = 0.20\%$ $\left(Peak\ axial\right)$

Next, we use the equation $\epsilon_1 = \epsilon_x cos^2\alpha + \epsilon_y sen^2\alpha$ to rotate the global strain components to the fiber frame. The maximum and the minimum strain components in the fiber direction are

$\varepsilon_{max} = 0.20 \times cos^2 55 + 0.30 \times sen^2 55 = 0.27\%$

$\varepsilon_{min} = 0.15 \times cos^2 55 + 0.20 \times sen^2 55 = 0.18\%$

The R ratio in the fiber direction is

$R = \dfrac{\epsilon_{min}}{\epsilon_{max}} = \dfrac{0.18}{0.27} = 0.67$

The long-term pure static strength of boron-free glass in 20 years is

$\log(S_S) = 0.400 - 0.077 \log(20 \times 365 \times 24)$

$Ss = 1.20\%$

The long-term pure cyclic strength of boron-free glass in Y = 10^8 cycles is

$\log(S_C) = 0.519 - 0.089 \log 10^8$

$Sc = 0.64\%$

The loading frequency is

$frequency = \dfrac{10^8}{20 \times 365 \times 24 \times 60 \times 60} = 0.15\ cycles/sec$

The interaction parameters Gsc for the load ratio R = 0.67 and the load path of 0.15 Hz in the fiber direction are not known. For illustration purposes, we assume Gsc = 10. The parameters Gs and Gc come from the pure static and the pure cyclic equations respectively. For a pipeline of boron-free glass the 20 years rupture safety factor SF is

$$\left(\frac{0.67\times0.27\times SF}{1.20}\right)^{1/0.077}_{static}+\left(\frac{(1-0.67)\times0.27\times SF}{0.64}\right)^{1/0.089}_{cyclic}$$

$$+\left(\frac{0.67\times(1-0.67)\times0.27^2\times SF^2}{1.20\times0.64}\right)^{1/10}=1.0$$

$SF = 5.0$

The unified equation predicts excellent performance against long-term rupture.

The short-term safety factor is

$$SF = \frac{3.0\%}{0.27\%} = 11$$

Example 18.3

Compute the 20 years weep safety factor for the pipeline of Example 18.2.

The computation protocol is similar, except that we now deal with transverse strains. The maximum global hoop and axial strains are the same as in Example 18.2.

$$\epsilon_y = 0.20 + 0.10 = 0.30\% \quad \left(Peak\ hoop\right)$$

$$\epsilon_x = 0.15 + 0.05 = 0.20\% \quad \left(Peak\ axial\right)$$

The minimum global strains are, as before

$$\epsilon_y = 0.20\% \quad \left(Minimum\ hoop\right)$$

$$\epsilon_x = 0.15\% \quad \left(Minimum\ axial\right)$$

Next, we use the equation $\epsilon_2 = \epsilon_x sen^2\alpha + \epsilon_y cos^2\alpha$ to compute the transverse strain components in the local ply frame. The maximum and the minimum transverse strain components are

$$\varepsilon_{max} = 0.20\times sen^2 55 + 0.30\times cos^2 55 = 0.23\%$$

$$\varepsilon_{min} = 0.15 \times sen^2 55 + 0.20 \times \cos^2 55 = 0.17\%$$

The R ratio in the transverse direction is

$$R = \frac{\varepsilon_{min}}{\varepsilon_{max}} = \frac{0.17}{0.23} = 0.74$$

The long-term pure static weep strength in the transverse direction is equal to the weep threshold. Assuming vinyl ester resin, we have

$$Tw = 0.40\%$$

The long-term transverse pure cyclic weep strength of vinyl ester UD plies at $Y = 10^8$ cycles is

$$\log S_C = -0.398 - 0.051 \log 10^8$$

$$S_C = 0.16\%$$

The loading frequency is not relevant in time-independent weep failures. The transverse interaction parameters Gsc for R = 0.74 is not known. For illustration purposes, we assume Gsc = 1 000. Entering this value in the simplified form of the unified equation, we have

$$\left(\frac{(1-0.74) \times 0.23 \times SF}{0.16} \right)^{1/0.051} + \left(\frac{0.74 \times (1-0.74) \times 0.23^2 \times SF^2}{0.40 \times 0.16} \right)^{1/1000} = 1.0$$

$$SF = 1.8$$

The unified equation predicts a reasonable weep performance in 20 years of continuous service for the vinyl ester pipeline. The weep short-term safety factor is

$$SF = \frac{0.40\%}{0.23\%} = 1.7$$

Example 18.4

Continuing with the analysis of the pipeline of Example 18.2, let us assume it has a weep barrier of vinyl ester resin. The mode of failure is weep and the target lifetime is X = 20 years.

The computation protocol is the same of Examples 18.2 and 18.3, except the critical ply is the weep barrier of chopped fibers. The maximum and minimum hoop strains are

$$\epsilon_y = 0.20 + 0.10 = 0.30\% \quad \left(Peak\ hoop\right)$$

$$\epsilon_y = 0.20\% \quad \left(Minimum\ hoop\right)$$

The corrosion barrier is isotropic and needs no strain rotation. The R ratio for the weep barrier is

$$R = \frac{\epsilon_{min}}{\epsilon_{max}} = \frac{0.20}{0.30} = 0.66$$

The long-term pure static weep strength is equal to the weep threshold. Assuming vinyl ester resin, we have

$$S_S = Tw = 1.0\%$$

The weep pure regression equation of vinyl ester chopped plies is in Chapter 9

$$log\epsilon = log\left(1,0\right) - 0.066 \times logN$$

The long-term pure cyclic weep strength of vinyl ester chopped plies in 20 years is

$$log\epsilon = -0.066 \times log10^8$$

$$S_C = 0.30\%$$

The loading frequency is not relevant in the computation of weep failures. The interaction parameters Gsc for vinyl ester chopped plies at the load ratio R = 0.66 is not known. For illustration purposes, we assume Gsc = 100. Entering this value in the simplified form of the unified equation, we have

$$\left(\frac{(1-0.66) \times 0.30 \times SF}{0.30}\right)^{1/0.066} + \left(\frac{0.66 \times (1-0.66) \times 0.30^2 \times SF^2}{1.0 \times 0.30}\right)^{1/100} = 1.0$$

$$SF = 2.2$$

The unified equation predicts good weep performance for the pipeline with a weep barrier of vinyl ester resin. The long-term safety factor for the pipeline without a weep barrier, from Example 18.3, is SF = 1.8. The weep short-term safety factor for the vinyl ester chopped ply is

$$SF = \frac{1.0\%}{0.30\%} = 3.3$$

The above results are illustrative and do not reflect the true pipe behavior. This is not because the unified equation is wrong, but because the assumed interaction parameters Gsc are certainly inaccurate. Not until the interaction parameters Gsc are accurately measured can the unified equation give reliable results. As explained in Chapters 9 and 17, the accurate determination of Gsc requires the development of many regression equations at several frequencies and load ratios R.

Just how many regression equations are required will depend on the desired accuracy. Perhaps, as a minimum, the following suffices:

- The fiber loaded UD plies need a minimum of five load paths (loading frequency) for each load ratio R. Taking five load ratios, we need $5 \times 5 = 25$ equations.
- The transverse loaded UD plies do not depend on the load path. Taking 5 load ratios gives $1 \times 5 = 5$ equations.
- The chopped fiber plies need an additional 25 equations.

The required number of experimental regression equations is not very large. Just 55 equations would allow the accurate solution of all load-dependent durability problem for any laminate and loading.

Example 18.5

Explain the conditions for applicability of the unified equation and the non-pure regression equations.

The non-pure regression equations are easy to use and contain all information needed to compute the ply long-term load-dependent failure. When available, the non-pure regression equations are the way to go. However, they are seldom available. The unified equation is general, and finds use in all situations. The unified equations are a major step forward in the analysis of load-dependent durability of composites.

Example 18.6

Explain the condition for applicability of the Goodman diagrams.

The Goodman diagrams are as good as the unified equation to compute the easy cases of time-independent durability. Both methods give excellent results. However, unlike the unified equation, the Goodman diagrams do not capture the time variable and cannot compute cases of time-dependent durability. The computation of time-dependent durability requires the unified equation.

APPENDIX 18.1

PRACTICAL RANGES OF THE LONG-TERM SAFETY FACTORS SF

The obstacle to the widespread use of the unified equation is, no doubt, the lack of experimental non-pure regression equations to determine the interaction parameter Gsc. In spite of the tremendous simplifications proposed in Chapter 9, the application of the unified equation still requires the experimental measurement of several expensive regression equations. The effort to determine such equations will probably not be undertaken in the near future.

The problem, then, is the lack of ply regression equations to determine the interaction parameter Gsc. The other parameters entering the unified equation, like the pure slopes Gs and Gc, and the long-term ply strengths Ss and Sc, derive directly from known and readily available pure regression equations. The pure equations presented in this book certainly need better and systematic measurement for improved accuracy. In spite of this shortcoming, these equations are good enough for computation purposes. The effort to refine the available pure equations, however, is minor compared with the huge experimental undertaking involved in the measurement of the non-pure regression lines for UD and chopped plies at several R ratios and load paths. The accurate measurement of the interaction parameter Gsc is a dream for the future. A design simplification to eliminate the need for the interaction parameter Gsc is highly desirable.

This appendix introduces such a simplification and provides an interesting shortcut to estimate the upper and lower bounds of the long-term safety factor SF independent of the interaction parameter Gsc. Let us explain how to do this extraordinarily important simplification. We begin with the full version of the unified equation:

$$\left(\frac{R \times \varepsilon \times SF}{S_S} \right)^{\frac{1}{G_S}} + \left(\frac{(1-R) \times \varepsilon \times SF}{S_C} \right)^{\frac{1}{G_C}} + \left(\frac{R \times (1-R) \times \varepsilon^2 \times SF^2}{S_S \times S_C} \right)^{\frac{1}{G_{SC}}} = 1.00 \quad (18.1)$$

Equation (18.1) finds use in the solution of time-dependent rupture durability problems involving fiber strain-corrosion. The time-independent failures, such as infiltration, weep, and loss of stiffness, use a time-independent simplified version of the unified equation

$$\left(\frac{(1-R) \times \varepsilon \times SF}{S_C} \right)^{\frac{1}{G_C}} + \left(\frac{R(1-R) \times \varepsilon^2 \times SF^2}{S_S \times S_C} \right)^{\frac{1}{G_{SC}}} = 1.00 \qquad (18.2)$$

Our goal is to establish the interval limited by the highest and lowest possible values of SF in Equations (18.1) and (18.2), independent of the elusive parameter Gsc. This simplification does not give the true SF, but an interval defined by its lower and upper bounds. The designer can be sure that the true SF value falls in the computed interval.

The Lower Bound The lower limit of the long-term safety factor SF comes from a close examination of Figure 18.1, which is identical to Figure 9.3.

From Figure 18.1, the pure cyclic loadings (R = 0) cause the most damage and produce the lowest possible safety factors. Therefore, setting R = 0 in the unified Equation (18.1), gives the lowest possible value of the long-term safety factor. This is the cycle-dominated lower limit of SF. Entering R = 0 in Equation (18.1), we have

$$\left[\frac{\epsilon \times SF}{S_C} \right]^{\frac{1}{G_C}} > 1.00$$

Which gives

$$SF > \frac{S_C}{\epsilon}$$

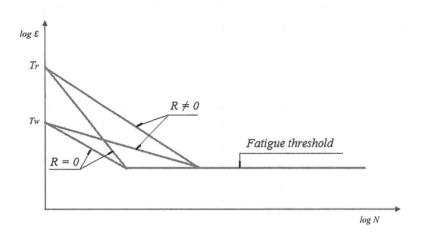

FIGURE 18.1   *This is the same as Figure 9.3. The slopes of the regression lines decrease with increments in the R ratio. The pure cyclic loadings (R = 0) are the most damaging and produce the least values of the safety factors SF.*

In the above equation, we recall, Sc is the ply long-term pure cyclic strength and ε is the peak strain. The computation of the cycle-dominated lower limit of SF assumes the entire load to be cyclic.

A similar analysis for the static load components leads to the static-dominated lower limit. Setting R = 1 in Equation (18.1), we have

$$\left[\frac{\epsilon \times SF}{S_S} \right]^{\frac{1}{G_S}} > 1.00$$

Which gives

$$SF > \frac{S_S}{\epsilon}$$

The computation of the static-dominated lower limit of SF assumes the entire load to be static.

We have computed two values for the lower limit of SF: one cyclic and one static. The lower bound is the least of the two.

The Upper Bound The highest possible value of the long-term cyclic safety factor SF comes from dropping the terms corresponding to the static and the interaction damages in Equation (18.1). The unified equation in this situation reduces to

$$\left[\frac{\left(1 - R\right) \times \epsilon \times SF}{S_C} \right]^{\frac{1}{G_C}} < 1.00$$

Which gives

$$SF < \frac{S_C}{\left(1 - R\right) \times \epsilon}$$

The computation of the cyclic upper limit neglects the static and the interaction damages and therefore produces the highest possible value of SF. This, of course, is the cyclic upper limit of SF.

A similar analysis for the static load component leads to the static upper limit

$$\left[\frac{R \times \epsilon \times SF}{S_S} \right]^{\frac{1}{G_S}} < 1.00$$

Which gives

$$SF < \frac{S_S}{R \times \epsilon}$$

We have computed two upper limits for the safety factor SF, one cyclic and another static. The upper bound is the least of the two.

The SF Interval for Time-Dependent Failures The foregoing analyses produced two lower limits and two upper limits for SF. We choose the lowest values of each as the lower and upper bounds of SF. By doing this we assure that, regardless of the interaction parameter Gsc, the long-term safety factor SF will always fall in the interval defined by the lower and upper bounds.

In the above equations, the ply long-term cyclic Sc and static Ss strengths come from known pure cyclic regression equation and the owner-defined lifetime X and number of cycles Y. The load ratio R and peak strain ε come from the computation protocols developed in the Part I of this book.

The development of a confidence interval for the long-term safety factor SF has tremendous practical implications. The confidence interval establishes the range of the long-term safety factor *independent of the interaction parameter Gsc*. This enormous simplification allows the application of the unified equation to all loadings, independent of the availability of expensive regression equations. The determination of the confidence interval for the long-term SF removes the current obstacle to the practical use of the unified equation.

The SF Interval for Cycle-Dependent Failures The static upper limit of cycle-dependent failures is defined only at R = 1. Therefore, for cycle-dependent failures, the upper and lower bounds of the SF interval are purely cyclic.

Example 18.7

Compute the ranges of the long-term safety factors SF for the numerical Examples 18.2, 18.3 and 18.4 discussed in this chapter.

In Example 18.2: $Ss = 1.20\%$, $Sc = 0.64\%$, $R = 0.67$ and $\varepsilon = 0.27\%$.

This is a time-dependent problem requiring the full version of the unified equation. The static and cyclic lower limits are

$$SF > \frac{S_C}{\epsilon} \quad \left(Cyclic\right)$$

$$SF > \frac{0.64}{0.27} = 2.40$$

$$SF > \frac{S_S}{\epsilon} \quad \left(Static\right)$$

$$SF > \frac{1.20}{0.27} = 4.40$$

The lower bound is the lowest of the two, $SF = 2.40$.

The cyclic and static upper limits are

$$SF < \frac{S_C}{\left(1-R\right) \times \epsilon} \quad \left(Cyclic\right)$$

$$SF < \frac{0.64}{\left(1-0.67\right) \times 0.27} = 7.20$$

$$SF < \frac{S_S}{R \times \epsilon} \quad \left(Static\right)$$

$$SF < \frac{1.20}{0.67 \times 0.27} = 6.60$$

The upper bound is the lowest of the two, $SF = 6.60$.

The confidence interval for SF is

$$2.40 < SF < 6.60$$

The actual long-term SF falls in the range 2.40 – 6.60. This is the best we can do at this time, given the current state of the art. The determination of the true SF requires knowledge of the Gsc values not available at this time. The $SF = 5.0$ computed in Example 18.2 assumed $Gsc = 10$.

In Example 18.3: $Ss = 0.40\%$, $Sc = 0.16\%$, $R = 0.74$ and $\epsilon = 0.23\%$.

This is a cycle-dependent problem requiring the simplified version of the unified equation. The upper and lower bounds of the SF interval in this case are.

The lower bound is

$$SF > \frac{S_C}{\epsilon} \quad \left(Cyclic\right)$$

$$SF > \frac{0.16}{0.23} = 0.70$$

The upper bound is

$$SF < \frac{S_C}{\left(1-R\right)\times \epsilon} \quad \left(Cyclic\right)$$

$$SF < \frac{0.16}{\left(1-0.74\right)\times 0.23} = 2.70$$

The confidence interval is

$$0.70 < SF < 2.70$$

The actual SF falls in the interval 0.70 – 2.70 and indicates a possibility of long-term weep failure. This is because in the computation of the lower bound we made the extremely conservative assumption that the entire loading is cyclic. The determination of the true SF requires knowledge of the Gsc values not available at this time. The $SF = 1.8$ computed in Example 18.3 assumed $Gsc = 1\ 000$.

In Example 18.4: $Ss = 1.00\%$, $Sc = 0.30\%$, $R = 0.66$ and $\epsilon = 0.30\%$.

Again, this is a cycle-dependent problem. The computation of the upper and lower bounds of SF for weeping failures takes into account only the cyclic limits

The lower bound is

$$SF > \frac{S_C}{\epsilon} \quad \left(Cyclic\right)$$

$$SF > \frac{0.30}{0.30} = 1.00$$

The upper bound is

$$SF < \frac{S_C}{\left(1-R\right)\times\epsilon} \quad \left(Cyclic\right)$$

$$SF < \frac{0.30}{\left(1-0.66\right)\times0.30} = 2.90$$

The confidence interval for SF is

$1.00 < SF < 2.90$

The actual SF falls in the range 1.00 – 2.90 and indicates a possibility of failure. This is because in the computation of the lower bound we made the extremely conservative assumption that the entire loading is cyclic. The determination of the true SF requires knowledge of the Gsc values not available at this time. The SF = 2.2 computed in Example 18.4 assumed Gsc = 100.

19 The Unified Equation Applied to API 15HR

19.1 INTRODUCTION

This chapter illustrates the application of the unified equation to solve the API 15HR long-term requirement for composite pipes. The requirement is....

> "...no pipe failure in 20 years of continuous operation under nominal pressure and $N = 10^9$ cycles at a strain ratio $R = 0.9$".

We address the problem from the perspectives of weep and rupture, which are the only possible modes of long-term failure in this case.

19.2 THE GLOBAL STRAINS

To simplify the discussions, we ignore the residual strain components in this simple and illustrative example. We start with the computation of the mechanical global strains from the average global stresses and the Hooke's relationship.

$$[\sigma] = [A] \times [\varepsilon]$$

The stiffness matrix [A] of standard ± 55 angle-ply laminates used in oil pipes is:

$$[A] = \begin{bmatrix} 132470 & 92200 & 0 \\ 92200 & 235070 & 0 \\ 0 & 0 & 101220 \end{bmatrix}$$

Entering this in the expanded form of the Hooke's equation, we have

$$\begin{bmatrix} \sigma_x \\ \sigma_y \\ \tau_{xy} \end{bmatrix} = \begin{bmatrix} 132470 & 92200 & 0 \\ 92200 & 235070 & 0 \\ 0 & 0 & 101220 \end{bmatrix} \times \begin{bmatrix} \varepsilon_x \\ \varepsilon_y \\ \gamma_{xy} \end{bmatrix} \qquad (19.1)$$

Assuming a 2:1 loading from internal pressure, and in the absence of external torque, Equation (19.1) reduces to

$$
\begin{bmatrix} \sigma_x \\ 2\sigma_x \\ 0 \end{bmatrix} = \begin{bmatrix} 132470 & 92200 & 0 \\ 92200 & 235070 & 0 \\ 0 & 0 & 101220 \end{bmatrix} \times \begin{bmatrix} \varepsilon_x \\ \varepsilon_y \\ \gamma_{xy} \end{bmatrix} \tag{19.2}
$$

Expanding the above, we obtain

$$
\varepsilon_y = (3.41)\varepsilon_x \tag{19.3}
$$

$$
\gamma_{xy} = 0 \tag{19.4}
$$

From Equation (19.3), we see that the mechanical hoop strain components in standard oil pipes submitted to 2:1 loadings are 3.41 times the axial strain components. Equation (19.4) indicates absence of shear strains in the global frame. The API 15HR standard recognizes the extrapolated HDB as the only valid long-term pipe strength. To obtain the extrapolated HDB, we make use of one of the many regression equations available in the literature for oil pipes. Assuming a $X = 20$ years lifetime we have

$$
\log\left(HDB\right) = -0.01 - 0.06\log(20 \times 365 \times 24) \tag{19.5}
$$

$$
HDB = 0.48\%
$$

The interpretation of the above HDB is as follows:

"A water-filled pipe subjected to a hoop static tensile strain of 0.48% will develop cracks that grow in time to allow the penetration of water. It takes 20 years for the cracks to reach sizes large enough to allow the passage of water. After 20 years under the sustained 0.48% hoop strain the water emerges as droplets on the outside surface".

This is the classical interpretation of the HDB. As discussed in Chapter 13, this interpretation ignores the pipe wall thickness and assumes crack growth under static strains. This is not correct. The ASTM D2992 B protocol tests pressurized pipe specimens filled with water that do not corrode the resin. The resin does not grow strain-corrosion cracks from water attack. Furthermore, the pressure remains constant throughout the test generating stationary cracks that do not grow with time. The observed water leakage does not come from any pipe deterioration or crack growth. The water leaks through stationary pathways developed by excessive pressures, higher than the weep threshold. The weep time of the failed pipe is in fact a "travel time", i.e., the time taken by the pressurized water to traverse the pathway formed by stationary cracks in the pipe wall.

The product standard API 16HR does not recognize the weep threshold. Therefore, we conduct the following analysis taking into account the incorrect classical HDB concept, instead of the correct weep threshold. The extrapolated long-term HDB is an average hoop strain, with a 50% probability of weeping the pipe after 20 years of continuous operation. From API 15HR, the allowable static hoop strain for a target lifetime of 20 years is

$$\varepsilon_y = f_1 f_2 (HDB) = 0.85 \times 0.67 \times 0.48 = 0.27\%$$

Where the factors f_1 and f_2 come from the API standard. Entering the above in Equation (19.3), we obtain the static axial strain component corresponding to the allowable hoop strain component.

$$\varepsilon_x = \frac{0.27}{3.41} = 0.08\%$$

We have just computed the global strain components in the axial and in the hoop directions of oil pipes under the nominal operating static conditions established in API 15HR. In the next section, we rotate these global strains to the local frame of the ±55 UD plies. The transverse strain component controls the weep failure. The strain component in the fiber direction controls the long-term rupture. The pipe operates under the conditions stipulated by API 15HR as stated in the introduction to this chapter.

19.3   THE PLY LOCAL STRAINS

The mechanical pipe strain components in the global frame are

$$\begin{bmatrix} \epsilon_x \\ \epsilon_y \\ \gamma_{xy} \end{bmatrix} = \begin{bmatrix} 0.08\% \\ 0.27\% \\ 0 \end{bmatrix}$$

As discussed in Chapter 3, the global strain components rotated to the UD ply frame are

$$
\begin{bmatrix} \epsilon_1 \\ \epsilon_2 \\ \frac{1}{2}\gamma_{12} \end{bmatrix} = \begin{bmatrix} 0.33 & 0.67 & \pm 0.94 \\ 0.67 & 0.33 & \mp 0.94 \\ \mp 0.47 & \pm 0.47 & -0.34 \end{bmatrix} \times \begin{bmatrix} \epsilon_x \\ \epsilon_y \\ \frac{1}{2}\gamma_{xy} \end{bmatrix}
$$

Entering the computed values of the global strain components in the above, we have

$$
\begin{bmatrix} \epsilon_1 \\ \epsilon_2 \\ \frac{1}{2}\gamma_{12} \end{bmatrix} = \begin{bmatrix} 0.33 & 0.67 & \pm 0.94 \\ 0.67 & 0.33 & \mp 0.94 \\ \mp 0.47 & \pm 0.47 & -0.34 \end{bmatrix} \times \begin{bmatrix} 0.08 \\ 0.27 \\ 0 \end{bmatrix}
$$

$\varepsilon_1 = 0.18\%$ (Static strain component in the fiber direction)

$\varepsilon_2 = 0.14\%$ (Static strain component transverse to the fibers)

$\gamma_{12} = \pm 0.18\%$ (Static shear strain in the local ply system)

The above are the static mechanical local strain components on the UD plies of standard oil pipes under the API 15HR nominal operating conditions. The residual thermal and hydric strains are not included in the analysis. An accurate analysis should include the residual strains.

19.4   LONG-TERM RUPTURE

We proceed to check the long-term rupture superimposing the prescribed loading (N = 10^9 cycles and R = 0.9) on the nominal static working conditions. The unified equation for rupture failure in the fiber direction is

$$
\left(\frac{R\epsilon \times SF}{S_S} \right)^{\frac{1}{G_S}} + \left(\frac{(1-R) \times \epsilon \times SF}{S_C} \right)^{\frac{1}{G_C}} + \left(\frac{R(1-R) \times \epsilon^2 \times SF^2}{S_S \times S_C} \right)^{\frac{1}{G_{SC}}} = 1.0 \qquad (19.6)
$$

Where
   ϵ is the peak strain in the fiber direction
   R = 0.9 is the strain ratio
   Ss = is the pure long-term (20 years) static strength in the fiber direction
   Sc = is the pure long-term (10^9 cycles) cyclic strength in the fiber direction
   Gs = 0.077 is the pure static slope of boron-free glass

Gs = 0.130 is the pure static slope of E glass

Gc = 0.089 is the pure cyclic slope, any glass

Gsc is the interaction parameter for the target lifetime of $X = 20$ years and $Y = N = 10^9$ cycles.

The peak strain in the fiber direction comes from R = 0.9 and the known minimum strain value 0.18%.

$$R = \frac{0.18}{\epsilon}$$

$\epsilon = 0.20\%$

The long-term pure static and pure cyclic strengths of UD plies in the fiber direction come from the appropriate pure regression lines. Let us compute these long-term ply strengths.

The 20 years pure static strength of boron-free UD plies is

$$\log(S_S) = 0.400 - 0.077 \log(20 \times 365 \times 24)$$

$S_S = 0.99\%$ $\left(\text{Boron-free glass}\right)$

The 20 years pure static strength of E glass UD plies is

$$\log(S_S) = 0.347 - 0.130 \log(20 \times 365 \times 24)$$

$S_S = 0.46\%$ $\left(\text{E glass}\right)$

The long-term pure cyclic strength of UD plies in the fiber direction for $N = Y = 10^9$ cycles is

$$\log(S_C) = 0.477 - 0.089 \log(N)$$

$$\log(S_C) = 0.477 - 0.089 \log(10^9)$$

$S_C = 0.474\%$ $\left(\text{Boron-free and E glass}\right)$

Now we come to the interaction parameter Gsc. The loading frequency prescribed by API 15HR is

$$frequency = \frac{10^9}{20 \times 365 \times 24 \times 60 \times 60} = 1.6 \text{ Hz}$$

The Gsc values for the load-path corresponding to such a frequency are not known. The only Gsc values for fiber loaded UD plies available at this time come from the Guangxu Wei tests conducted at 20 Hz. Unfortunately we cannot solve the API 15HR problem, because we do not have the Gsc value for the load path 1.6 Hz and R = 0.9. However, for illustration purposes, we can try a solution with Gsc = 9.0, taken from Table 18.8 for N = 10^9 cycles and R = 0.1 at the test path 20 Hz. See Table 18.8 of Chapter 18. The chosen Gsc = 9.0 is good enough for illustration purposes.

Example 19.1

Compute the long-term rupture durability of boron-free oil pipes operating under the stated API 15HR conditions.

We enter the computed and supplied data on the unified Equation (19.6).

$$\left(\frac{0.9 \times 0.20 \times SF}{0.99} \right)^{\frac{1}{0.077}} + \left(\frac{0.1 \times 0.20 \times SF}{0.474} \right)^{\frac{1}{0.089}} + \left(\frac{0.9(1-0.9) \times 0.2^2 \times SF^2}{0.99 \times 0.474} \right)^{\frac{1}{9.0}} = 1.0$$

Solving for SF we obtain SF = 4.8

The unified equation predicts that pipes of boron-free glass under the combined cyclic and static loads mandated by API 15HR have a long-term (20 years) safety factor SF = 4.8 against rupture. This SF assumes Gsc = 9.0. Let us compute the upper and lower bounds of the interval for the long-term safety factor. This is a time-dependent problem.

The lower limits of the SF are

$$SF > \frac{S_C}{\epsilon} \quad \left(Cyclic \right)$$

$$SF > \frac{0.474}{0.20} = 2.37$$

$$SF > \frac{S_S}{\epsilon} \quad \left(Static \right)$$

$$SF > \frac{0.99}{0.20} = 4.95$$

The lower bound is the least of the values computed above. Therefore, the lower bound is SF = 2.37.

The upper limits of the SF are

$$SF < \frac{S_C}{(1-R)\times\epsilon} \quad (Cyclic)$$

$$SF < \frac{0.474}{(1-0.90)\times0.20} = 23.70$$

$$SF < \frac{S_S}{R\times\epsilon} \quad (Static)$$

$$SF < \frac{0.99}{0.90\times0.20} = 5.50$$

The upper bound is the least of the above values. Therefore, the upper bound is SF = 5.50.

The interval for the SF is

$$\left[2.37 < SF < 5.50\right]$$

The actual long-term SF falls in the range [2.37 – 5.50]. This is the best we can do in the absence of the correct value of the interaction parameter Gsc. The determination of the true SF requires knowledge of the Gsc values, not available at this time. The SF = 4.8 computed in this example assumed Gsc = 9.0.

Example 19.2

We next solve the API 15HR problem for pipes of standard E glass. We enter the known inputs in the unified equation.

$$\left(\frac{0.9\times0.20\times SF}{0.46}\right)^{\frac{1}{0.130}} + \left(\frac{0.1\times0.20\times SF}{0.474}\right)^{\frac{1}{0.089}} + \left(\frac{0.9(1-0.9)\times0.2^2\times SF^2}{0.46\times0.474}\right)^{\frac{1}{9.0}} = 1.0$$

$$SF = 2.2$$

The foregoing analysis indicates that pipes of regular E glass meet the API 15HR requirements for long-term rupture with a safety factor SF = 2.2. However, the superior performance of the boron-free glass (SF = 4.8) versus the regular E glass (SF = 2.2) is clear. The long-term safety factors computed in this example are illustrative, since we have assumed a fictitious value for the interaction parameter Gsc.

The best we can do at this time, in the absence of the true Gsc values, is to compute the confidence interval for the long-term safety factor. Let us compute this interval for the pipes of E glass.

The lower limits of the SF are

$$SF > \frac{S_C}{\epsilon} \quad \left(Cyclic\right)$$

$$SF > \frac{0.474}{0.20} = 2.37$$

$$SF > \frac{S_S}{\epsilon} \quad \left(Static\right)$$

$$SF > \frac{0.46}{0.20} = 2.30$$

The lower bound is the lowest of the above limits, SF = 2.30.

The upper limits of the SF are

$$SF < \frac{S_C}{\left(1-R\right) \times \epsilon} \quad \left(Cyclic\right)$$

$$SF < \frac{0.474}{\left(1-0.90\right) \times 0.20} = 23.70$$

$$SF < \frac{S_S}{R \times \epsilon} \quad \left(Static\right)$$

$$SF < \frac{0.46}{0.90 \times 0.20} = 2.56$$

The upper bound is the lowest of the above limits, SF = 2.56.

The confidence interval for the long-term SF is

$$\left[2.30 < SF < 2.56 \right]$$

The long-term SF falls in the range [2.30 – 2.56]. The lower bound 2.30 in this case is higher than the SF = 2.2 computed by assuming Gsc = 9.0. This unexpected result comes from the very low hydrolytic stability of the E glass over magnifying the static contribution to failure at R values close to R = 1.0. The unified equation captures the low hydrolytic stability of E glass, which the lower bound does not. A similar anomaly occurs at R values close to R = 0.0 in fibers of low cyclic resistance.

The preceding examples deal with the long-term time-dependent ruptures involving strain-corrosion of the fibers. We next address the long-term cycle-dependent weep failure.

19.5   WEEP FAILURE

The following facts are undisputed in relation to the weep failure of oil pipes:

- Under static pressure, the pipe develops many small stationary cracks parallel to the UD fibers.
- Under cyclic pressure, the otherwise stationary cracks grow in number and in length.
- The crack density grows under cyclic pressure. Static pressures do not grow the crack density.
- Under cyclic pressure, the initially small cracks become macro-cracks parallel to the UD fibers. The coalescence of such macro-cracks provides the pathway for the passage of water.

We also keep in mind the following.

- The crack openings in the UD plies are determined by the pipe stiffness and the applied pressure, and do not change under cyclic loads. The cyclic loads increase the length and density, not the opening of the cracks.
- Under pure static loads the cracks are stationary and do not grow in number, in length or in opening.

In the weeping process, the water moves along many narrow and long macro-cracks parallel to the UD fibers. The cracks in the innermost+55 ply cross with similar cracks in the –55 ply immediately above. The water migrates from the inner to the outer plies at these crossing points and eventually weeps out. The macro-cracks are narrow, the pathways are tortuous and the travel distances are relatively large. The travel time from the innermost to the outermost ply may be very long. It is obvious that the weep time depends on the pipe thickness. Other things being equal, the pipes with larger wall thickness have longer weep times.

The weep time is not a fundamental material property. For one thing, it depends on the pipe thickness. One very simple way to boost the weep time and obtain higher HDB values in the ASTM D 2992 B protocol is to increase the wall thickness of the test specimens. Not being a fundamental material property, the HDB should be not be used in pipe design. The fundamental pipe parameter in control of the weep process is the weep threshold, discussed in Chapter 8.

Example 19.3

Compute the short-term weep safety factor of oil pipes operating under the specified API 15HR conditions. This is a cycle-dependent problem independent of the glass composition. We solve the problem assuming pipes of polyester and vinyl ester resins.

The mode of failure is weep and the focus is on the transverse strains. The peak strain in the transverse direction under the rated operating conditions and R = 0.9 is 0.14/0.9 = 0.16%.

The short-term weep safety factors are

$$short-term\ safety\ factor = \frac{0.50}{0.16} = 3.13 \quad \left(Vinyl\ ester\right)$$

$$short-term\ safety\ factor = \frac{0.25}{0.16} = 1.56 \quad \left(Polyester\right)$$

There is no risk of the peak strains exceeding the weep thresholds.

Example 19.4

Compute the long-term weep SF of oil pipes under the API 15HR loading. This is a time-independent problem requiring the simplified version of the unified equation.

$$\left(\frac{(1-R)\varepsilon \times SF}{S_C}\right)^{\frac{1}{0,040}} + \left(\frac{R(1-R) \times \varepsilon^2 \times SF^2}{S_S \times S_C}\right)^{\frac{1}{G_{SC}}} = 1.0 \qquad (19.7)$$

Where
    $\varepsilon = 0.16\%$ (peak transverse strain)
    S_C = Long-term cyclic transverse strength ($N = 10^9$ cycles).
    S_S = Long-term static strength, taken as equal to the weep threshold
    $R = 0.9$

G_{SC} = Interaction parameter for R = 0.9 and N = 10^9 cycles

The peak transverse strain component is

$$\epsilon = \frac{0.14}{0.9} = 0.16\%$$

The Gsc values entering the simplified version of the unified equation are time-independent, which is a blessing to the analyst. The values of Gsc to use in Equation (19.7) come from Tables 18.8 and 18.9 for R = 0.9. Unfortunately, we do not have these values. We propose to use the values for N = 10^9 and R = 0.1 in this case. The tabulated values are Gsc = 25 for vinyl ester and Gsc = 46 for polyester. See Tables 18.8 and 18.9 in Chapter 18.

The long-term cyclic weep strengths S_C comes from the pure cyclic weep regression line for transverse loaded UD plies.

$$\log \epsilon = \log(Tw) - G_C \log N$$

These equations are in Chapter 9. The weep equation for UD plies of brittle polyester is

$$\log \varepsilon = \log 0.25 - 0.039 \log N$$

For vinyl ester resins, the weep regression equation is

$$\log \varepsilon = \log 0.40 - 0.051 \log N$$

In the above, 0.25% and 0.40% are the weep thresholds of transverse loaded polyester and vinyl ester UD plies respectively. The API requirement calls for N = 10^9 cycles. The long-term cyclic strength S_c for UD plies of vinyl ester resins is

$$\log S_C = \log 0.40 - 0.051 \log 10^9$$

$$S_C = 0.14\%$$

The long-term transverse static strength S_s in cycle-dependent failures is the same as the weep threshold, since the resin matrix is not subject to strain-corrosion. See Figure 9.2 in Chapter 9. Therefore, for vinyl ester UD plies

$$S_S = 0.40\%$$

Entering the above in the simplified unified equation

$$\left(\frac{(1-0.9)\times0.16\times SF}{0.14}\right)^{\frac{1}{0.051}} + \left(\frac{0.9\times(1-0.9)\times0.16^2\times SF^2}{0.40\times0.14}\right)^{\frac{1}{25}} = 1.0 \qquad (19.8)$$

Solving for SF we obtain

$$SF = 4.5$$

The unified equation predicts an excellent weep life for oil pipes of vinyl ester resins under the specified API loading. We can be sure that, for a target lifetime of $X = 20$ years and $Y = 10^9$ cycles, the vinyl ester pipes under the loading prescribed by API 15HR will not weep.

From Appendix 18.1, the long-term safety factor falls in a confidence interval defined by an upper and a lower bound. The computation of the upper and lower bounds in this case involve only the cyclic load component, since the weeping failure is cycle-dominated and time-independent.

The cyclic lower bound is

$$SF > \frac{S_C}{\epsilon} \quad \left(Cyclic\right)$$

$$SF > \frac{0.14}{0.16} = 0.90$$

The cyclic upper bound is

$$SF < \frac{S_C}{\left(1-R\right)\times\epsilon} \quad \left(Cyclic\right)$$

$$SF < \frac{0.14}{\left(1-0.90\right)\times0.16} = 8.75$$

The confidence interval for the weep SF is

$$\left[0.90 < SF < 8.75\right]$$

The actual SF falls in the range [0.90 – 8.75]. The lower bound 0.90 is obviously too low. This is because high load ratios, like $R = 0.9$, depress the lower bound.

Example 19.5

Repeat the above computations for polyester oil pipes.

The weep threshold for transverse loaded polyester UD plies is 0.25%. The long-term transverse cyclic strength for polyester UD plies in this case is

$$\log S_C = \log 0.25 - 0.039 \log N$$

$$\log S_C = \log 0.25 - 0.039 \log 10^9$$

$$Sc = 0.11\%$$

Entering the above in the simplified unified equation, we have

$$\left(\frac{(1-0.9)\times 0.16 \times SF}{0.11}\right)^{\frac{1}{0.039}} + \left(\frac{0.9 \times (1-0.9)\times 0.16^2 \times SF^2}{0.25 \times 0.11}\right)^{\frac{1}{46}} = 1.0$$

Solving for SF we obtain

$$SF = 3.0$$

The long-term safety factor is satisfactory and the polyester oil pipes seem to work well.

The reader will note the long-term safety factors higher than the short-term ones. This situation is not a cause of concern. The short-term and the long-term safety factors are not related.

From Appendix 18.1, the long-term safety factor falls in a confidence interval defined by an upper and a lower bound.

The cyclic lower bound is

$$SF > \frac{S_C}{\epsilon} \quad \left(Cyclic\right)$$

$$SF > \frac{0.11}{0.16} = 0.69$$

The cyclic upper bound is

$$SF < \frac{S_C}{(1-R)\times \epsilon} \quad (Cyclic)$$

$$SF < \frac{0.11}{(1-0.90)\times 0.16} = 6.88$$

The confidence interval for the weep SF of polyester oil pipes is

$$[0.69 < SF < 6.88]$$

The actual SF falls in the range [0.69 – 6.88]. The lower bound 0.69 is too low, because the high load ratio R = 0.9 depresses the lower bound.

This completes our analysis of the API 15HR cyclic requirement.

20 Short-Term Strengths of ± 55 Oil Pipes

20.1 INTRODUCTION

The failure envelopes proposed by ISO 14692 for the long-term analysis of composite oil pipes are cumbersome, artificial and inaccurate. In this chapter, we discuss an easier and more accurate analytical tool than those failure envelopes.

20.2 THE GENERAL EQUATION

The general Equation (20.1) computes the mechanical global strains on composite circular cylinders under the simultaneous action of pressure, torque, hoop and axial loads

$$
\begin{bmatrix}
\dfrac{P \times \Phi}{4} + N_x \\[2ex]
\dfrac{P \times \Phi}{2} + N_y \\[2ex]
\dfrac{2 \times T}{\pi \times \Phi^2}
\end{bmatrix}
=
\begin{bmatrix}
A_{11} & A_{12} & A_{13} \\
A_{21} & A_{22} & A_{23} \\
A_{31} & A_{32} & A_{33}
\end{bmatrix}
\times
\begin{bmatrix}
\varepsilon_x \\
\varepsilon_y \\
\gamma_{xy}
\end{bmatrix}
\tag{20.1}
$$

Where
  x designates the axial direction.
  y designates the hoop direction.
  P is the internal fluid pressure.
  Φ is the pipe diameter.
  N_x is the axial force per unit length.
  N_y is the hoop force per unit length.
  T is the torque.
  [A] is the laminate tensile stiffness matrix.
  [ε] is the matrix of mechanical global strains.

Equation (20.1) is general and applies to any circular pipe. The protocol to compute the laminate stiffness matrix [A] is in the Part I of this book. Equation (20.1) gives the pipe mechanical global strain components for the most general membrane loading possible. To facilitate the presentation, the discussion that follows excludes the bending loads.

20.3   ANGLE-PLY LAMINATES

The angle-ply and balanced laminates of commercial pipes reduce Equation (20.1) to

$$
\begin{bmatrix}
\dfrac{P \times \Phi}{4} + N_x \\[3mm]
\dfrac{P \times \Phi}{2} + N_y \\[3mm]
\dfrac{2 \times T}{\pi \times \Phi^2}
\end{bmatrix}
=
\begin{bmatrix}
A_{11} & A_{12} & 0 \\
A_{21} & A_{22} & 0 \\
0 & 0 & A_{33}
\end{bmatrix}_{55}
\times
\begin{bmatrix}
\varepsilon_x \\
\varepsilon_y \\
\gamma_{xy}
\end{bmatrix}
\tag{20.2}
$$

This chapter will illustrate the computation protocol for oil and sanitation pipes. From Chapter 4, the stiffness matrix [A] of standard ± 55 angle-ply laminates with a 70% glass loading is

$$
[A] =
\begin{bmatrix}
132470 & 96220 & 0 \\
96220 & 235070 & 0 \\
0 & 0 & 101220
\end{bmatrix}
\times [t] \quad \text{kg/cm}
$$

In the above, "t" is the total laminate thickness. Entering this matrix [A] in Equation (20.2), we have

$$
\begin{bmatrix}
\dfrac{P \times \Phi}{4} + N_x \\[3mm]
\dfrac{P \times \Phi}{2} + N_y \\[3mm]
\dfrac{2 \times T}{\pi \times \Phi^2}
\end{bmatrix}
=
\begin{bmatrix}
132470 & 96220 & 0 \\
96220 & 235070 & 0 \\
0 & 0 & 101220
\end{bmatrix}
\times [t] \times
\begin{bmatrix}
\varepsilon_x \\
\varepsilon_y \\
\gamma_{xy}
\end{bmatrix}
\tag{20.3}
$$

Dividing through by the laminate thickness "t" we obtain Equation (20.4), which is Equation (20.3) expressed in terms of the stress components

$$
\begin{bmatrix}
\dfrac{P \times \Phi}{4 \times t} + \sigma_x \\[3mm]
\dfrac{P \times \Phi}{2 \times t} + \sigma_y \\[3mm]
\dfrac{2 \times T}{\pi \times \Phi^2 \times t}
\end{bmatrix}
=
\begin{bmatrix}
132470 & 96220 & 0 \\
96220 & 235070 & 0 \\
0 & 0 & 101220
\end{bmatrix}
\times
\begin{bmatrix}
\varepsilon_x \\
\varepsilon_y \\
\gamma_{xy}
\end{bmatrix}
\qquad (20.4)
$$

Equation (20.4) links the pressure, torque and average stresses to the global mechanical strains. The stiffness matrix in Equation (20.4) is valid for 70% UD glass loading and ± 55 angle-ply circular pipes.

20.4 STRAIN ANALYSIS

The analysis of composite pipes is best done in terms of strains, instead of average stresses. The reasons for the use of strains in place of stresses are:

1. The failure strains are ply properties not affected by Poisson effects. The strain failure envelope of plies are rectangular, not elliptical as those of stresses.
2. Unlike the stresses, which change from ply to ply, the global strains have the same value for all plies of circular pipes.

Solving Equation (20.4) for the global strain components gives

$$
\begin{bmatrix}
\varepsilon_x \\
\varepsilon_y \\
\gamma_{xy}
\end{bmatrix}
=
\begin{bmatrix}
132470 & 96220 & 0 \\
96220 & 235070 & 0 \\
0 & 0 & 101220
\end{bmatrix}^{-1}
\times
\begin{bmatrix}
\dfrac{P \times \Phi}{4 \times t} + \sigma_x \\[3mm]
\dfrac{P \times \Phi}{2 \times t} + \sigma_y \\[3mm]
\dfrac{2 \times T}{\pi \times \Phi^2 \times t}
\end{bmatrix}
\qquad (20.5)
$$

Which, upon inversion of matrix [A], becomes

$$
\begin{bmatrix}
\varepsilon_x \\
\varepsilon_y \\
\gamma_{xy}
\end{bmatrix}
=
\begin{bmatrix}
1.07 & -0.44 & 0 \\
-0.44 & 0.61 & 0 \\
0 & 0 & 0.99
\end{bmatrix}
\times \begin{bmatrix} 10^{-5} \end{bmatrix} \times
\begin{bmatrix}
\dfrac{P \times \Phi}{4 \times t} + \sigma_x \\[3mm]
\dfrac{P \times \Phi}{2 \times t} + \sigma_y \\[3mm]
\dfrac{2 \times T}{\pi \times \Phi^2 \times t}
\end{bmatrix}
\qquad (20.6)
$$

The global strain components in Equation (20.6) rotated to the local ply frames gives

$$
\begin{bmatrix} \epsilon_1 \\ \epsilon_2 \\ \frac{1}{2}\gamma_{12} \end{bmatrix} = \begin{bmatrix} 0.33 & 0.67 & \pm0.94 \\ 0.67 & 0.33 & \mp0.94 \\ \mp0.47 & \pm0.47 & -0.34 \end{bmatrix} \times \begin{bmatrix} \epsilon_x \\ \epsilon_y \\ \frac{1}{2}\gamma_{xy} \end{bmatrix}
$$

$$
\begin{bmatrix} \epsilon_1 \\ \epsilon_2 \\ \frac{1}{2}\gamma_{12} \end{bmatrix} = \begin{bmatrix} 0.33 & 0.67 & \pm0.94 \\ 0.67 & 0.33 & \mp0.94 \\ \mp0.47 & \pm0.47 & -0.34 \end{bmatrix} \times \begin{bmatrix} 1.07 & -0.44 & 0 \\ -0.44 & 0.61 & 0 \\ 0 & 0 & 0.50 \end{bmatrix}
$$

$$
\times \begin{bmatrix} 10^{-5} \end{bmatrix} \times \begin{bmatrix} \dfrac{P\times\varnothing}{4\times t}+\sigma_x \\ \dfrac{P\times\varnothing}{2\times t}+\sigma_y \\ \dfrac{2\times T}{\pi\times\varnothing^2\times t} \end{bmatrix}
$$

$$
\begin{bmatrix} \epsilon_1 \\ \epsilon_2 \\ \gamma_{12} \end{bmatrix} = \begin{bmatrix} 0.06 & 0.26 & \pm0.47 \\ 0.57 & -0.09 & \mp0.47 \\ \mp1,42 & \pm1,00 & -0.34 \end{bmatrix} \times \begin{bmatrix} 10^{-5} \end{bmatrix} \times \begin{bmatrix} \dfrac{P\times\varnothing}{4\times t}+\sigma_x \\ \dfrac{P\times\varnothing}{2\times t}+\sigma_y \\ \dfrac{2\times T}{\pi\times\varnothing^2\times t} \end{bmatrix} \qquad (20.7)
$$

Equation (20.7) gives the UD ply mechanical strain components of standard ± 55 oil laminates as a function of the applied loading. We recall that

P is the internal pressure.
Φ is the pipe diameter.
t is the pipe wall thickness.
σ_x is the average axial stress.
σ_y is the average hoop stress.
T is the torque.

Equation (20.7) computes all mechanical strains on the UD plies of standard cylindrical ± 55 oil pipes. Equation (20.7) is easy to use, accurate (in fact exact) and requires no arbitrary and cumbersome failure envelope. The ply total strain components come from adding the residual thermal and hydric strains to the mechanical strains computed in Equation (20.7). The load-dependent durability analysis readily follows from the unified equation and the total ply strains.

In the following examples, we apply Equation (20.7) to compute the short-term strengths of pipes.

Example 20.1

AXIAL SHORT-TERM TENSILE STRENGTH

The external loading in this case consists of just the axial tensile axial stress. Taking $P = T = \sigma_y = 0$ in Equation (20.7) gives

$$
\begin{bmatrix} \epsilon_1 \\ \epsilon_2 \\ \gamma_{12} \end{bmatrix} = \begin{bmatrix} 0.06 & 0.26 & \pm 0.47 \\ 0.57 & -0.09 & \mp 0.47 \\ \mp 1.42 & \pm 1.00 & -0.34 \end{bmatrix} \times \begin{bmatrix} 10^{-5} \end{bmatrix} \times \begin{bmatrix} \sigma_x \\ 0 \\ 0 \end{bmatrix} \tag{20.8}
$$

Expanding the above, we obtain three values for the short-term axial tensile strength of the pipe.

$$\varepsilon_1 = 0.06 \times 10^{-5} \times \sigma_x$$

$$\varepsilon_2 = 0.57 \times 10^{-5} \times \sigma_x$$

$$\gamma_{12} = \mp 1.42 \times 10^{-5} \times \sigma_x$$

The first equation refers to fiber rupture. Assuming the short-term tensile rupture strain in the fiber direction as $\varepsilon_1 = 0.03$ we have

$$\sigma_x = \frac{0.03}{0.06 \times 10^{-5}} = 50\,000 \text{ kg/cm}^2$$

The second equation refers to transverse ply failure. There are three scenarios to consider.

1. Axial strength to infiltration. This is the short-term axial strength to the onset of cracking. Entering the infiltration Ti values listed in Table 8.4, Chapter 8, in place of the strain component ε_2, we have

$$\sigma_x = \frac{0.002}{0.57 \times 10^{-5}} = 350 \text{ kg/cm}^2 \quad (\textit{Polyester pipe})$$

$$\sigma_x = \frac{0.003}{0.57 \times 10^{-5}} = 520 \text{ kg/cm}^2 \quad (\textit{Vinyl ester pipe})$$

2. Axial strength to weep. This is the short-term axial strength to the onset of weeping. Entering the weep Tw values listed in Table 8.4, Chapter 8, in place of the strain component ε_2, we have

$$\sigma_x = \frac{0.0025}{0.57 \times 10^{-5}} = 435 \text{ kg/cm}^2 \quad (\textit{Polyester pipe})$$

$$\sigma_x = \frac{0.004}{0.57 \times 10^{-5}} = 700 \text{ kg/cm}^2 \quad (\textit{Vinyl ester pipe})$$

3. Axial strength to rupture. This is the short-term axial strength of the pipe. Entering the rupture Tr values listed in Table 8.4, Chapter 8, in place of the strain component ε_2, we have

$$\sigma_x = \frac{0.004}{0.57 \times 10^{-5}} = 700 \text{ kg/cm}^2 \quad (\textit{Polyester pipe})$$

$$\sigma_x = \frac{0.006}{0.57 \times 10^{-5}} = 1050 \text{ kg/cm}^2 \quad (\textit{Vinyl ester pipe})$$

The third equation refers to shear failure. Assuming $\gamma_{12} = 0.02$ we have

$$\sigma_x = \frac{0.02}{1.42 \times 10^{-5}} = 1400 \text{ kg/cm}^2$$

As we see, the transverse failures of the UD ply governs the short-term axial strength of ± 55 oil pipes. This is a simple and direct computation. We invite the reader to compare it with the failure envelope proposed by ISO 14692.

Example 20.2

HOOP SHORT-TERM TENSILE STRENGTH

The loading in this case is the tensile hoop stress. All other stresses are zero.

$$\begin{bmatrix} \epsilon_1 \\ \epsilon_2 \\ \gamma_{12} \end{bmatrix} = \begin{bmatrix} 0.06 & 0.26 & \pm0.47 \\ 0.57 & -0.09 & \mp0.47 \\ \mp1,42 & \pm1,00 & -0.34 \end{bmatrix} \times \begin{bmatrix} 10^{-5} \end{bmatrix} \times \begin{bmatrix} 0 \\ \sigma_y \\ 0 \end{bmatrix} \qquad (20.9)$$

Expanding the above, we obtain three values for the short-term hoop tensile strength.

$$\varepsilon_1 = 0.26 \times 10^{-5} \times \sigma_y$$

$$\varepsilon_2 = -0.09 \times 10^{-5} \times \sigma_y$$

$$\gamma_{12} = 1.00 \times 10^{-5} \times \sigma_y$$

The first equation refers to fiber rupture. Assuming as before $\varepsilon_1 = 0.03$ we have

$$\sigma_y = \frac{0.03}{0.26 \times 10^{-5}} = 11\ 500\ \text{kg/cm}^2$$

The second equation refers to transverse ply failure. The negative – compressive – transverse strain do not cause failure.

The third equation refers to shear failure. Assuming $\gamma_{12} = 0.02$, we have

$$\sigma_y = \frac{0.02}{1.00 \times 10^{-5}} = 2000\ \text{kg/cm}^2$$

The short-term hoop strength of oil pipes is shear-dominated and resin-independent.

Example 20.3

TORQUE SHORT-TERM STRENGTH

The short-term torque strength comes from Equation (20.7) with the axial stress and the internal pressure equal to zero.

$$\begin{bmatrix} \epsilon_1 \\ \epsilon_2 \\ \gamma_{12} \end{bmatrix} = \begin{bmatrix} 0.06 & 0.26 & \pm0.47 \\ 0.57 & -0.09 & \mp0.47 \\ \mp1,42 & \pm1,00 & -0.34 \end{bmatrix} \times \begin{bmatrix} 10^{-5} \end{bmatrix} \times \begin{bmatrix} 0 \\ 0 \\ \dfrac{2 \times T}{\pi \times \varnothing^2 \times t} \end{bmatrix} \qquad (20.10)$$

The computation of the torque short-term strengths is straightforward.

Example 20.4

COMPLETE LOADING

By setting $\sigma_y = 0$ and rearranging terms, Equation (20.7) takes the following form.

$$
\begin{bmatrix} \epsilon_1 \\ \epsilon_2 \\ \gamma_{12} \end{bmatrix} = \begin{bmatrix} 0.15 & 0.06 & \pm 0.30 \\ 0.00 & 0.57 & \mp 0.30 \\ \pm 0.15 & \mp 1,42 & -0.22 \end{bmatrix} \times \begin{bmatrix} 10^{-5} \end{bmatrix} \times \begin{bmatrix} \dfrac{P \times \varnothing}{t} \\ \sigma_x \\ \dfrac{T}{\varnothing^2 \times t} \end{bmatrix}
\tag{20.11}
$$

Equation (20.11) computes all strain components on the local frame of the ± 55 UD standard plies from a complete set of external loads.

Example 20.5

Suppose we choose to ignore the torques T producing strains that are less than 5.0% of those from the internal pressure P. Compute the maximum torque T meeting this criterion.

Expanding Equation (20.11), we obtain

$$
\begin{bmatrix} \epsilon_1 \\ \epsilon_2 \\ \gamma_{12} \end{bmatrix} = \begin{bmatrix} 0.15 \dfrac{P\varnothing}{t} + 0.06\sigma_x \pm 0.30 \dfrac{T}{\varnothing^2 \times t} \\ 0.10 \dfrac{P\varnothing}{t} + 0.57\sigma_x \mp 0.30 \dfrac{T}{\varnothing^2 \times t} \\ \pm 0.15 \dfrac{P\varnothing}{t} \mp 1.42\sigma_x - 0.22 \dfrac{T}{\varnothing^2 \times t} \end{bmatrix} \times \begin{bmatrix} 10^{-5} \end{bmatrix}
\tag{20.12}
$$

For strains in the fiber direction, the above criterion establishes that we can ignore the torques if

$$
0.30 \frac{T}{\Phi^2 t} < 0.05 \times 0.15 \frac{P\Phi}{t}
$$

$$
T < 0.025 \times P\Phi^3
$$

For strains transverse to the fibers, our criterion gives

$$
0.30 \frac{T}{\Phi^2 t} < 0.05 \times 0.10 \frac{P\Phi}{t}
$$

$$
T < 0.017 \times P\Phi^3
$$

For shear strains our criterion gives

$$
0.22 \frac{T}{\Phi^2 t} < 0.05 \times 0.15 \frac{P\Phi}{t}
$$

$T < 0.034 \times P\Phi^3$

The transverse strain gives the least torque and therefore governs our choice. Therefore, we can ignore as irrelevant all torques less than

$T < 0.017 \times P\Phi^3$

Example 20.6

Compute the mechanical strains in the local ply frame of an oil pipe operating under the following conditions.

P = 1000 psi
t = 0.3 in (7.6 mm)
T = 100.0 kg.m
Φ = 30 cm
σ_x = 100 kg/cm²

We can use Equations (20.11) or (20.12). Let us use Equation (20.12).

$$
\begin{bmatrix} \epsilon_1 \\ \epsilon_2 \\ \gamma_{12} \end{bmatrix} = \begin{bmatrix} 0.15\dfrac{P\varnothing}{t} + 0.06\sigma_x \pm 0.30\dfrac{T}{\varnothing^2 \times t} \\[2ex] 0.10\dfrac{P\varnothing}{t} + 0.57\sigma_x \mp 0.30\dfrac{T}{\varnothing^2 \times t} \\[2ex] \pm 0.15\dfrac{P\varnothing}{t} \mp 1.42\sigma_x - 0.22\dfrac{T}{\varnothing^2 \times t} \end{bmatrix} \times \begin{bmatrix} 10^{-5} \end{bmatrix}
$$

The first thing to do is to convert all quantities into units of kg and cm. This is required, since the matrix [A] is in units of kg and cm.

$$
\begin{bmatrix} \epsilon_1 \\ \epsilon_2 \\ \gamma_{12} \end{bmatrix} = \begin{bmatrix} 0.15\dfrac{70 \times 30}{0.76} + 0.06 \times 100 \pm 0.30\dfrac{100 \times 100}{30^2 \times 0.76} \\[2ex] 0.10\dfrac{70 \times 30}{0.76} + 0.57 \times 100 \mp 0.30\dfrac{100 \times 100}{30^2 \times 0.76} \\[2ex] \pm 0.15\dfrac{70 \times 30}{0.76} \mp 1.42 \times 100 - 0.22\dfrac{100 \times 100}{30^2 \times 0.76} \end{bmatrix} \times \begin{bmatrix} 10^{-5} \end{bmatrix}
$$

$$
\begin{bmatrix} \epsilon_1 \\ \epsilon_2 \\ \gamma_{12} \end{bmatrix} = \begin{bmatrix} 420.5 \pm 4.4 \\ 280.3 \mp 4.4 \\ \pm 278.5 - 3.2 \end{bmatrix} \times \begin{bmatrix} 10^{-5} \end{bmatrix}
$$

Ignoring the small torque strains, the mechanical strains are

$$
\begin{bmatrix} \varepsilon_1 \\ \varepsilon_2 \\ \gamma_{12} \end{bmatrix} = \begin{bmatrix} 425 \\ 285 \\ \pm 280 \end{bmatrix} \times \begin{bmatrix} 10^{-5} \end{bmatrix}
$$

To obtain the total strain components we add the residual thermal and hydric components to the above mechanical strains.

21 Impermeable Pipes

21.1 INTRODUCTION

Two issues require attention in the composite pipes currently used in underground service. The first issue is their high permeability, which has been a barrier to acceptance in solvent and gas service. The second is their low tolerance to damage from impact loads, which has been a concern for contractors unwilling to exercise extra care in handling, shipping and installing composite pipelines. It would be desirable to solve or at least improve the composite pipes in relation to these issues. The improvement proposed in this chapter involves the insertion of a thin impermeable aluminum foil between the pipe's resin-rich liner and its corrosion/weep barrier.

The impermeable aluminum barrier certainly solves the pipe's low permeability issue, while improving their tolerance to damage from impact loads and careless handling. Furthermore, the aluminum foil eliminates the expensive hydrostatic pressure test currently mandated on all sanitation and oil pipes. The thin foil chemically bonds to the resin and become an integral part of the pipe wall.

This chapter develops compelling arguments showing how the low cost aluminum foil opens new and unprecedented opportunities for composite pipelines. The impermeable pipes find new and undreamed of applications in the transmission of gases and solvents, while improving their performance in well-established markets like oil and sanitation.

21.2 PERMEABILITY, DIFFUSIVITY AND SOLUBILITY

The permeation of chemicals through the pipe wall is a cause of concern in underground pipes for two reasons.

- Fear of soil contamination from chemicals carried in the pipes.
- Fear of product contamination from soil pollutants.

The risk of soil or product contamination has always been a concern in applications involving underground plastic pipes. Furthermore, the high permeability of composite pipes may promote the occurrence of a new and unsuspected type of long-term anomalous failure, discussed later in this chapter. For now, let us say a few words about the permeability, diffusivity and solubility of systems involving composite laminates.

The permeability of a system consisting of a laminate and a diffusing material is

$$P = DS$$

Where P is the permeability, D is the diffusivity and S is the solubility of the diffusing material in the laminate. Each combination of diffusing material and laminate construction has its own P, D and S, which change in each case with the resin matrix. It is quite possible that the multiplicity of values taken by P, D and S has discouraged the efforts to measure them. Be it as it may, the values of D, S and P for systems involving composite materials are scarce in the published literature. The rest of this section is dedicated to a brief discussion of P, D and S. Table 21.1 shows a few values of the water diffusivity D in laminates at 37 C.

The solubility S designates the maximum amount of penetrant that a resin matrix can accept at saturation. The solubility S controls the maximum working temperature and the load bearing capability of laminates in contact with solvents. High values of S means high solvent intake and high resin swelling, which reduces the laminate structural and temperature capabilities. The literature carries ample documentation showing the decreased laminate performance in the presence of solvents.

There are three categories of organic solvents.

- *Low solubility* - The low solubility solvents cause little harm to composites and are not a deterrent to structural service. The most notorious representative in this category is water, with a solubility $S = 1.0\%$. The low swelling caused by such a small water absorption does little harm to composite laminates, allowing their use in load-bearing applications. The high pressure pipes used in the oil industry is a good example. Low solubility solvents like water, therefore, are not a deterrent to composites in load-bearing structural service.

TABLE 21.1

The diffusivity D of water is higher in polyester than in vinyl ester novolac laminates.

Laminate	$D\left(\times 10^{-8}\ \frac{m^2}{s}\right)$	
	Polyester	Vinyl ester novolac
Resin casting	4.0–6.0	1.0
Chopped strands	2.0–3.0	1.0
Woven roving	2.0–3.0	1.0

TABLE 21.2

The low water uptake (1.0%) allows the use of composite pipes in aqueous structural service. The water pickup causes a small decrease (usually 10^0C) in the maximum allowed operating temperature. The higher solubility of methanol and toluene preclude the use of composites in structural or high temperature services.

Solubility – S	Water	Methanol	Toluene
High reactivity polyester	1.0%	10.0%	18.0%
Vinyl ester novolac	1.2%	13.0%	10.0%

- *Medium solubility* - The swelling from the solvents in this category may restrict the use of composite laminates to non-load bearing and low temperature services. There is a variety of organic solvents falling in the category. The most notorious are, maybe, ethanol and methanol, with a solubility S = 14.0%.
- *High solubility* - The high swelling from these solvents may cause resin cracking and laminate failure even in non-structural service. Acetone is a good example of a solvent in this category.

Table 21.2 illustrates the solubility S of water, ethanol and toluene in castings of two premium resins. The reported solubility values are valid for castings at 37 C.

In conclusion:

1. The small solubility S = 1% of water has no appreciable effect on composites pipes, other than a minor, say 10^0C, decrease in the maximum allowed operating temperature.
2. The medium solubility of some organic solvents may reduce the working temperature/pressure rating of composite pipes to very low values.
3. Composite pipes are not good for service in high solubility solvents.

The traditional composite pipelines are not adequate for service in high solubility solvents. Their use is questionable even in medium solubility solvents, such as ethanol, where the service is limited to low temperatures and pressures. The new impermeable composite pipes do not have these restrictions. The aluminum foil bars the ingress of solvents in the structural plies and the pipelines perform as in water service. As an example, we mention the transmission of methanol or ethanol, where the traditional permeable pipelines are limited to room temperature and gravity service. The impermeable pipelines, however, are good in ethanol or methanol at any temperature and pressure service, just as if they carried water.

The following sections describe some applications of the impermeable pipes.

21.3   IMPERMEABLE PIPES

An apparent major benefit from the use of impermeable aluminum foil in oil and sanitation pipes would be the elimination of the weep failures. This, however, never happens as the thin aluminum foil is not strong enough to stop the large crack densities that develop in these pipes above the weep threshold. The formation of large crack densities in the composite laminate may rupture the thin foil and cause the impermeable pipes to weep just like any other. The benefit of the impermeable foil lies in the improved damage tolerance of the pipes, which display a higher weep threshold than the regular ones. For details on this important topic, see the appendix of this chapter.

The aluminum foil certainly serves the important function of preventing premature pipe weeping from manufacturing defects. The manufacturing defects from entrapped air, dry fibers and encapsulated foreign objects are quite distinct and should not be confused with the pipe cracks from mechanical or impact loads. Our concern at this time relates to the manufacturing defects. The cracks from impact loads are treated in the appendix of this chapter. The large manufacturing defects may provide direct pathways in the pipe wall causing weep failures below the rated pressures. These defects are a major concern in high-pressure applications involving sanitation and oil pipes. Their detection involves expensive and time-consuming short-term (30 seconds) quality control hydrostatic pressure testing of all manufactured pipes. Let us expand on this.

The current pipe standards for sanitation and oil pipes require two quality control pressure tests.

- A high-pressure test to ascertain the pipe rating. This is a destructive test conducted on representative samples at 4 times the rated pressure for 10 minutes with no weep. The aluminum foil will not affect this test and will not interfere in the assessment of the rated pressure.
- A 30 seconds test at 1.5 times the rated pressure, to ascertain the absence of manufacturing defects. The rationale here is that all defective pipes weep when pressurized at 1.5 times the rated pressure for 30 seconds. This is a non-destructive test mandated on all pipes used in sanitation or oil service. The aluminum foil does not rupture at pressures less than 1.5 the rated pressure, and prevents

the pipe weeping, even in the presence of gross manufacturing defects. The impermeable pipes do not weep at pressures less than weep threshold and do not require the quality control hydrostatic test.

The prevention of premature pipe weeping from gross manufacturing defects is an undisputed benefit provided by the aluminum foil. This benefit may raise the argument that the impermeable foil masks the presence of major manufacturing defects and force the acceptance of faulty pipes. This clever argument is not valid, however, since the gross manufacturing defects hidden by the foil do no harm other than weeping the pipe. Since the foil stops the weeping, the defective pipe will live a normal life, as if the gross manufacturing defects were not there.

The impermeable aluminum foil is a welcome low cost solution that eliminates the expensive quality control hydrostatic pressure testing currently required on all sanitation and oil pipes. Furthermore, the foil presence improves the weep threshold and the damage tolerance of all pipes. For details, see the appendix of this chapter.

21.4   APPLICATIONS OF THE ALUMINUM FOIL

The following sections describe the benefits of the aluminum foil in some applications of composite materials.

21.4.1   ANOMALOUS FAILURE IN SANITATION PIPES

The anomalous failure involves an unsuspected mode of deterioration just recently described. This section explains the causes and prevention of the anomalous failure.

The water diffusion through the wall of composite pipes derives from Fick's first equation

$$[flow] = \frac{D \times S \times C}{t}$$

Where [flow] is the amount of water passing across a unit area of the wall per unit time, D is the coefficient of diffusion, S is the solubility, C is the water concentration and t is the wall thickness. The solubility S and the diffusivity D of water in a few laminates are in Tables 21.1 and 21.2. The permeability is the product $P = D.S$.

Example 21.1

Compute the flow of water through the wall of a polyester pipe at 37C. Assume the following conditions

- Wall thickness t = 5.0 mm.
- $D = 3.0 \times 10^{-8} \frac{m^2}{s}$
- S = 0.01 (Solubility of water in polyester).
- C = 1 g/cm³ (the pipe carries water 100% pure).

Entering the above in Fick's equation, we obtain

$$[flow] = \frac{3.0 \times 10^{-8} \frac{m^2}{s} \times 0.01 \times 1 \, g/cm^3}{5.0 \, mm}$$

$[flow]$ = 190 g/m².year

It is obvious that the impermeable aluminum foil would stop this water flow through the pipe wall. The flow of solvents and gases would be higher than the flow of water. The aluminum foil stops all flows. The impermeable pipelines can carry solvents and gases, which the current pipe technology cannot do.

The amount of water lost by diffusion computed in Example 21.1 is not high and amounts to just about 3 liters per square meter in 15 years. As a rule, the pipeline owner is not aware of such a loss, since the water molecules that permeate the pipe dissipate harmlessly into the atmosphere. However, the permeating water first fills all cavities, cracks and voids in the pipe wall before proceeding in their path. The 3 liters of permeated water mentioned above may be very harmful in the long-term, if they accumulate in cracks and cause the anomalous failure. Let us discuss the mechanism leading to the anomalous failure. The water pressure in the crack/void drops linearly from a maximum near the inner surface, to a minimum near the outer surface. Under steady operating conditions, there is nothing unusual or damaging with the pressurized water accumulated in the crack/void. The problem arises when the pressure starts to oscillate. Let us explain this.

Figure 21.1 illustrates the discussion that follows. Suppose a a crack in the wall of a pressurized pipe filled with water. The water pressure in the pipe equilibrates the pressure in the crack. A sudden pressure drop causes the crack to expand and grow by a small amount, just enough to balance the lower pressure in the pipe. As the pipe pressure rises to its previous value, the crack fills once again with water. The process continues, with small crack expansions following each pressure drop. After many cycles of pressure

FIGURE 21.1A *Anomalous blisters of composite pipes in water. Note the cracked sand core that triggers the onset of the anomalous blister.*

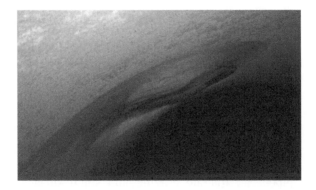

FIGURE 21.1B *The anomalous blisters grow to large sizes, like 500 mm in diameter and 100 mm in height.*

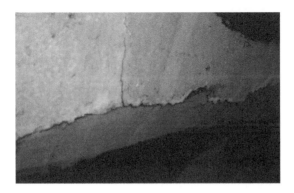

FIGURE 21.1C *There is no mechanism to check the growth of the anomalous blisters, which increase in size to the point of rupture.*

oscillation, the crack accumulates a considerable enlargement. This process continues without stopping and eventually the initial small wall crack develops into an enormous blister filled with pressurized water.

The anomalous blister continues its unchecked growth until the pipe eventually ruptures. This mode of failure was recognized just recently (mid 2012) and came as a surprise. In fact, the surprise was so great that we named this mode of failure "anomalous failure", to emphasize its abstruse origin and emphasize its fundamental difference from the better-known osmotic blisters.

The sand core of underground pipes is particularly susceptible to delamination from mishandling or impact loads, which provide the cracks that start the anomalous blisters. The following concerns are always in the minds of contractors and owners involved in the installation of composites sanitation pipes with sand core.

- Have the unavoidable impacts from shipping and receiving damaged the pipes?
- Has that pipe drop caused any undetectable damage?
- Have we at some point inadvertently mishandled the pipe?
- What about the sand core? Is it cracked or delaminated? Will anomalous blisters develop at some point?

There are no simple answers to these questions. A visual inspection may indicate the presence of surface crazing and hairline cracks, which increase the owner's concern. The aluminum foil eliminates all these worries. As explained in the appendix of this chapter, the high weep threshold of the impermeable pipes can tolerate crazing and small surface cracks from impact, transportation and mishandling. Furthermore, the aluminum foil bars the ingress of water and prevents the formation of anomalous blisters in pipes with sand core.

21.4.2 UNDERGROUND STORAGE TANKS

The underground tanks to store ethanol, methanol, gasoline and mixtures thereof, require the use of special and expensive resins. The impermeable aluminum foil allows the use of low cost resins in place of the expensive premium products. Furthermore, the impermeable tanks could operate at high temperatures and pressures.

21.4.3 SOLVENT STORAGE AND TRANSMISSION

One of the shortcomings of composite pipes is their inability to work in solvent service at high pressure/temperature. The aluminum foil prevents the solvent penetration and allows the use of low cost pipes in solvent service at high pressures and temperatures.

21.4.4   INDUSTRIAL EFFLUENTS

The aluminum foil allows the use of regular sanitation pipelines in services involving industrial effluents at high pressures and temperatures. The only concern would be the chemical attack on the aluminum.

21.4.5   GAS TRANSMISSION

The impermeable aluminum foil eliminates the anomalous failure and permeation that has prevented the full acceptance of composite pipelines in gas transmission services. The underground impermeable pipelines would be fully gas-tight and immune to any type of corrosion, both internal and external.

21.4.6   CHEMICAL SERVICE

The original development of the impermeable pipes contemplated services in high-pressure pipelines for ethanol transmission. Later the concept evolved to include underground storage tanks, oil and sanitation pipes. We excluded the aluminum foil from chemical service for two reasons:

- The laminates used in non-penetrating chemicals do not need impermeable liners.
- The corrosion resistance of the aluminum foil in chemical service is unknown.

However, the insertion of an aluminum foil between the resin-rich liner and the corrosion barrier of chemical laminates might not be a bad idea. The impermeable foil would prevent the passage of molecules and bring the chemical concentration to zero in the corrosion barrier. This extremely desirable condition would fully protect the corrosion barrier for as long as the foil is able to maintain its integrity. The corrosion resistance of the thin aluminum foil in such conditions can range from very long in sanitation, oil and solvent service to very low in other chemicals. However, whatever the durability, the foil would do some good and have no bad effects. At worse, assuming instant chemical destruction of the foil, the original corrosion barrier would still be there, intact and ready to do its job.

The best results should occur in chemicals that are harmful to the corrosion barrier and not quite so aggressive to aluminum. In such cases, the foil would certainly give full protection to the corrosion barrier for some time, until its destruction. This would certainly extend the laminate service life. The inclusion of two foils would give even better protection. The idea of using impermeable aluminum foils in chemical service, although not yet quantified, may be a wise decision to improve the service lives of corrosion barrier in especially nasty chemicals.

21.5   MARKET ACCEPTANCE

The proposed impermeable pipes are the old conventional pipes with a thin aluminum foil behind the liner. The foil chemically bonds to the resin to become an integral part of the pipe. The impermeable foil is an important innovation to the traditional composite technology, allowing considerable cost reductions and opening new opportunities in applications like solvents, industrial effluents, gas transmission and perhaps chemical service.

The Impermeable pipe solution is simple, straightforward and market-ready. No extensive and expensive qualification testing is required. In fact, we can eliminate the current qualification routine for new pipes since these are not "new pipes", but the same old regular, pre-qualified products in the manufacturer's portfolio, augmented by the presence of the impermeable foil.

Figures 21.2, 21.3 and 21.4 illustrate the manufacture of impermeable pipes. The helically wound aluminum foil goes directly on top of the resin-rich liner. The regular structural laminate goes on top of the foil. To improve the pipe's impact resistance and eliminate the risk of damage from careless

FIGURE 21.2   *Helically winding the impermeable foil on top of a resin-rich liner prior to the structural plies.*

FIGURE 21.3 *The aluminum foil over the resin-rich liner.*

FIGURE 21.4 *The pipe in the picture has the following layup*
- *Dark graphite-filled conductive inner liner for use in hydrocarbon solvents*
- *Aluminum foil*
- *Filament wound ± 55 angle-plies*

handling and transportation, we suggest the insertion of a ply of chopped fibers on top of the foil, before the filament wound structure. The ply of chopped fibers protects the pipe from cracks originating in the UD plies. The foil will have perfect adhesion to the resin and become an integral part of the pipe.

The end-users will realize that the "new impermeable pipes" are in essence the "old standard pipes", with the added feature of impermeability. Market acceptance should not be a major problem. The impermeable pipes bring the following confirmed benefits to the composites industry:

- Zero contamination in underground service
- Solvent service at high pressure/temperature
- Industrial effluent service at high pressure/temperature
- Gas pipeline service
- Elimination of pressure weep tests
- Elimination of anomalous failure
- No need of extra careful handling

In addition to the above, the following are very likely benefits:

- Improved service life in chemical service
- Elimination of strain-corrosion in underground sewer pipelines

APPENDIX 21.1

WEEP THRESHOLDS OF IMPERMEABLE PIPES

As discussed in Appendix 16.2, the weep regression lines are pipe features that depend on the resin toughness, the operating temperature and the wall thickness. Figure 21.5 shows the weep line of a thin-walled pipe plotting below that of a thick-walled pipe and producing a low extrapolated HDB value. The pipe's weep thresholds, however, are the same in both cases, independent of the wall thickness. As we have discussed in this entire book, the weep thresholds are ply properties independent of the operating temperature and wall thickness. The weep thresholds are the true determinants of the weep failure.

This appendix explores the effect of tough and impermeable aluminum foils on the weep thresholds and damage tolerance of composite pipes.

The motivation for the following discussion is the amazing 30.0% elongation at break of the aluminum foil, which is 10 times higher than the 3.0% rupture strain of commercial composite pipes. Such an overwhelming disparity in toughness suggests that pipes lined with aluminum membranes would pressure test to burst without weeping. This, however, is not true since the thin foil may rupture at the crack tips of highly strained critical plies. The passage of water through the ruptured foil would weep the impermeable pipes before they burst. The weep lines of impermeable pipes are the same as those of regular pipes.

However, the inherent toughness of the aluminum foil increases the damage tolerance of the critical ply, and substantially raises the pipe's weep threshold.

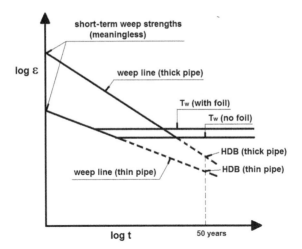

FIGURE 21.5 *The weep regression lines are pipe – not ply – features that predict the travel times of pressurized water in traversing cracked walls. The weep lines depend on the resin toughness, the operating temperature and the pipe's wall thickness. They predict the weep times of failed pipes and have little engineering interest. The meaningful parameter in the analysis of pipe failure is the weep threshold, which is a fundamental ply property independent of the operating temperature and pipe thickness. The tough aluminum foil raises the weep threshold and improves the damage tolerance of the pipes.*

Figure 21.5 shows the weep threshold Tw of foiled pipes plotted above that of regular permeable pipes. The improved weep performance afforded by the foil presence suggests the possibility of alleviating – and perhaps eliminating – the low tolerance of composite pipes to impact damage. We can imagine three damage scenarios:

1. **No visible impact damage.** The small and invisible damages from low impact energies do not exceed the weep threshold and the pipe remains essentially pristine. The critical ply does not fail and the damaged pipe operates as discussed in the numerical Examples 9.7 and 9.8. See Chapter 9. The aluminum foil is not necessary in these cases, which are representative of the vast majority of the situations found in practice.
2. **Visible impact crazing and hairline cracks.** The internal damages from impact blows high enough to produce visible surface crazing and hairline cracks may exceed the weep threshold of regular pipes and make them unfit for pressure service. However, the improved damage tolerance afforded by the aluminum foil allows the use of

crazed and hairline cracked pipes in pressure service. The impact damages in these cases do not exceed the high weep threshold of the impermeable pipes, which are safe to use in pressure service, as discussed in the numerical Examples 9.7 and 9.8. See Chapter 9.

3. **Visible Impact delamination and large cracks.** The internal damages from impact energies high enough to produce delamination or large cracks possibly exceed the weep threshold of the aluminum foil and fail the impermeable pipes. The aluminum foil may not be of help and all damaged pipes in this situation, impermeable or otherwise, are unfit for pressure service.

In conclusion, the tough aluminum foil raises the pipe's weep threshold and improves its tolerance to impact loads. The question is…. are the impacted pipes good enough for pressure service? We propose the following decision rules:

1. Impact damages with no visible surface crazing or hairline cracks do not exceed the weep threshold and do not disqualify any pipe for pressure service. This is the situation normally found in practice.
2. Impact damages with visible surface crazing or hairline cracks may exceed the weep threshold of permeable pipes – those with no foil. The pipes with no foil may not tolerate visible surface crazing or hairline cracks. The safe and conservative recommendation is to discard such pipes for pressure service.
3. The impermeable pipes lined with tough aluminum foils can tolerate impact loads producing visible surface crazing and hairline cracks. Such pipes are safe to use in pressure service.
4. Impact damages causing delamination and large cracking quite possibly rupture the aluminum foil and exceed the pipe's weep threshold. Delaminated and cracked pipes, even those with aluminum foil, may be unfit for pressure service.

The experimental qualification of cracked pipes for pressure service requires expensive and unjustifiable long-term pressure tests. This is so because of the large weep times of pipes cracked at the weep threshold. In the absence of such testing, the conservative decision rules proposed above should suffice.

Appendix: The Fatigue Mechanism

A.1 INTRODUCTION

The published literature has little information on the mechanism of fatigue crack growth. My search did not produce a single article or even a passing mention of the mechanism explaining how a subcritical, repetitive load, would cause the growth of otherwise stationary cracks. This appendix is an attempt to explain this.

We approach the subject from the viewpoint of fracture mechanics. Our efforts will focus on the derivation of an expression for the crack extension per cycle, with a semblance to the well-known Paris law. In its simplest form, the Paris law states that the crack growth per cycle can be estimated as a power function of the range of the stress intensity factor $\Delta K = K_{max} - K_{min}$.

$$\frac{da}{dN} = Y\left(\Delta K\right)^{Z} \tag{A.1}$$

In the above, Y and Z are empirical constants that depend on the material as well as on the loading parameters, such as the frequency, the R ratio and others. The crack extension per cycle is da/dN, where "a" is the crack length and N is the number of cycles.

To my knowledge, there is no theoretical justification for the Paris law. In fact, this empirical law saw strong objections when first proposed in 1961, and has survived all these years awaiting justification. This appendix is an attempt to explain the fatigue mechanism leading to the Paris law. The discussion assumes the validity of one of fracture mechanics basic tenets, which prohibits crack growth if the stress intensity factor K_I is less than the critical K_{IC}. Thus, according to fracture mechanics, crack growth occurs only if at some point in the loading cycle the stress intensity factor K_I exceeds the critical K_{IC}. In mathematical terms, no crack ever grows (da/dN = 0) unless the stress intensity facto $K_I > K_{IC}$, or the energy release rate in crack growth $G_I > G_{IC}$. Both K_{IC} and G_{IC} are well-known material properties.

The classic fatigue model does not meet the preceding condition. Indeed, the classic model assumes crack growth driven by subcritical loadings in which $K_I < K_{IC}$. This situation cannot be correct. According to fracture mechanics, the cyclically loaded subcritical cracks should remain stationary and never grow. However, they do grow. An explanation is required as to why cyclical subcritical loadings, far lower than the minimum static requirement, cause crack growth.

A.2   RUPTURE FAILURE

The reader is referred to Chapter 8 for a discussion of the four load-dependent long-term failure modes of composite laminates. This appendix focus on just one of those four modes, specifically the rupture failure. This appendix will not discuss the infiltration, weep and stiffness failures. Such simplification facilitates the explanation with no loss of generality, since, in the final analysis all load-dependent failures reduce to rupture of one kind or another.

The questions pursued in this appendix are:

- How is it possible that subcritical cyclic loadings in which $K_I < K_{IC}$ cause crack growth while static loadings of equal K_I do not?
- What drives the cyclic crack growth in the fatigue process?

To answer the above questions, we must first understand the rupture mechanism of cracked materials. The rupture of cracked materials can be described in terms of the critical stress intensity factor K_{IC} or the critical energy release rate G_{IC}. This appendix will make use of these material properties to describe the crack growth in the fatigue process.

The stress intensity factor defines the magnitude of the stress field near the crack tips. Its mathematical expression is

$$K_I = \sigma\sqrt{\pi a} \tag{A.2}$$

Given the crack size "2a" and the global stress "σ", the K_I factor comes from Equation (A.2). K_I defines the magnitude of the stress field near the crack tips. The rupture process is initiated – the crack starts to grow – when the stress at the crack tips exceed the material strength. This occurs when the stress intensity factor exceeds a critical value, i.e., when $K_I > K_{IC}$.

Cracks under static loads grow in two ways.

- The "static" load increases monotonically, i.e., it is not "exactly" static. This is the case observed in short-term rupture tests.
- The static load is "exactly" static, and the crack growth comes from strain-corrosion.

The above is valid for static loads. Under cyclic loads, both the global stress and the crack size vary with time. However, since the crack growth per cycle is small, the fatigue analysis per cycle assumes constant crack sizes. Therefore, the discussion of the fatigue phenomenon assumes variable loads and constant crack sizes, as in the case of quasi-stationary, monotonic static failure.

Next, we discuss another approach frequently used in the study of brittle failure of ductile materials. This is the energy release approach, which is similar to the stress case just discussed. The elastic energy stored at the crack tips of any crack of size "2a" subjected to a stress "σ" is

$$Energy = \frac{\pi\sigma^2 a^2}{2E}$$

For any given material, the elastic energy stored at the crack tips varies with the square of both the global stress and the crack size. The energy released per unit area when a crack grows under constant global stresses is

$$\frac{d(energy)}{da} = G_I = \frac{\pi\sigma^2 a}{E} \tag{A.3}$$

Equation (A.3) quantifies the energy released per unit area at the onset of crack growth, assuming static loads. In analogy with the stress intensity case, rupture will occur only if the energy release rate from Equation (A.3) is more than the critical level G_{IC}. Therefore, the two conditions for crack growth are:

$$K_I > K_{IC} \text{ or } G_I > G_{IC}$$

Comparing the expressions (A.2) and (A.3), we have

$$K_I^2 = EG_I \tag{A.4}$$

From Equation (A.4) we see that the crack growth phenomenon can be approached from the perspectives of either G_I or K_I.

- The energy approach is convenient when the stress is constant, as in strain-corrosion. Under constant stress, the released energy plots as straight lines against the crack size "a", which simplifies the analysis.
- The stress intensity approach is convenient when the crack size is constant and the stress varies with time, as in the case of fatigue. As explained in the next section, in such cases K_I plots like the stress.

The study of fatigue crack growth focus one load cycle, in which the stresses are variable and the crack sizes are constant. Therefore, in our study of fatigue, we make use of the stress intensity factor K_I.

A.3 FATIGUE FAILURE

The fatigue events start with a crack of size "2a" subjected to a cyclic loading. In one fatigue cycle, the applied stress varies from a minimum value σ_{min} to a maximum value σ_{max} in a time interval ΔT. The crack size "2a" is constant in one cycle. As the stress grows, the stress intensity factor K_I also grows. The crack growth starts when $K_I > K_{IC}$ The growth continues as long as $K_I > K_{IC}$ and ceases when K_I falls below the critical value on the descending part of the load cycle. This is obvious and clear. The crack will grow only if and when $K_I > K_{IC}$. The duration of the time interval in which $K_I > K_{IC}$ is short and the crack extension per cycle is small.

There is nothing special in the mechanism of crack growth in a fatigue cycle. Crack growth occurs in the short time interval when $K_I > K_{IC}$. Furthermore, the crack extension per cycle is small given the short duration of this time interval.

A.4 ACTUAL AND NOMINAL STRESS WAVES

Figure A.1 shows the nominal stress waveform. To facilitate the analysis, we have assumed a triangular loading. The nominal waveform parameters are in upper case letters, as usual in this type of study. We will see next, when we come to Figure A.3, that the nominal stress wave depicted in Figure A.1 should not grow cracks, since the maximum K_I that it gives is less than K_{IC}.

$$\sigma_{max} \sqrt{\pi a} < K_{IC}$$

From fracture mechanics, we would expect the loading depicted in Figure A.1, in which $K_I < K_{IC}$, not to grow cracks of sizes less than 2a. Such a loading could cycle forever with absolutely no crack growth. However, things do not happen that way. We know from experience that subcritical cyclic loadings, like the one in Figure A.1, do grow cracks of sizes less than 2a. The question is... what makes the cyclic loads so special and different from the static ones. This appendix proposes and develop the hypothesis that the nominal waveform of Figure A.1 misrepresents reality.

Figure A.2 shows the actual shape of the stress wave, with the wave parameters in lower case letters. To facilitate the comparison, we have plotted the actual and nominal waveforms together, in a single figure. Figures A.1 and

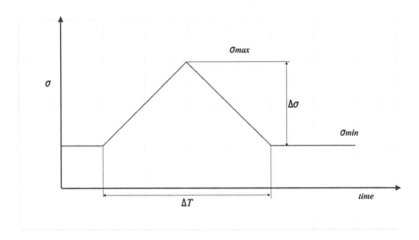

FIGURE A.1 *The nominal waveform and the wave parameters.*

A.2 plot the stress waves as a function of time. The stress intensity factors "K_I" are in Figure A.3.

In Figure A.2, the ascending part of the actual waveform is steeper than the nominal waveform. Such a steeper rise is justifiable, since the material near the crack accelerates from a closed crack position to a maximum open crack in the ascending part of the wave. The descending part of the actual waveform is also steeper than the actual wave, since the open crack quickly returns to the closed position. The actual stress becomes very small towards the end of the cycle. To facilitate the analysis, Figure A.2 assumes a triangular shape for the actual stress waveform. This approximation should not differ much from the real waveform.

The actual stress waveform is narrower and taller than the nominal. Compared with the nominal, the actual wave has a shorter duration δt and a higher peak. The actual peak stress may attain values high enough to make $k_I > K_{IC}$ and promote crack growth. The relation between the actual and nominal waveforms derives by equating their delivered energy or momentum. Figure A.2 help us equate the momentum delivered by each waveform. The reader is reminded that the minimum stress σ_{min} remains active during the entire time interval ΔT in both cases.

$$\int_{t}^{t+\Delta T} \left(\sigma_{min} + \delta\sigma \right) dt = \int_{t}^{t+\Delta T} \left(\sigma_{min} + \Delta\sigma \right) dt$$

$$\sigma_{min}\Delta T + \delta\sigma \times \frac{\delta t}{2} = \sigma_{min}\Delta T + \Delta\sigma \times \frac{\Delta T}{2}$$

The contribution from σ_{min} cancels out for both waveforms. From the above equation, we have

$$\delta\sigma = \Delta\sigma \times \frac{\Delta T}{\delta t} \tag{A.5}$$

The actual stress range is the nominal stress range multiplied by the impact factor

$$Impact\ factor = \frac{\Delta T}{\delta t}$$

The impact factor compensates the decrease in the nominal load duration by increasing the actual stress magnitude.

The same reasoning applies to the stress intensity factor. Figure A.3 shows a diagram similar to that in Figure A.2, showing the actual stress intensity factors K_I. The K_I diagram is identical to the stress diagram. This is a direct consequence of our choice to develop the discussions based on the stress intensity factor, instead of the released energy. Incidentally, Figure A.3 also shows K_{IC} and the crossover points, t_0 and $t_0 + \Delta\tau$, marking the short active time interval $\Delta\tau$ of crack growth.

A quick inspection of Figure A.3 gives the time dependence of the actual stress intensity factor.

$$k_I = K_{Imin} + \frac{2\delta k}{\delta t} t \tag{A.6}$$

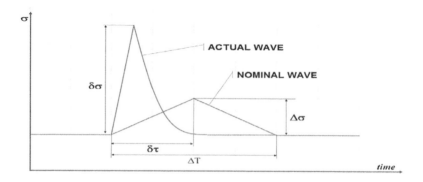

FIGURE A.2 *The actual waveform versus the nominal. The actual wave parameters are in lower case letters.*

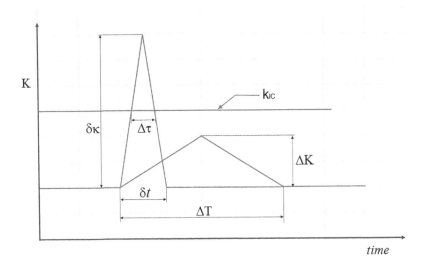

FIGURE A.3 *The actual and the nominal stress intensity factors plotted side by side. Note the active interval $\Delta\tau$ in which the actual stress intensity factor exceeds the critical $k_I > K_{IC}$*

In Equation (A.6), the actual stress intensity factor is in lower case font, as we have explained. The actual stress intensity factor in Equation (A.6) drives the crack growth per cycle.

A.5   THE ENERGY BALANCE

According to Irwin, the elastic energy accumulated near the crack tips is

$$energy \; at \; crack \; tips = \frac{\pi\sigma^2 a^2}{2E}$$

The energy change near the crack tips when the applied load cycles back and forth is

$$d\left(energy\right) = \frac{\pi a^2\sigma}{E}\,d\sigma + \frac{\pi\sigma^2 a}{E}\,da - G_{IC}da$$

Integrating the above over one cycle, we obtain

$$\oint d\left(energy\right) = 0$$

$$\oint \frac{\pi a^2 \sigma}{E} d\sigma = 0$$

The above equations are identically zero over one cycle, since the energy supplied in the ascending part of the cycle returns to the system in the descending part. The equation for the remaining part of the energy is

$$\oint \frac{\pi \sigma^2 a}{E} da = \oint G_{IC} da \tag{A.7}$$

From Equation (A.7), the crack growth results from the released elastic energy. Since G_{IC} is constant, the integration of Equation (A.7) gives

$$\Delta a = \frac{1}{G_{IC}} \int_a^{a+\Delta a} \frac{\pi \sigma^2 a}{E} da \tag{A.8}$$

Recalling Equations (A.2) and (A.4)

$$K_I = \sigma \sqrt{\pi a}$$

$$G_{IC} E = K_{IC}^2$$

Entering the above in (8), we have

$$\Delta a = \frac{1}{K_{IC}^2} \int_a^{a+\Delta a} k^2 da \tag{A.9}$$

Equation (A.9) gives the crack extension Δa in one cycle. The reader will note the lower case stress intensity factor "k" representative of the actual, not the nominal, waveform. Integrating over the time variable "t", instead of the crack size "a", we have

$$\Delta a = \frac{1}{K_{IC}^2} \int_{t_0}^{t_0+\Delta \tau} k^2 \frac{da}{dt} dt \tag{A.10}$$

In Equation (A.10), da/dt is the speed of crack growth. We next derive a tentative expression for such speed. Figure (A.4) plots a qualitative representation of the crack speed over one cycle. The crack speed starts from zero at t_0, grows to a maximum at $t_0 + \Delta \tau/2$ and drops to zero at $t_0 + \Delta \tau$. The driver of the

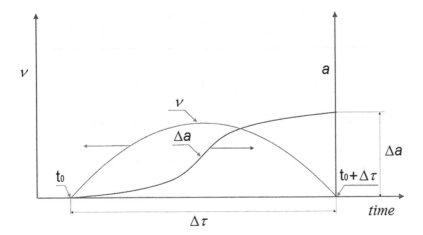

FIGURE A.4 *The speed of crack growth v = da/dt is driven by the difference between the actual and the critical stress intensity factor. This representation gives zero values for the speed at the start and finish of the cycle, and shows a maximum speed at mid-cycle.*

crack growth is the stress at its tips, which relates to the difference between the actual and the critical stress intensity factors $k - K_{IC}$. We may postulate the following expression for the crack speed.

$$\frac{da}{dt} = Y\left(k - K_{IC}\right)^m \tag{A.11}$$

Equation (A.11) is the same Equation (A.1) in the Paris law, applied to the speed of crack growth instead of the crack growth per cycle.

Entering (A.11) in (A.10), we obtain

$$\Delta a = \frac{Y}{K_{IC}^2} \int_{t_0}^{t_0+\Delta\tau} \left(k\right)^2 \left(k - K_{IC}\right)^m dt \tag{A.12}$$

It is more convenient to integrate over the stress intensity factor "k" instead of the time variable "t". Recalling from Figure A.3 that $k = K_{min} + \dfrac{2\delta k}{\delta t} t$ we have $dk = \dfrac{2\delta k}{\delta t} dt$ which, upon entering Equation (A.12), gives

$$\Delta a = \frac{A}{K_{IC}^2} \times \frac{\delta t}{\delta k} \int_{K_{IC}}^{k_{max}} k^2 \left(k - K_{IC}\right)^m dk \tag{A.13}$$

Equation (A.13) computes the crack extension per cycle in terms of the lower case actual, not the upper case nominal, loading. To simplify the above integration, we assume a constant velocity of crack growth. Taking m = 0 in Equation (A.13), we have

$$\Delta a = \frac{A}{K_{IC}^2} \frac{\delta t}{\delta k} \int\limits_{K_{IC}}^{k_{max}} k^2 dk \tag{A.14}$$

$$\Delta a = \frac{A}{K_{IC}^2} \frac{\delta t}{\delta k} \left(k_{max}^3 - K_{IC}^3 \right) \tag{A.15}$$

We can simplify Equation (A.15) by entering the relation

$$\frac{\delta t}{\delta k} = \frac{\Delta \tau}{k_{max} - K_{IC}}$$

$$\Delta a = \frac{A}{K_{IC}^2} \times \Delta \tau \times \frac{k_{max}^3 - K_{IC}^3}{k_{max} - K_{IC}} \tag{A.16}$$

Equation (A.16) gives the crack extension per cycle in terms of the active interval $\Delta \tau$ and the lower case actual stress intensity factor. The Paris law, however, relates the crack extension to the upper case nominal – not the actual – stress wave. The conversion comes from the relation $k_{max} = \dfrac{\Delta T}{\delta t} \Delta K + K_{min}$ and a few mathematical tricks.

$$\Delta a = C_1 \times \Delta K^2 + C_2 \times \Delta K + C_3$$

Where C_1, C_2 and C_3 are constants that vanish when the active interval $\Delta \tau = 0$.

Since Δa is the crack growth per cycle, the above equation is in fact

$$\frac{da}{dN} = C_1 \times \Delta K^2 + C_2 \times \Delta K + C_3 \tag{A.17}$$

We now take a long look at Equation (A.17) and ask ourselves how close it is to Equation (A.1). Not much, I am afraid. Our result apparently is not in agreement with the Paris law. However, for $\Delta K \gg 1$, we can write Equation (A.17) as

$$\frac{da}{dN} = C_1 \times \Delta K^2 \tag{A.18}$$

Our analysis seems to agree with the Paris law (A.1), with the parameter $Z = 2$ when $\Delta K \gg 1$.

A.6 THE UNIFIED EQUATION

The effect of the nominal K_{min} on the crack extension rate is particularly interesting. Suppose a fatigue situation in which everything is constant, except K_{min}. From Figure A.3, increments in K_{min} increase the active interval $\Delta\tau$ and the rate of crack growth. Therefore, changes in K_{min} – everything else fixed – modulate the rate of crack growth. Such an interesting conclusion suggests the possibility of separating the damages from complex loadings that combine cyclic and static components.

- We compute the damage from the pure cyclic load component acting alone.
- We compute the damage from the pure static component acting alone.
- The K_{min} created by the presence of the static load component, modulates the process and accounts for the interaction damage.
- We compute the total damage by adding the partial pure static, pure cyclic and interaction contributions.

The possibility of adding the damages from pure load components justifies the unified equation discussed in Chapter 16 of this book.

A.7 THE FATIGUE LIMIT

We proceed now to discuss the existence of a fatigue limit for homogeneous materials. Later we discuss the composites case.

Homogeneous materials like metals, ceramics, glass fibers and plastics, rupture from fatigue when a single growing crack reaches a critical size. The fatigue durability of homogeneous materials is the number of cycles to grow a single subcritical crack to its critical size. The fatigue durability of homogeneous materials comes from the concept of crack extension per cycle. From fracture mechanics, we know that the cracks in homogeneous materials do not grow unless $K_I > K_{IC}$. From our discussion in this appendix, this statement leads to

$$K_{min} + \frac{\Delta T}{\delta t} \Delta K > K_{IC} \tag{A.19}$$

The above inequality suggests the existence of a fatigue limit, or endurance limit, for any homogeneous material under cyclic loadings. Equation (A.18) allows crack growth only if

$$\Delta K > \left(K_{IC} - K_{min} \right) \times \frac{\delta t}{\Delta T} \tag{A.20}$$

This is a very interesting conclusion. Our analysis predicts the existence of a fatigue limit for homogeneous materials.

The above fatigue scenario holds for homogeneous materials and does not apply to composites. One important feature of composite materials is their ability to arrest crack growth. This property derives from the fibers presence, which arrest the resin cracks and keep them small. The fatigue rupture of composites comes from the coalescence of many small cracks instead of the growth of one large crack. Our Equation (A.17), as well as the Paris Equation (A.1), are applicable locally to predict the crack growth in the homogeneous fibers and the matrix, but cannot describe the global behavior of composite plies or laminates.

The ability of composite materials to develop many small cracks – the composite effect – explains their outstanding fatigue performance. This ability is not present in homogeneous materials, like metals, that develop a single large and very harmful crack. The density of small cracks in composites is so high, and their size is so varied, as to render useless any attempt to predict their failure by the growth of one single crack, as described in Equations (A.17) and (A.1). In engineering practice, the fatigue analysis of composites uses regression equations linking the applied stresses (or strains) to the number of cycles to failure. The concepts of crack growth and crack size have little use in the analysis of composites. This book provides a comprehensive discussion of this topic.

We can imagine a situation in composites, in which all large cracks grow until arrested, and the small cracks, those not satisfying Equation (A.20), do not grow at all. The applied load can cycle forever and the material will not fail. Therefore, there must be a fatigue limit for composites as well.

The fatigue limit of composites has nothing to do with the fatigue threshold discussed in Chapter 9. See Chapter 9.

Bibliography

1. *Stress Analysis of Fiber Reinforced Materials.* Michael W. Hyer, McGraw-Hill, 1998.
2. *Engineering Mechanics of Composite Materials.* Isaac Daniel and Ori Ishai. Oxford University Press, 1994.
3. Pultruded composites durability. A key value. Mark Greenwood. *CFA (ACMA) Composites Conference*, 2001.
4. Stress-corrosion of GRP. P. J. Hogg, J. N. Price, D. Hull. *39th Annual Conference of the SPI Composites Institute*, 1984.
5. Strain limited design criteria for reinforced plastic process equipment. L. Norwood and A. Millman. *SPI Composites Institute Conference*, 1980.
6. Chemical resistance of GRP pipes measured over 30 years. Hogni Jonsson. *TUV's FRP Conference in Munich*, 2009.
7. Mode of failure of hydrostatically overstressed reinforced plastic pipe. W. G. Gottenberg, R. C. Allen, W. V. Breitigan, C. T. Dickerson. *34th Annual Conference of the SPI Composites Institute*, 1979.
8. Design of high pressure fiberglass downhole tubing. A proposed new ASTM specification. Frank Pickering. *38th Annual Conference of the SPI Composites Institute*, 1983.
9. Surface cracking of RP by repeated bending. J. Malmo. *36th Annual Conference of the SPI Composites Institute*, 1981.
10. Historical background of the interface; studies and theories. Porter Erickson. *SPI Conference of the Composites Institute*, 1970.
11. *Effect of Winding Angle on the Failure of Filament Wound Pipes.* B. Spencer, D. Hull. University of Liverpool, UK, 1978.
12. Damage initiation and development in chopped strand mat composites. P. E. Boudan, W. J. Cantwell, H. H. Kausch, S. J. Youd. *International Conference on Composites Materials, Madrid*, 1993.
13. Corrosion resistance of glass fiber materials. A crucial property for reliability and durability of FRP structures in aggressive environments. Stefanie Romhild, Gunnar Bergman. *Nace Corrosion 2014*, paper 04612.
14. Effect of internal stress on corrosion behavior of glass fibers. A. Schmiermann, M. Gehde, G. Ehrestein. *44th Annual Conference of the SPI Composites Institute*, 1989.
15. Fatigue behavior of fiber-resin composites. J. Mandell. *Developments in Reinforced Plastics*, vol. 2, edited by G. Prichard. Applied Sciences Publisher, 1980.
16. Behavior of fiber reinforced plastic materials in chemical service. R. F. Regester. *NACE Corrosion Journal*, November 1968.
17. Behavior of glass fiber composite pipes under internal pressure as a function of composite cohesion parameters. J. Pabiot, P. Krawzak, C. Monnier. *49th Annual Conference of the SPI Composites Institute*, 1994.
18. Effect of seawater on the interfacial strength of reinforced plastic. Catherine Wood, Walter Bradley. Elsevier Science Limited, 1996.
19. *Effect of Fiber Coating on Properties of Vinyl Ester Composites.* B. A. Sjogren, R. Joffe, L. Berglund, E. Mader. Elsevier Science Limited, 1999.
20. *The Effects of Matrix and Interface on Damage in FRP Cross-ply Laminates.* B. A. Sjogren, L. A. Berglund. Elsevier Science Limited, 1999.
21. *Transverse properties of UD composites. Influence of fiber surface treatments.* K. Benzarti, L. Cangemi, F. dal Maso. Elsevier Science Limited, 2000.

22. *Effects of Glass Fiber Size Composition on Transverse Cracking in Cross-ply Laminates.* S. P. Fernberg, L. A. Berglund. Elsevier Science Limited, 2000.
23. *Performance of Epoxy Vinyl Ester Resins in Dilute Solvents. Synergistic Effects with HCl.* Don Kelley, Michael Jaeger. Ashland Specialty Chemicals.
24. Fatigue and failure mechanisms in GRP with special reference to random reinforcements. M. J. Owen, R. Dukes, T. R. Smith. *33rd Annual Conference of the SPI Composites Institute,* 1968.
25. High cycle fatigue of UD glass/polyester performed at high frequency. Guangxu Wei. Master's thesis. Montana State University, 1995.
26. Solvent resistance of fiber reinforced plastics. W. McClellan, Thomas Anderson, R. Stavinoha. *30th Annual Conference of the SPI Composites Institute,* 1975.
27. *A Molecular Mechanism for Stress Corrosion in Vitreous Silica.* Terry Michaelske, Stephen Freimann. Sandia Laboratories, 87115.
28. *Stress Corrosion Mechanism in Silicate Glasses.* Matteo Ciccoti. Universite Montpellier, France.
29. *Strength of High Performance Glass Reinforcement Fiber.* Michelle Korwin, Edson, Douglas Hofmann, Peter McGiniss. Owens Corning, Granville, Ohio.
30. *Finite Element Analysis of Composite Materials.* Ever J. Barbero. Department of mechanical and aerospace engineering, West Virginia University.
31. *Selection of Wind Turbine Blade Materials for Fatigue Resistance.* John Mandell. The American Ceramics Society, Cocoa Beach, FL, February, 24, 2010.

Index